Multivariate Statistical Methods in Quality Management

Multivariate Statistical Methods in Quality Management

Kai Yang

Jayant Trewn

McGraw-Hill

New York Chicago San Francisco Lisbon London Madrid
Mexico City Milan New Delhi San Juan Seoul
Singapore Sydney Toronto

Library of Congress Cataloging-in-Publication Data

Yang Kai,
  Multivariate statistical methods in quality management/Kai Yang, Jayant Trewn.
      p.   cm.
  Includes bibliographical references and index.
  ISBN 0-07-143208-6
  1. Quality control—Statistical methods.   2. Multivariate analysis.
  I. Trewn, Jayant. II. Title.

  TS156.Y34   2003
  658.5'62—dc22
                                                                    2003066522

Copyright © 2004 by The McGraw-Hill Companies, Inc. All rights reserved. Printed in the United States of America. Except as permitted under the United States Copyright Act of 1976, no part of this publication may be reproduced or distributed in any form or by any means, or stored in a data base or retrieval system, without the prior written permission of the publisher.

1 2 3 4 5 6 7 8 9 0 DOC/DOC 0 1 0 9 8 7 6 5 4

ISBN 0-07-143208-6

The sponsoring editor for this book was Kenneth S. McCombs and the production supervisor was Sherri Souffrance. It was set in Century Schoolbook by International Typesetting and Composition. The art director for the cover was Margaret Webster-Shapiro.

Printed and bound by RR Donnelley.

This book was printed on recycled, acid-free paper containing a minimum of 50% recycled, de-inked fiber.

McGraw-Hill books are available at special quantity discounts to use as premiums and sales promotions, or for use in corporate training programs. For more information, please write to the Director of Special Sales, Professional Publishing, McGraw-Hill, Two Penn Plaza, New York, NY 10121-2298. Or contact your local bookstore.

Information contained in this work has been obtained by The McGraw-Hill Companies, Inc. ("McGraw-Hill") from sources believed to be reliable. However, neither McGraw-Hill nor its authors guarantee the accuracy or completeness of any information published herein, and neither McGraw-Hill nor its authors shall be responsible for any errors, omissions, or damages arising out of use of this information. This work is published with the understanding that McGraw-Hill and its authors are supplying information but are not attempting to render engineering or other professional services. If such services are required, the assistance of an appropriate professional should be sought.

To our parents, families, and friends for their continuous support

Contents

Preface   xiii

Chapter 1. Multivariate Statistical Methods and Quality   1

1.1   Overview of Multivariate Statistical Methods   1
      1.1.1   Graphical multivariate data display and data stratification   3
      1.1.2   Multivariate normal distribution and multivariate sampling distribution   3
      1.1.3   Multivariate analysis of variance   4
      1.1.4   Principal component analysis and factor analysis   5
      1.1.5   Discriminant analysis   6
      1.1.6   Cluster analysis   7
      1.1.7   Mahalanobis Taguchi system (MTS)   7
      1.1.8   Path analysis and structural model   8
      1.1.9   Multivariate process control   10
1.2   Applications of Multivariate Statistical Methods in Business and Industry   10
      1.2.1   Data mining   11
      1.2.2   Chemometrics   12
      1.2.3   Other applications   13
1.3   Overview of Quality Assurance and Possible Roles of Multivariate Statistical Methods   13
      1.3.1   Stage 0: Impetus/ideation   13
      1.3.2   Stage 1: Customer and business requirements study   15
      1.3.3   Stage 2: Concept development   15
      1.3.4   Stage 3: Product/service design/prototyping   15
      1.3.5   Stage 4: Manufacturing process preparation/product launch   16
      1.3.6   Stage 5: Production   16
      1.3.7   Stage 6: Product/service consumption   17
      1.3.8   Stage 7: Disposal   17
1.4   Overview of Six Sigma and Possible Roles of Multivariate Statistical Methods   18
      1.4.1   Stage 1: Define the project and customer requirements (D or define step)   20
      1.4.2   Stage 2: Measuring process performance   21
      1.4.3   Stage 3: Analyze data and discover causes of the problem   21
      1.4.4   Stage 4: Improve the process   22
      1.4.5   Stage 5: Control the process   23

viii   Contents

Chapter 2. Graphical Multivariate Data Display and Data Stratification   25

2.1 Introduction   25
2.2 Graphical Templates for Multivariate Data   26
    2.2.1 Charts and graphs   26
    2.2.2 Templates for displaying multivariate data   29
2.3 Data Visualization and Animation   33
    2.3.1 Introduction to data visualization   33
2.4 Multivariate Data Stratification   38
    2.4.1 Multi-vari chart technique   39
    2.4.2 Graphical analysis of multivariate variation pattern   41

Chapter 3. Introduction to Multivariate Random Variables, Normal Distribution, and Sampling Properties   47

3.1 Overview of Multivariate Random Variables   47
3.2 Multivariate Data Sets and Descriptive Statistics   50
    3.2.1 Multivariate data sets   50
    3.2.2 Multivariate descriptive statistics   51
3.3 Multivariate Normal Distributions   55
    3.3.1 Some properties of the multivariate normal distribution   56
3.4 Multivariate Sampling Distribution   57
    3.4.1 Sampling distribution of $\overline{X}$ 57
    3.4.2 Sampling distribution of S   58
    3.4.3 Central limit theorem applied to multivariate samples   58
    3.4.4 Hotelling's T^2 distribution   59
    3.4.5 Summary   60
3.5 Multivariate Statistical Inferences on Mean Vectors   60
    3.5.1 Small sample multivariate hypothesis testing on a mean vector   62
    3.5.2 Large sample multivariate hypothesis testing on a mean vector   63
    3.5.3 Small sample multivariate hypothesis testing on the equality of two mean vectors   64
    3.5.4 Large sample multivariate hypothesis testing on the equality of two mean vectors   66
    3.5.5 Overview of confidence intervals and confidence regions in multivariate statistical inferences   67
    3.5.6 Confidence regions and intervals for a single mean vector with small sample size   68
    3.5.7 Confidence regions and intervals for a single mean vector with large sample size   70
    3.5.8 Confidence regions and intervals for the difference in two population mean vectors for small samples   71
    3.5.9 Confidence regions and intervals for the difference in two population mean vectors for large samples   72
    3.5.10 Other Cases   73
Appendix 3A: Matrix Algebra Refresher   73
    A.1 Introduction   73
    A.2 Notations and basic operations   73
    A.3 Matrix operations   75

Chapter 4. Multivariate Analysis of Variance   81

4.1 Introduction   81
4.2 Univariate Analysis of Variance (ANOVA)   82
    4.2.1 The ANOVA table   85

4.3	Multivariate Analysis of Variance	86
	4.3.1	MANOVA model	86
	4.3.2	The decomposition of total variation under MANOVA model	88
4.4	MANOVA Case Study	95

Chapter 5. Principal Component Analysis and Factor Analysis   97

5.1	Introduction	97
5.2	Principal Component Analysis Based on Covariance Matrices	99
	5.2.1	Two mathematical representations of principal component analysis	100
	5.2.2	Properties of principal component analysis	101
	5.2.3	Covariance and correlation between X and principal components Y	103
	5.2.4	Principal component analysis on sample covariance matrix	103
5.3	Principal Component Analysis Based on Correlation Matrices	108
	5.3.1	Principal component scores and score plots	111
5.4	Principal Component Analysis of Dimensional Measurement Data	114
	5.4.1	Properties of the geometrical variation mode	117
	5.4.2	Variation mode chart	118
	5.4.3	Visual display and animation of principal component analysis	121
	5.4.4	Applications for other multivariate data	122
5.5	Principal Component Analysis Case Studies	124
	5.5.1	Improving automotive dimensional quality by using principal component analysis	124
	5.5.2	Performance degradation analysis for IRLEDs (Yang and Yang, 2000)	131
5.6	Factor Analysis	141
	5.6.1	Common factor analysis	143
	5.6.2	Properties of common factor analysis	143
	5.6.3	Parameter estimation in common factor analysis	147
5.7	Factor Rotation	148
	5.7.1	Factor rotation for simple structure	149
	5.7.2	Procrustes rotation	152
5.8	Factor Analysis Case Studies	152
	5.8.1.	Characterization of texture and mechanical properties of heat-induced soy protein gels (Kang, Matsumura, and Mori, 1991)	152
	5.8.2	Procrustes factor analysis for automobile body assembly process	154
	5.8.3	Hinge variation study using procrustes factor analysis	156

Chapter 6. Discriminant Analysis   161

6.1	Introduction	161
	6.1.1	Discriminant analysis steps	162
6.2	Linear Discriminant Analysis for Two Normal Populations with Known Covariance Matrix	163
6.3	Linear Discriminant Analysis for Two Normal Population with Equal Covariance Matrices	167
6.4	Discriminant Analysis for Two Normal Population with Unequal Covariance Matrices	169
6.5	Discriminant Analysis for Several Normal Populations	170

	6.5.1	Linear discriminant classification	170
	6.5.2	Discriminant classification based on the Mahalanobis squared distances	171
6.6	Case Study: Discriminant Analysis of Vegetable Oil by Near-Infrared Reflectance Spectroscopy		175

Chapter 7. Cluster Analysis — 181

7.1	Introduction		181
7.2	Distance and Similarity Measures		183
	7.2.1	Euclidean distance	183
	7.2.2	Standardized euclidean distance	183
	7.2.3	Manhattan distance (city block distance)	184
	7.2.4	Distance between clusters and linkage method	185
	7.2.5	Similarity	189
7.3	Hierarchical Clustering Method		190
7.4	Nonhierarchical Clustering Method (K-Mean Method)		195
7.5	Cereal Brand Case Study		197

Chapter 8. Mahalanobis Distance and Taguchi Method — 201

8.1	Introduction		201
8.2	Overview of the Mahalanobis-Taguchi System (MTS)		202
	8.2.1	Stage 1: Creation of a baseline Mahalanobis space	203
	8.2.2	Stage 2: Test and analysis of the Mahalanobis measure for abnormal samples	205
	8.2.3	Stage 3 variable screening by using Taguchi orthogonal array experiments	206
	8.2.4	Stage 4: Establish a threshold value (a cutoff MD) based on Taguchi's quality loss function and maintain a multivariate monitoring system	214
8.3	Features of the Mahalanobis-Taguchi System		216
8.4	The Mahalanobis-Taguchi System Case Study		216
	8.4.1	Clutch disc inspection	217
8.5	Comments on the Mahalanobis-Taguchi System by Other Researchers and Proposed Alternative Approaches		221
	8.5.1	Alternative approaches	221

Chapter 9. Path Analysis and the Structural Model — 223

9.1	Introduction		223
9.2	Path Analysis and the Structural Model		225
	9.2.1	How to use the path diagram and structural model	228
9.3	Advantages and Disadvantages of Path Analysis and the Structural Model		235
	9.3.1	Advantages	235
	9.3.2	Disadvantages	236
9.4	Path Analysis Case Studies		237
	9.4.1	Path analysis model relating plastic fuel tank characteristics with its hydrocarbon permeation (Hamade, 1996)	237
	9.4.2	Path analysis of a foundry process (Price and Barth, 1995)	241

Chapter 10. Multivariate Statistical Process Control   243

  10.1 Introduction   243
  10.2 Multivariate Control Charts for Given Targets   245
        10.2.1 Decomposition of the Hotelling T^2 248
  10.3 Two-Phase T^2 Multivariate Control Charts with Subgroups   251
        10.3.1 Reference sample and new observations   251
        10.3.2 Two-phase T^2 multivariate process control for subgroups   255
  10.4 T^2 Control Chart for Individual Observations   259
        10.4.1 Phase I reference sample preparation   260
        10.4.2 Phase II: Process control for new observations   264
  10.5 Principal Component Chart   265

Appendix Probability Distribution Tables   271

References   291

Index   295

Preface

Multivariate statistical methods are a collection of methods and procedures that analyze, interpret, display, and make decisions based on multiple related random variables. Some well-known multivariate statistical methods include principal component analysis, factor analysis, discriminant analysis, cluster analysis, and so on. The foundation of multivariate statistical methods was gradually developed starting from the beginning of the twentieth century. Multivariate statistical methods are playing important roles in psychology, sociology, and many other areas.

In quality engineering and Six Sigma practice, statistical methods are playing important roles. Statistical process control and sampling inspection methods are among the first applied statistical methods developed for quality assurance practice. Statistically designed experiments (design of experiments) and regression analysis are among the most important workhorses in the Six Sigma movement, and their applications contributed a great deal to improving the quality and profitability of many prominent multinational corporations in the world.

However, almost all the popular statistical methods that are used in quality engineering practice and Six Sigma are univariate statistical methods. Currently, the certification requirements of certified quality engineers and certified Six Sigma black belts do not include multivariate statistical methods in the body of knowledge.

On the other hand, with the development of computer information technology and sensor technology, gigabytes upon gigabytes of data are available in business and industry, and most of them are multivariate in nature. For example, in the service industry again, huge amounts of business data are produced each day. Transaction records and customer survey data are multivariate data. Data mining techniques are developed to dig important information from these huge data sources and gain valuable clues to guide sales and promotion efforts. In the chemical and automobile industries, huge amounts of online process data are measured and recorded in real time. In the semiconductor manufacturing industry, because the production process involves multiple layers, many subprocesses, and process steps, a large number of process variables and process testing results are produced continuously. Almost all these data are multivariate in nature. Multivariate statistical methods are very powerful methods that can analyze multivariate data and extract valuable information for decision makers. It is the authors' strong belief that multivariate statistical methods are vastly underutilized methods in quality engineering. They can provide great benefits to quality engineering and Six Sigma if they are used properly.

The major barriers to applying multivariate statistical methods in quality engineering practices and Six Sigma are mostly psychological. For a long time, multivariate statistical methods bore the image of being ultracomplex, ultraclumsy, difficult-to-use, and difficult-to-interpret with regard to analysis results. However, with the development of computer power, most complicated computations of multivariate statistical analysis can be accomplished within a split second. With the advancement of data visualization and animation techniques, presenting the results of multivariate analysis becomes easier and easier.

The Objectives of This Book

There are many books that discuss multivariate statistical methods, but very few books cover material in the context of business and industry. Almost all the examples are related to such areas as psychology and education.

This book is trying to serve as the introductory book on multivariate statistical methods for quality professionals. In this book we are trying to accomplish the following objectives:

1. To provide an in-depth and clear coverage of all important and relevant multivariate statistical methods for quality professionals
2. To discuss the theory and background of each method clearly with examples and illustrations
3. To illustrate how to apply these powerful multivariate statistical methods to solve real world problems by using case studies in business and industry
4. To provide a roadmap on how to integrate multivariate statistical methods in quality assurance practice and Six Sigma projects

Background Needed

The background required to study this book is some familiarity with univariate statistical concepts such as normal distribution, mean, variance and simple data analysis techniques, as well as matrix algebra.

Summary of Chapter Contents

Chapter 1 begins with an overview of multivariate statistical methods and their applications in business and industry. Then it discusses how multivariate statistical methods can benefit quality assurance practice and Six Sigma projects.

Chapter 2 discusses graphical multivariate data display, visualization and stratification. These graphical display techniques can play important roles in communicating multivariate data analysis results and in helping problem solving.

Chapter 3 discusses multivariate random variables, multivariate normal distribution, and sampling properties. We discuss these challenging topics by using comparisons between similar concepts in univariate statistics and multivariate statistics.

Chapter 4 discusses multivariate analysis of variance (MANOVA). We also discuss MANOVA by comparing univariate analysis of variance (ANOVA). An automotive industrial case study is provided to illustrate how MANOVA is used in business and industry.

Chapter 5 provides very detailed discussions of two very important multivariate statistical methods, the principal component analysis (PCA) and factor analysis (FA). We discuss the relevant mathematical backgrounds and numerical computational procedure. Most of the examples are related to business and industry. Several case studies are presented.

Chapter 6 provides a comprehensive coverage of discriminant analysis for various cases. A food industry case study is also provided.

Chapter 7 provides a concise and complete coverage of cluster analysis aided by many graphs and examples.

Chapter 8 discusses the Mahalanobis distance and the Taguchi method. The Taguchi method was developed by Japanese quality expert Genechi Taguchi and his coworkers. Many prominent Japanese and American companies are embracing this method and have developed many case studies. However, this method also gets mixed reviews from many researchers. In this chapter, we also cover the alternative approaches proposed by other researchers.

Chapter 9 presents path analysis and the structural model. This method is relatively unknown to quality professionals. However, it can be very useful in using messy data to establish cause and effect models. Two industrial case studies are also presented.

Chapter 10 discusses multivariate statistical process control. We discuss several of the most frequently used multivariate control charts and their setup procedures. Many illustrative examples are given in this chapter.

What Distinguishes This Book from Others in the Area?

This book's main distinguishing feature is its orientation toward readers in business and industry and toward quality professionals. Specifically, the selection of contents, the writing style, the selection of examples and case studies are all based on the needs of readers in business and industry. There are many books available on multivariate statistical methods but very few of them (if any) are authored toward the needs of quality professionals in business and industry.

Acknowledgments

In preparing this book we received advice and encouragement from several people. For this we are thankful to Dr. G. Taguchi, Dr. K. Murty, Dr. K. Kapur, Dr. Way Kuo, Dr. S. Albin, Dr. O. Mejabi, Dr. H. Pham, and Dr. Rajesh Jugulum. We are appreciative of the help of many individuals. We are very thankful to the efforts of Kenneth McCoombs of McGraw-Hill and Waseem Andrabi and Jane Stark of International Typesetting and Composition. We want to acknowledge and express our gratitude to Robert Frutiger, Joseph Langhauser, Hemanth Munipalli, Yinzhong Jiang, Weimin Xie of General Motors, and Khi-Young Jang of Ford Motor Company for their help and encouragement.

Kai Yang
Jayant Trewn

Multivariate Statistical Methods in Quality Management

Chapter 1

Multivariate Statistical Methods and Quality

1.1 Overview of Multivariate Statistical Methods

The word *multivariate* not only means many variables, but also means that these variables might be correlated. *Statistical method* refers to a range of techniques and procedures for analyzing data, interpreting data, displaying data, and making decisions based on data. Therefore, *multivariate statistical methods* are collections of methods and procedures that analyze, interpret, display, and make decisions based on multivariate data. The foundation of multivariate statistical methods was gradually developed, starting from the beginning of the twentieth century. Multivariate statistical methods are playing important roles in the areas of psychology, sociology, and many others.

Statistical methods are playing important roles in quality engineering practice. Statistical process control and sampling inspection methods are among the first applied statistical methods developed for quality assurance practice. Statistically designed experiments (design of experiments) and regression analysis are among the most important "work horses" in Six Sigma movement and their applications contributed a great deal to improving the quality and profitability of many prominent multinational corporations in the world.

However, almost all the popular statistical methods that are used in quality engineering practice are univariate statistical methods. Very few industrial engineering departments in the United States offer multivariate statistical methods courses. Currently, the certification requirements of certified quality engineers and certified Six Sigma black belts do not include multivariate statistical methods in the body of knowledge.

On the other hand, with the development of sensor technology and computer information technology, gigabytes upon gigabytes of data are available in business and industry, and most of these data are multivariate in nature. For example, in the chemical industry, after 1970, due to the growing applications of spectroscopy and chromatography, huge amount of online measurement data or signals on chemical processes are produced in real time. These data provide real-time information about process variables such as temperature, pressure, chemical and physical properties of the semifinished product or incoming materials, etc. A near infrared reflectance spectrophotometer usually produces 1050 variables in the wavelength range of 400 to 2500 nm on a single scan. A new research area, called chemometrics, has been formed to provide tools to make good use of measured data (Workman et al., 1996), enabling practitioners to make sense of measurements and to quantitatively model and produce visual representations of information. As the production process in the semiconductor manufacturing industry is featured by multiple layers, many subprocesses and process steps, large numbers of process variables and process testing results are produced continuously, and these variables and results are definitely multivariate in nature. In the service industry, again, huge amounts of business data are produced each day. Transaction records and customer survey data are multivariate data. Data mining techniques (Berry and Linoff, 2000; Edelstein, 1999) are developed to dig out important information from these huge data sources and gain valuable clues to guide sales and promotion efforts. Both data mining and chemometrics involve extensive use of multivariate statistical methods.

The major barriers in the application of multivariate statistical methods to quality engineering practices are mostly psychological. For a long time, multivariate statistical methods were seen as ultracomplex, ultraclumsy, difficult-to-use, and difficult-to-interpret with regard to analysis results. However, with the development of computer power, most of the complicated computation of multivariate statistical analysis can be accomplished within a split second. With the advancement of data visualization and animation technique, presenting the results of multivariate analysis becomes easier and easier. Multivariate statistical methods are very powerful methods. For example, some of the multivariate statistical methods, such as principal component analysis or cluster analysis, are very powerful "data reduction" and "data stratification" techniques, and they have some distinct features that univariate statistical methods do not possess. It is the authors' strong belief that multivariate statistical methods are vastly underutilized methods in quality engineering. They can provide great benefits to quality engineering if they are used properly.

Before we discuss the roles of multivariate statistical methods in quality engineering practice, we provide a brief overview of the multivariate statistical methods that are going to be discussed in this book.

1.1.1 Graphical multivariate data display and data stratification

It is well known that "a picture is worth a thousand words." Actually, it is even more so that "a picture is worth thousands of data." The bulk of raw multivariate data or data analysis results can easily overwhelm users. Graphical multivariate data display includes many methods, ranging from simple charts, graphs, and graphical templates to sophisticated computer graphics based data visualization and animation (Post et al., 2003; Rosenblum et al., 1994; Fortner, 1995). However, the overall objective of graphical data display is to create visual images so that either the raw multivariate data or the data analysis results can be readily understood by the users. Graphical data display can also help communicate raw data or analysis results to people with different backgrounds.

Graphical data stratification is a "graphical analysis of variance." One of the graphical data stratification methods is *multi-vari chart* (Seder, 1950; Breyfogle, 2003). In this method, multivariate data are displayed graphically. The variation in multivariate data sets are partitioned according to different sources such as time related, batch related, part-to-part related, group-to-group related, etc. The major contributor(s) to the variation could provide clues for finding the root cause for the excessive variation. The multi-vari chart is one of the major root cause analysis tools in the Six Sigma approach in quality improvement.

Graphical multivariate data display and data stratification will be discussed in detail in Chap. 2.

1.1.2 Multivariate normal distribution and multivariate sampling distribution

In univariate statistics, normal distribution is one of the most important distributions, especially in quality engineering application. For multiple (and possibly correlated) random variables, multivariate normal distribution can also be defined. There are similarities and differences between univariate normal distribution and multivariate normal distribution. Here we list a few comparisons:

1. A univariate normal random variable is a single variable X; multivariate normal random variables are expressed by $\mathbf{X} = (X_1, X_2, ..., X_p)^T$, which is a vector.

2. In univariate normal distribution, the mean (or expected value) is a scalar μ; for multivariate normal distribution, the mean is a vector $\boldsymbol{\mu}$.
3. In univariate normal distribution, the variance is a scalar σ^2. It is well known that σ^2 is a measure of dispersion or variation. For multivariate normal distribution, the measure of dispersion is a matrix, called covariance matrix Σ. Actually, the covariance matrix not only gives the measure of dispersion for all variables $X_1, X_2,..., X_p$, but also gives the mutual variation relationship among variables. For example, if X_1 increases by one standard deviation, how much change will $X_2,...,X_p$ have?

Actually, the above three comparisons will hold true for all distributions, not just for normal distributions.

4. When analyzing samples from univariate normal population, many sampling distributions, such as the t distribution, are developed. In multivariate normal distribution, similar sampling distributions, such as the T^2 distribution, are also developed. In univariate statistical hypothesis testing, we can use the t statistic to test the equality of means, and in multivariate statistical hypothesis testing, we can use the T^2 statistic to test the equality of mean vectors. The T^2 statistic and other measures, such as Mahalanobis squared distance (similar to univariate z score or t score), are playing important roles in many other multivariate statistical methods.

Multivariate normal distribution and multivariate sampling distributions are discussed in detail in Chap. 3.

1.1.3 Multivariate analysis of variance

In univariate statistical methods, analysis of variance (ANOVA) is a general technique that can be used to test the hypothesis that the means among two or more groups are equal, assuming that the sampled populations are normally distributed.

In an analysis of variance, the variation in the response measurements is partitioned into components that correspond to different sources of variation. The aim of this procedure is to split the total variation in the data into a portion due to random error and portions due to changes in the values of the independent variable(s).

If we have a set of multiple random variables, univariate analysis of variance will become a multivariate analysis of variance (MNOVA). It is a general technique that can be used to test the hypothesis that the mean *vectors* among two or more groups are equal, under the assumption that the sampled populations are multivariate normally distributed.

In MNOVA, the variation in the response measurements is also partitioned into components that correspond to different sources of variation. The computation in MNOVA involves a lot of matrix computation and different hypothesis testing procedure, other than the F test used in univariate analysis of variance.

1.1.4 Principal component analysis and factor analysis

Principal component analysis (PCA) and factor analysis (FA) are unique multivariate statistical methods. There is nothing like these methods in univariate statistical methods.

Principal component analysis and factor analysis are multivariate statistical methods that can be used for data reduction. By *data reduction* we mean that these methods are very powerful in extracting a small number of hidden factors from a massive amount of multivariate data, and these factors account for most of the variation in the data. The structures of these hidden factors can help a great deal in searching the root causes for the variation. For example, in automobile assembly operation, the dimensional accuracies of body panels are very important in "fit and finish" for automobiles. Many sensors, such as optical coordinate measurement machines (OCMM), are used to measure dozens of actual dimensional locations, for every automobile, at selected key points (see Fig. 1.1). These measurements are definitely multivariate data, and they are correlated to each other because the body structure is

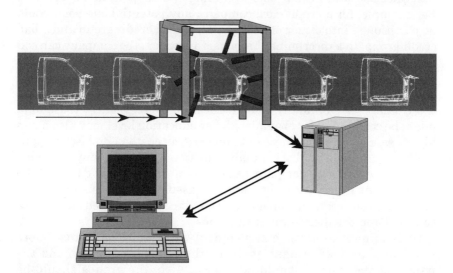

Figure 1.1 Automobile body subassembly dimensional measurement process.

physically connected and the dimensional variation of one location often affects the dimensional variation of other locations.

By looking at dimensional measurement raw data, you will find that there will be piece-to-piece variation at each measurement point. If there are 50 measurement locations, the raw data will overwhelm any operator or process engineer. If we do some simple data analysis, such as mean and standard deviation of each point, we will have 50 pairs of means and standard deviations. We will know how bad the dimensional variation is, but by using only this kind of information it is really difficult to find the root cause of the variation. By using principal component analysis, we may find that most of the variations (usually more than 80 percent) at these 50 locations are caused by a small number (maybe 2 to 3) of principal components. Each principal component may display a special feature such as "metals are twisting"; "subassemblies are varying around an axis" (maybe a locating pin). These features can provide valuable clues about what causes the variation. If we can successfully find the root causes of the features or symptoms indicated by these principal components and take actions to eliminate the root causes, then we can eliminate a big percentage of variation, and the quality of production will be greatly improved.

Principal component analysis and factor analysis will be discussed in detail in Chap. 5.

1.1.5 Discriminant analysis

Discriminant analysis is a powerful multivariate statistical method that can classify individual objects or units into uniquely defined groups. For example, for a credit card company, any potential customer could be partitioned into either of two groups, a good credit group and a bad credit group. In a discriminant analysis approach, the company can collect enough data on existing customers. Some of the existing customers belong to the good credit group and some of them belong to the bad credit group. The multivariate data collected may include educational level, salary, size of family, indebtedness, age, etc. Clearly, these data might be correlated; for example, higher education level may also mean higher salary. By using these data on existing customers, discriminant analysis can develop discriminant functions. Now, for any new customer, the credit card company can collect the same set of data and the discriminant function can be used to classify the new customer as a potential good credit customer or bad credit customer based on the data collected and the discriminant functions.

Discriminant analysis can also deal with multiple groups (more than two). In discriminant analysis, the number of groups and the names of the groups should be known. Also, the groups should be exclusive and able to partition the whole population. For example, if we partition the people into three categories—underweight, normal weight, and overweight—then the number of groups and name of groups are known. These three groups are also mutually exclusive (you cannot be both overweight and normal weight), and are able to partition the whole population (everyone can only belong to one of the three groups; there is no room for a fourth group).

Discriminant analysis is also a unique multivariate statistical method. There is no similar method in univariate statistics. Actually, the cluster analysis that we will be discussing next is also a unique multivariate statistical method.

Discriminant analysis will be discussed in detail in Chap. 6.

1.1.6 Cluster analysis

Cluster analysis is a multivariate statistical method that identifies hidden groups in a large number of objects based on their characteristics. Similar to discriminant analysis, each object has multiple characteristics, the values of which vary from object to object. The primary objective of cluster analysis is to identify similar objects by the characteristics they possess. Cluster analysis clusters similar objects into groups so that the objects within a group are very similar and objects from different groups are significantly different in their characteristics. Unlike discriminant analysis, in which the number of groups and group names are known before the analysis, in cluster analysis the number of groups and group features are unknown before the analysis and are determined in the analysis.

In taxonomy, cluster analysis is used to partition similar animals (or species) into groups based on their characteristics. Unlike discriminant analysis, the number of groups, the name and definition of each group for animals are not predetermined. They are determined during the analysis and after the analysis is done. In marketing analysis, cluster analysis is used to partition the customers into several groups based on customer data such as education level, salary, past purchasing history, size of family, etc. For each group, different marketing strategy will be used depending on the features of each group.

Cluster analysis will be discussed in Chap. 7.

1.1.7 Mahalanobis Taguchi system (MTS)

Mahalanobis Taguchi system (Taguchi et al., 2001; Taguchi and Jugulum, 2002) is a multivariate data based pattern recognition and diagnosis system promoted by the Japanese quality expert Genichi Taguchi. In this approach, a large number of multivariate data sets are collected. Then the data sets for a "healthy" or normal group and other data sets for an abnormal or unhealthy group are partitioned. For example, in liver disease diagnosis, for each potential patient, many kinds of medical tests are conducted. Therefore, for each patient, a complete set of test records is a multivariate data set. In the MTS approach, the test records of a large number of people are collected. The people who are known to be healthy are called healthy groups and their test data sets are used to create a baseline metric for the healthy population. This baseline metric for the healthy population is based on the Mahalanobis distance. As we previously mentioned, the role of the Mahalanobis distance in multivariate statistics is similar to z score or t score in univariate statistics. In MTS, this Mahalanobis distance is scaled so that the average length for the Mahalanobis distance for the known healthy group is approximately equal to 1, which Taguchi called a "unit space." For the objects in the abnormal group, the scaled Mahalanobis distances are supposed to be significantly larger than 1. In the MTS approach, it is recommended that people should start with a large number of variables in each multivariate data set so that the chance for the multivariate data set to contain important variables is good. The important variables in this case are such that they make the scaled Mahalanobis distance large for abnormal objects. Then Taguchi's orthogonal array experiment is used to screen this initial variable set for important variables. After the orthogonal experiment, a smaller set of important variables are selected as the monitoring variables to detect future abnormal conditions.

Unlike discriminant analysis, MTS does not assume that "abnormal group" is a separate population because Dr. Taguchi thinks that each abnormal situation is a different case. No particular probability distribution is used to determine the cutoff line dividing normal and abnormal conditions. Partially because of these assumptions, the Mahalanobis Taguchi system gets mixed reviews from researchers (Woodall 2003, Kenett 2003).

1.1.8 Path analysis and structural model

In many quality improvement projects, the linear regression method and the design of experiment method are very important tools to model relationships between dependent variables and independent variables. In industrial applications, the dependent variables are often key products or process performance characteristics. The independent variables are often design parameters or process variables.

Linear regression models, such as the one specified by Eq. (1.1), are often used as the empirical relationship between dependent variables and independent variables.

$$Y = \beta_1 X_1 + \beta_2 X_2 + \beta_3 X_3 + \varepsilon \qquad (1.1)$$

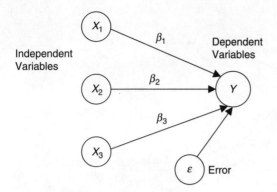

Figure 1.2 Path diagram for linear regression.

Equation (1.1) is a special case of linear regression, in which there is one dependent variable Y, and three independent variables X_1, X_2, and X_3.

We can use the path diagram shown in Fig. 1.2 to represent the linear regression (1.1).

In the path diagram, the value by the arc means coefficient to be multiplied with. For example, if the circle before the arc is X_1, the value by the arc is β_1, then at the end of arc is $\beta_1 X_1$. However, one of the major weaknesses in linear regression is its limitations on multicolinearity, that is, the potential high correlations among independent variables and dependent variables.

Path analysis with a structural model is an alternative multivariate model, which can deal with correlated independent variables and correlated dependent variables. In many industrial processes, it is very easy to collect large amount of process data. Some of these data are related to independent variables and some to dependent variables. Traditional linear regression models will have lots of difficulties in dealing with this kind of data and establishing good mathematical models relating independent variables with dependent variables. Path analysis and structural model can work very well in this situation. Figure 1.3 shows an example of a path diagram in a structural model.

In Fig. 1.3, there are two dependent variables, Y_1 and Y_2, and there are three independent variables, X_1, X_2, and X_3. In this case, X_1, X_2, and X_3 are all correlated, so in the path diagram, they are connected by double arrows. Y_1 and Y_2 are also mutually related, so they are connected by a double arrow.

The structural model approach is able to design and estimate this kind of mathematical model with clear interpretations relating independent variables with dependent variables despite the presence of correlations.

Path analysis and structural model are discussed in Chap. 9.

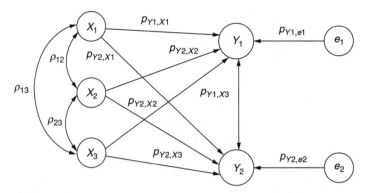

Figure 1.3 Path diagram for three independent variables and two dependent variables.

1.1.9 Multivariate process control

Statistical process control charts are well developed for univariate cases. By univariate case we mean treating one variable at a time. However, in many real industrial situations, the variables to be controlled in the process are multivariate in nature. For example, in an automobile body assembly operation, the dimensions of subassemblies and body-in-white are multivariate and highly correlated. In the chemical industry (Mason and Young, 2002), many process variables such as temperature, pressure, and concentration are also multivariate and highly correlated.

A multivariate control chart that is based on the T^2 statistic, which is a multivariate statistic similar to the t statistic in the univariate case, can be used to develop control charts for process control in multivariate cases.

Multivariate process control is discussed in Chap. 10.

In Sec. 1.2, we will review the current successful stories in applying multivariate statistical methods in business and industry. Section 1.3 gives an overview of quality engineering and discusses the possible roles of multivariate statistical methods in quality engineering. Section 1.4 briefly gives an overview of the most popular quality improvement strategy in business world today, Six Sigma, and discusses the possible roles of multivariate statistical methods in Six Sigma.

1.2 Applications of Multivariate Statistical Methods in Business and Industry

Though the quantity of literature available on applying multivariate statistical methods in quality engineering area is limited, multivariate statistical methods have been playing important roles in many areas of business and industrial practice. The most noticeable include data mining, chemometrics, variation reduction in automobile manufacturing process, Mahalanobis Taguchi system applications, and multivariate process control in chemical industry. In this section, brief reviews will be given to applications in these areas.

1.2.1 Data mining

Data mining is a process of analyzing data and summarizing it into useful information. In business application of data mining, useful information derived from data mining can be used to increase revenue, cut costs, or both. Data mining is primarily used by companies with strong customer focus such as retail, financial, communication, and marketing organizations. It enables these companies to determine relationships among internal factors such as price, product positioning, or staff skills, and external factors such as economic indicators, competition, and customer demographics. Data mining enables companies to determine the impact of these factors on sales, customer satisfaction, and corporate profitability, and to develop marketing and sales strategy to enhance corporate performance and cut down losses.

Many organizations are using data mining to help manage all phases of the customer life cycle, including acquiring new customers, increasing revenue from existing customers, and retaining good customers. For example, Blockbuster Entertainment mines its video rental history database to recommend rentals to individual customers. American Express can suggest products to its cardholders based on analysis of their monthly expenditure. Telecommunications and credit card companies have been applying data mining to detect fraudulent uses of their services.

The goals of data mining include prediction, often called predictive data mining. In this case, data mining uses a variety of data analysis tools to discover patterns and relationships in data that are used to make valid predictions. For example, a credit card company may use the data set from known cheaters to derive a model or set of models that can quickly identify transactions which have a high probability of being fraudulent.

The goal of data mining could also be data reduction. Data reduction is used to aggregate or amalgamate the information contained in large data sets into manageable (smaller) information nuggets. Data reduction methods can include simple tabulation, aggregation (computing descriptive statistics), or more sophisticated techniques such as cluster analysis and principal component analysis.

The technological basis for data mining is based on two fields, the field of artificial intelligence and the field of statistics. Many multivariate statistical methods, notably cluster analysis, discriminant analysis, and principal component analysis, as well as data visualization techniques, are playing very important roles in data mining.

1.2.2 Chemometrics

Chemometrics is the branch of chemistry that provides tools to make good use of measured data, enabling practitioners to make sense of measurements and to quantitatively model and produce visual representations of information. Chemometrics also provides a means of collecting relevant information through experimental designs. Therefore, chemometrics can be defined as the information aspect of chemistry.

Around 1970, chemical analysis and instrumentation changed from wet chemistry (e.g., precipitation and specific reagents), which provides direct information (e.g., color change and precipitation), to instrumental chemistry (e.g., spectroscopy and chromatography), which provides information in the form of numbers, signals, and data, often in large amounts. For example, a near infrared reflectance spectrophotometer usually produces 1050 variables in the wavelength range 400 to 2500 nm on a single scan. The traditional tools for analyzing chemical data were not well suited to this "data explosion" and to the type of data the instrument produced. The realization that the unused information in these data masses may be substantial and could be handled using computers was the launch pad for the field of chemometrics. Using methods from electrical engineering and psychology, chemists were able to interpret data sets considered large at the time, with 20 to 50 variables (measurements, properties) and 20 to 200 observations (cases, samples, compounds). Early chemometric methods tended towards classification and discriminant analysis. However, the demand for quantitative modeling leads to the inclusion of regression methods, thus allowing the development of quantitative prediction models.

Thirty years later (in 2004), chemometrics has grown into a well-established data analysis tool in areas such as multivariate calibration, quantitative structure-activity modeling, pattern recognition, and multivariate statistical process monitoring and control.

The goals of chemometrics include improving and controlling the chemical process and assisting in chemical product design. In chemical product design, multivariate statistical methods and response surface experimental design techniques are used to help in developing new molecular structures. This practice is called statistical molecular design.

The technological basis for chemometrics is again based on two fields—artificial intelligence and statistics. In the field of statistics, experimental design technique and multivariate statistical methods are very widely used in chemometrics. Cluster analysis, discriminant analysis, principal component analysis, and factor analysis are playing very important roles in chemometrics. In a broader sense, we can say that chemometrics is data mining in the chemical process industry.

1.2.3 Other applications

Multivariate statistical methods, notably principal component analysis, correlation analysis, factor analysis, and discriminant analysis, have been used in automobile body assembly dimensional variation reduction practice. The objective for variation reduction is to reduce the variation in dimension for the automobile assembly process and to improve "fit and finish" for automobile body panels (Hu and Wu, 1990, 1992; Roan et al., 1993; Ceglarek, 1994; Yang, 1996). Data mining techniques are also used in improving and controlling the semiconductor manufacturing process (Forrest, 2002).

1.3 Overview of Quality Assurance and Possible Roles of Multivariate Statistical Methods

To deliver quality to a product or service, we need a system of methods and activities, called quality assurance. Quality assurance is defined as all the planned and systematic activities implemented within the quality system that can be demonstrated to provide confidence that a product or service will fulfill requirements for quality.

Because quality is a way of doing business, it must be related to a specific product or service. For any product and service, its life span includes its creation, development, usage, and disposal. We call this whole life span the product/service life cycle. A good quality assurance program should act on all stages of the life cycle. Figure 1.4 illustrates a typical product/service life cycle (Yang and El-Haik, 2003).

The earlier stages of the cycle are often called "upstream" and the later stages are often called "downstream." We will briefly review each stage of the cycle and find what are the possible roles of multivariate statistical methods in supporting quality assurance at each stage.

1.3.1 Stage 0: Impetus/ideation

The product or service life cycle begins with impetus/ideation. The impetus for a new product/service could be the discovery of a new technology, such as the invention of semiconductors, with or without a clear idea in advance as to how it might be commercialized. Some form of market research can identify a great market opportunity, an obvious need to retire an existing product that has been eclipsed by competition (as in the annual redesign of automobile models), or a new idea using existing technologies to sell books via internet. Once the impetus is identified and it is determined that a viable product/service can be subsequently developed, the ideation phase follows. The ideation phase focuses on stating the possible product/service and setting a general direction, including identifying plausible options for a new product/service.

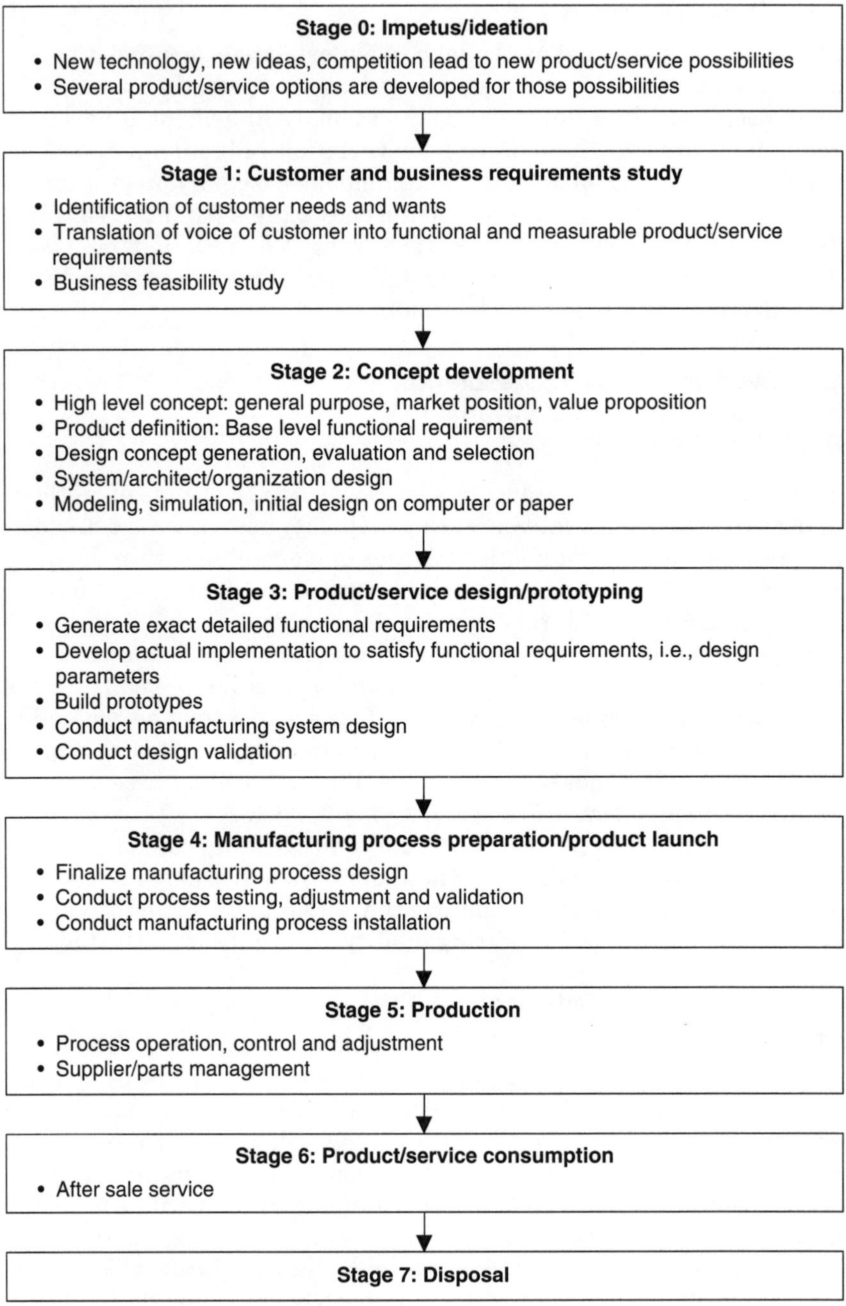

Stage 0: Impetus/ideation
- New technology, new ideas, competition lead to new product/service possibilities
- Several product/service options are developed for those possibilities

Stage 1: Customer and business requirements study
- Identification of customer needs and wants
- Translation of voice of customer into functional and measurable product/service requirements
- Business feasibility study

Stage 2: Concept development
- High level concept: general purpose, market position, value proposition
- Product definition: Base level functional requirement
- Design concept generation, evaluation and selection
- System/architect/organization design
- Modeling, simulation, initial design on computer or paper

Stage 3: Product/service design/prototyping
- Generate exact detailed functional requirements
- Develop actual implementation to satisfy functional requirements, i.e., design parameters
- Build prototypes
- Conduct manufacturing system design
- Conduct design validation

Stage 4: Manufacturing process preparation/product launch
- Finalize manufacturing process design
- Conduct process testing, adjustment and validation
- Conduct manufacturing process installation

Stage 5: Production
- Process operation, control and adjustment
- Supplier/parts management

Stage 6: Product/service consumption
- After sale service

Stage 7: Disposal

Figure 1.4 A typical product/service life cycle (Stage 0–5: Product/service development cycle).

There are several keys to the success of this phase. Two very important ones are the lead time to discover the possible new product/service idea and determine its viability, and the correct assessment of market acceptance of the new product/service.

Multivariate statistical methods can provide valuable support at this stage. Data mining study on the marketplace, similar products, and the service database may provide crucial information for the success of the new product or service.

1.3.2 Stage 1: Customer and business requirements study

Customer and business requirements study is the first stage. During both initial concept development and product definition stages, customer research, feasibility studies, and cost/value research should be performed. The purpose of customer research is to develop the key functional elements which will satisfy potential customers and therefore eventually succeed in the market. The purpose of a feasibility study and cost/value study is to ensure that the new product/service is competitive in the future market.

As we discussed in the last section, data mining is an excellent tool to dig out valuable information about customer needs, market segments, and market acceptability of the new product. It can provide tremendous help at this stage.

1.3.3 Stage 2: Concept development

Product/service concept development is the second stage. This stage starts with the initial concept development phase. It involves converting one or more options developed in the previous stage into a high-level product concept, describing the product's purpose, its general use, and its value proposition. Next is the product definition phase. It clarifies product requirements, which are the base-level functional elements necessary for the product to deliver its intended results.

Multivariate statistical method can also provide help at this stage. In the chemical industry, multivariate statistical methods, together with experimental design methods, provide support for chemical product design. This practice is called statistical molecular design. Data mining can also be used to procure good information from the warranty database to provide valuable inputs for designs.

1.3.4 Stage 3: Product/service design/prototyping

The third stage is product design/prototyping. In this stage, product/service scenarios are modeled and design principles are applied to generate exact detailed functional requirements and their actual implementation and design parameters. For product design, the design parameters could be dimension, material properties, and part specifications. For service design, the design parameters could be detailed organization layout and specifications. The design parameters should be able to provide all the detail necessary to begin construction or production. For product development, after product design, prototypes are built to test and validate the design. If the test results are not satisfactory, the designs are often revised. Sometimes, this build-test-fix cycle is iterated until satisfactory results are achieved. During this stage, manufacturing system design for the product is also produced to ensure that the product can be manufactured economically.

Multivariate statistical method can also provide support at this stage. In the product design stage, though computer aided engineering can provide valuable support in design analysis, as the computational power is increasing, it still takes a lot of time and money to run analytical models. Principal component analysis can help to achieve data reduction at this stage (Yang and Younis, 2003).

1.3.5 Stage 4: Manufacturing process preparation/product launch

The fourth stage is the manufacturing process preparation/product launch. During this stage, the manufacturing process design will be finalized. The process will undergo testing and adjustment and so there is another set of build-test-fix cycles for the manufacturing process. After iterations of cycles, the manufacturing process will be validated and accepted and installed for production.

Multivariate statistical methods can provide great help at this stage. Graphical data display and visualization and multi-vari charts have proved to be valuable tools for trouble-shooting (Breyfogle, 2003). Principal component analysis, factor analysis, and discriminant analysis can also provide great help in root cause analysis (Hu and Wu, 1990, 1992; Ceglarek, 1994; Yang, 1996).

1.3.6 Stage 5: Production

The fifth stage is the full scale production. In this stage, the product will be produced and shipped to the market. Some parts or subassemblies might be produced by suppliers. During production, it is very important that the manufacturing process is able to function consistently and free of defect and that all parts and subassemblies supplied by suppliers are consistent with quality requirements.

For quality assurance at this stage, the key task is to ensure that the final product is in conformance with product requirements. That is, all products, their parts, and subassemblies should conform to their design requirements; they should be interchangeable and consistent. The quality methods used in this stage include statistical process control (SPC), troubleshooting, and diagnostic methods.

Multivariate process control, the Mahalanobis Taguchi system, as well as root cause analysis tools based on graphical multivariate data display, multi-vari chart, principal component analysis, factor analysis, and discriminant analysis can all make valuable contributions at this stage. The monitoring and analysis of the production process can also be aided by path analysis and structural modeling.

The combined activities from stage 1 to stage 5 are also called the product development cycle.

1.3.7 Stage 6: Product/service consumption

The sixth stage is product consumption and service. During this stage, the products are consumed by customers. This stage is the most important to the consumers because it is the consumer who will form the opinion of the product and brand name. When customers encounter problems in using the product during consumption, such as defects, warranty and service are important to keep product in use and keep the customer satisfied.

For quality assurance at this stage, it is impossible to improve the quality level for the products already in use because they are already out of the hands of the producer. However, a good warranty and service program will certainly help to keep the product in use by repairing the defective units and providing other after sales services. The warranty and service program can also provide valuable information for the quality improvement of future production and product design.

Data mining on the customer survey and warranty database can provide a lot of decision support for this stage.

1.3.8 Stage 7: Disposal

The seventh stage is product disposal. With increasing concern over the environment, this stage is getting more and more attention. Once a product has been in the market for a while, a variety of techniques can be used to assess if the product is measuring up to expectations or if opportunities exist to take the product in new directions. Executives and product managers can then determine whether to stay put, perform minor design refinements, commence a major renovation or move forward to ideation and beginning the cycle for a new product. The ability to determine the right time to make the leap from an old product to a new one is an important skill. Data mining again will provide valuable decision support.

Table 1.1 summarizes the stages of product/service life cycles, quality assurance tasks for each stage, and the possible roles of multivariate statistical methods.

1.4 Overview of Six Sigma and Possible Roles of Multivariate Statistical Methods

Six Sigma is a methodology that provides businesses with the tools to improve the capability of their business processes (Pande et al., 2000). For Six Sigma, a process is the basic unit for improvement. A process could be a product or a service process that a company provides to outside customers, or it could also be an internal process within the company such as a billing process, a production process, etc. In Six Sigma, the purpose of process improvement is to increase the process's performance and decrease its performance variation. This increase in performance and decrease in process variation will lead to defect reduction and improvement in profits, employee morale, and quality of product, and eventually to business excellence.

Six Sigma is the fastest growing business management system in industry today. It has been credited with saving billions of dollars for companies over the past 10 years. Developed by Motorola in the mid-1980s, the methodology became well known only after Jack Welch at GE made it a central focus of his business strategy in 1995.

The name "Six Sigma" came from a statistical terminology, "sigma," or σ, meaning "standard deviation." For a normal distribution, the probability of falling within a ± 6 sigma range around the mean is 0.9999966. In a production process, the "Six Sigma standard" means that the process will produce defective products at the rate of 3.4 defects per million units. Clearly, Six Sigma indicates a degree of extremely high consistency and extremely low variability. In statistical terms, the purpose of Six Sigma is to reduce variation and thus to achieve very small standard deviations.

Six Sigma is a strategy that combines organizational support, professional training, and a system of statistical quality improvement methods. Six Sigma takes on problems on a project-to-project basis. The goal for any Six Sigma project is to enable the process to accomplish all key requirements with a high degree of consistency.

In a Six Sigma project, if the Six Sigma team selects the regular Six Sigma process improvement strategy, then a five stage process will be used to improve an existing process. These five stages are the following:

- Define the problem and customer requirements
- Measure the defects and process operation
- Analyze the data and discover causes of the problem
- Improve the process to remove causes of defects
- Control the process to make sure defects do not recur

TABLE 1.1 Product Life Cycle and Possible Roles of Multivariate Statistical Methods

Product/service life cycle stages	Quality assurance tasks	Multivariate statistical support
0. Impetus/ideation	• Ensure new technology/ideas are robust for downstream development	• Data mining
1. Customer and business requirements study	• Ensure new product/service concept has right functional requirements which satisfy customer needs	• Data mining
2. Concept development	• Ensure the new concept can lead to sound design and be free of design vulnerabilities • Ensure the new concept is robust for downstream development	• Data mining • Statistical molecular design for chemical product • Data reduction • Warranty data mining
3. Product/service design/prototyping	• Ensure designed product (design parameters) delivers desired product functions over its useful life • Ensure the product design is robust for variations from manufacturing, consumption and disposal stages	• Data reduction • Warranty data mining
4. Manufacturing process preparation/product launch	• Ensure the manufacturing process is able to deliver the designed product consistently	• Graphical data display • Multi-vari chart • Principal component analysis, factor analysis, discriminant analysis • Mahanalobis Taguchi system
5. Production	• Produce designed product with high degree of consistency and free of defect	• Multivariate process control • Graphical data display • Multi-vari chart • Principal component analysis, factor analysis, discriminant analysis • Path analysis and structural modeling • Mahanalobis Taguchi system
6. Product/service consumption	• Ensure customer has a satisfactory experience of consumption	• Data mining
7. Disposal	• Ensure customer is troublefree in disposing off the used product/service	• Data mining

This five step strategy is also called DMAIC. We will briefly describe what are the five steps.

1.4.1 Stage 1: Define the project and customer requirements (D or define step)

We need to launch a Six Sigma process improvement project when the process under improvement usually does not perform satisfactorily. At least we believe this process has a lot of room for improvement. Usually the define stage can be carried out in the following three steps:

Step 1. **Draft project charter.**

Step 2. **Identify and document the process.**

1. *Identify the process.* In a Six Sigma process improvement project, usually a team works on one process at a time. The process being identified is usually
   - A core process in the company such as product development, marketing, or customer service; so it is a very important process for the company
   - A support process such as human resource and information system because this process becomes a bottleneck or a waste center of the company
2. *Document the process.* After the process is identified, an appropriate flowchart-based process model will be used to model and analyze the process. After the process model is determined, major elements of the process model, suppliers, inputs, process map, process output, and customers should be defined.

Step 3. Identify, analyze, and prioritize customer requirements

1. *Identify customer requirements.* There are two kinds of customer requirements.
   - *Output requirements.* These are the features of the final product and service delivered to the customer at the end of the process.
   - *Service requirements.* These are the more subjective ways in which the customer expects to be treated and served during the process itself. Service requirements are usually difficult to define precisely.
2. *Analyze and Prioritize customer requirements.* The list of requirements can be long but not all requirements are equal in the eyes of the customer. We need to analyze and prioritize those requirements.

Possible roles for multivariate statistical methods. From the description of the "define" stage, it is clear that customer requirement research is very important in defining the objective of the Six Sigma project.

Clearly, data mining can provide valuable inputs for this stage, especially for Six Sigma projects of service industries. Unfortunately, in most of the Six Sigma articles and books, data mining has been totally ignored.

1.4.2 Stage 2: Measuring process performance

Measure is a very important step. This step involves collection of data to evaluate the current performance level of the process, and provide information for the analysis and improvement stages.

This stage usually includes the following steps:

1. *Select what to measure.* Usually, we measure the following:
   - *Input measures*
   - *Output measures*
   - *Data stratification.* This means that together with the collection of output measures, **Y**, we need to collect corresponding information about the variables which may have cause-and-effect relationship with **Y**, that is **X**. Sometimes, we do not know what **X** is and then we may collect other information which may relate to **X**, such as stratification factors, region, time, and unit. By analyzing the difference in performance levels at different stratification factors, we might be able to locate the critical **X** which may influence **Y**.
2. Develop data collection plan.
   - We will determine such issues as sampling frequency, who is going to do the measurement, format of data collection form, and measurement instruments.
3. Calculate process capability.

Possible roles for multivariate statistical methods. From the simple multivariate methods such as graphical multivariate data display, multi-vari chart to more sophisticated ones such as cluster analysis, principal component analysis or data mining can help a great deal in data stratification. Unfortunately, in most of the Six Sigma literature and books, only very simple and very basic techniques, such as histograms and scatter plots, and at most multi-vari charts, are discussed. Data reduction and data mining techniques can add much more to this stage.

1.4.3 Stage 3: Analyze data and discover causes of the problem

After data collection, we need to analyze the data and process in order to find how to improve the process. There are two main tasks at this stage:

Data analysis. Using collected data to find patterns, trends and other differences that could suggest, support, or reject theories about the cause and effect, the methods frequently used include the following:

Root cause analysis
Cause-effect diagram
Failure modes and effects analysis (FMEA)
Pareto chart
Validate root cause
Design of experiment
Shainin method

Process analysis. A detailed look at existing key processes that supply customer requirements in order to identify cycle time, rework, downtime, and other steps that do not add value for the customer.

Possible roles for multivariate statistical methods. Again, from simple graphical tools such as graphical multivariate data display or multi-vari chart to more sophisticated multivariate statistical methods based on root cause analysis tools such as principal component analysis, factor analysis or discriminant analysis can enhance the arsenals in root cause analysis to a totally new level. Linear regression and analysis of variance are major statistical tools in the "analyze" stage in current Six Sigma practice. It is well known that these methods do not work well for correlated X and Y variables. Path analysis and structural model can provide great help on this.

1.4.4 Stage 4: Improve the process

We should have identified the root causes for the process performance problem after we went through stage 3. If the root causes of process performance problem are identified by process analysis, the solutions are often found by process simplification, parallel processing, bottleneck elimination, etc.

Possible roles for multivariate statistical methods. In current Six Sigma practice, design of experiments (DOE), response surface methods (RSM), and Taguchi's robust engineering are major tools for this stage. As we discussed earlier, in chemometrics, multivariate statistical methods are used side by side with DOE and RSM to improve chemical product design. Similar approaches should also be developed to achieve improvement in other industries.

TABLE 1.2 Six Sigma Project Stages and Applicable Multivariate Statistical Tools

Six Sigma project stages	Applicable multivariate statistical tools
Define	Data mining
Measure	Graphical multivariate data display Multi-vari chart Data reduction
Analyze	Graphical multivariate data display Multi-vari chart Data reduction Principal component analysis, factor analysis, discriminant analysis, cluster analysis Path analysis and structural model
Improve	Data reduction Path analysis and structural model
Control	Multivariate process control Mahalanobis Taguchi system

1.4.5 Stage 5: Control the process

The purpose of this stage is to hold on to the improvement achieved from the last stage. We need to document the change made in the improvement stage. If the improvement is made by process management methods such as process simplification, we need to establish new process standards. If the improvement is made by eliminating the root causes of low performance, we need to keep track of process performance after improvement and control the critical variables relating to the performance by using control charts.

Possible roles for multivariate statistical methods. Multivariate process control and Mahalanobis Taguchi system can be useful at this stage. In many production/service processes, the variables that need to be controlled are inherently multivariate and correlated.

Table 1.2 illustrates the five stages of Six Sigma project and applicable multivariate statistical tools in each stage.

Chapter 2

Graphical Multivariate Data Display and Data Stratification

2.1 Introduction

Multivariate data sets are readily available in many industries such as manufacturing, service, and business administration. For example, in an automobile body assembly operation, the optical coordinate measurement machine can simultaneously measure dozens of dimensional locations on each automobile in the assembly line. In a process industry, many process variables such as temperature and flow rate, are constantly monitored by sensors and enormous amounts of multivariate data are generated. On the other hand, multivariate statistical methods are very powerful analytical methods, especially with the development of computing power. Multivariate statistical methods are able to analyze huge amounts of data and extract valuable information. However, compared to other popular quality methods, such as design of experiments (DOE), linear regression, and response surface methods, multivariate statistical methods are much less known to quality professionals. One of the major concerns in applying multivariate statistical methods is the difficulty involved in communicating multivariate data and analysis results to nonstatisticians. We believe that graphical multivariate data display and visualization are excellent tools to bridge the communication gap between multivariate statistical methods and their end users. To understand a large volume of multivariate data and analysis results and to make sense out of them is a real challenge for end users or subject matter experts. However, translating data back to graphics or images relating to the end user's discipline will usually be of great help.

This chapter will describe several techniques to display multivariate data. Section 2.2 will discuss the use of graphical templates to display multivariate data. Section 2.3 will introduce computerized data visualization and animation techniques to display multivariate data. Section 2.4 will discuss the use of visual analysis techniques to detect special patterns in multivariate data.

2.2 Graphical Templates for Multivariate Data

2.2.1 Charts and graphs

Graphical representation of data has been practiced for hundreds of years. However, graphical representation of data as an academic discipline started only in the middle of the twentieth century (Chernoff, 1978).

Charts and graphs are among the first graphical tools to display data. The advantages of graphs and charts are summarized by Schmid (1954) as follows:

1. In comparison to other types of presentations, well-designed charts are more effective in creating interest and in appealing to the attention of the reader.
2. Visual relationships, as portrayed by charts and graphs, are more clearly grasped and more easily remembered.
3. The use of charts and graphs saves time since the essential meaning of large masses of statistical data can be visualized at a glance.
4. Charts and graphs can provide a comprehensive picture of a problem and that makes possible a more complete and better balanced understanding than could be derived from tabular or textual forms of presentation.
5. Charts and graphs can bring out hidden facts and relationships and can stimulate as well as aid analytical thinking and investigation.

Following are categories of charts and graphs that are commonly used:

1. Charts that display original data such as bar, line, and pie charts.
2. Graphs depicting theoretical relationships such as probability density function graphs, contours of multivariate densities, and hazard rate plots.
3. Graphs intended to display data and results of analysis such as histograms, scatter plots, and time series charts.

4. Analytical graphs such as residual plots and half normal plots.
5. Nonnumerical graphs and charts such as maps, Venn diagrams, and flowcharts.

The objectives of using charts and graphs include the following:

1. Illustration or communication, that is, charts and graphs that are designed in order to make readers/audience understand easily.
2. Analysis or comprehension, that is, it is often useful to find a representation which permits an analyst to develop an understanding of what conclusions may be drawn and what relations and regularities exist.

Charts and graphs commonly used to display multivariate data sets. Given a multivariate data set, $\mathbf{X}_i = (x_{i1}, x_{i2}, ..., x_{ip})$, the following charts are often used:

Bar chart. It represents a multivariate data set by using the bar heights corresponding to the values of variables.

Line chart. It represents a multivariate data set by using a polygonal line which connects various heights corresponding to the values of the variables.

Circular profile. It is a variation of the line chart, in which the polygonal line connects points located on equally spaced rays, where the distance from the center represents the value for each of the variables.

Example 2.1: Multivariate Data Display Three students, Kevin, Cathy, and Josh, have taken the achievement tests in four areas, math, science, reading, and writing. Their scores are as follows (where 100 is the highest possible score in each area):

	Math	Science	Reading	Writing
Kevin	98	95	71	65
Cathy	62	58	86	82
Josh	82	78	77	80

Here we can display the multivariate data set by bar chart as in Fig. 2.1. We can also display this data set by the three line charts in Fig. 2.2. Finally, we can display this data set by the circular profiles shown in Fig. 2.3. All the three data displays can show the features of this data set. Kevin is very good at math and science but weak at reading and writing; Cathy is the opposite; Josh is a well-rounded student.

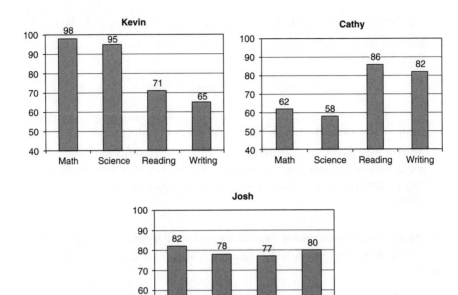

Figure 2.1 Bar charts for Example 2.1.

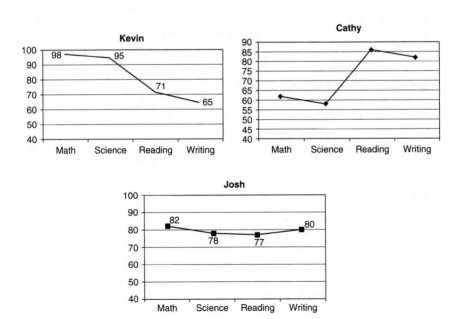

Figure 2.2 Line charts for Example 2.1.

Graphical Multivariate Data Display and Data Stratification 29

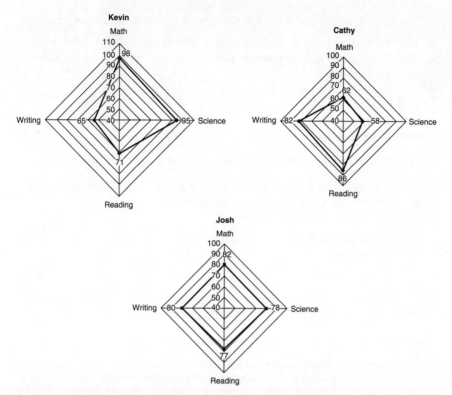

Figure 2.3 Circular profiles for Example 2.1.

2.2.2 Templates for displaying multivariate data

Multivariate data can also be displayed in formats other than graphs and charts. We will define a category of display as a template. By definition, a *template* is a system that helps you arrange information, or something that establishes or serves as a pattern. Therefore, the graphs and charts that we previously discussed are also templates. A good template that displays multivariate data should have the following properties:

1. *Accuracy.* It should not distort the original data and the original relationships within the data set.
2. *Informative.* It should provide all important information in the data.
3. *Self explanatory.* For the end users, it should be very easy to understand and follow. It should use some symbol/logo that end users are familiar with.

4. *Compactness.* It should not contain too much redundant information.
5. *Contrast or sensitivity.* If the data changes, the template should also change. The template should be able to detect any significant change in data.

Here we will provide several examples of templates.

Example 2.2 The following is a drawing of a machined part (see Fig. 2.4). Four measurements of thickness of the plate are denoted as (x_1, x_2, x_3, x_4). Assume that the nominal dimension for the plate thickness is 0.1 in and specification limits are 0.0095 and 0.105 in.

As a quality control procedure, three pieces of plates are randomly selected every 30 min in production for these four thickness readings.

The following is a sample of data set:

Thickness readings	x_1	x_2	x_3	x_4
Plate 1	0.092	0.095	0.099	0.105
Plate 2	0.109	0.097	0.096	0.107
Plate 3	0.102	0.104	0.106	0.110

Several templates can be designed to display this set of data:

Template design 1 This template design looks like an exaggerated cross section drawing for each plate (see Fig. 2.5). It gives a lot of information. The dotted lines indicate the specification limits, so plate 1 is out of lower specification. Plate 2 and plate 3 are out of upper specification. The shape of the templates for plate 1 and plate 3 indicates that they are "tapered." The shape of template for plate 2 indicates that plate 2 is "thin at the middle." This type of information may help

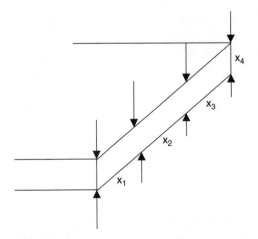

Figure 2.4 Dimensional measurements of a part.

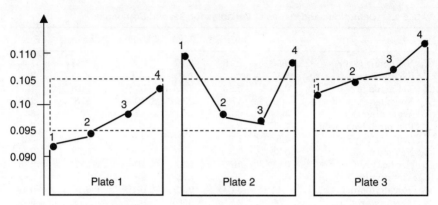

Figure 2.5 Template design 1.

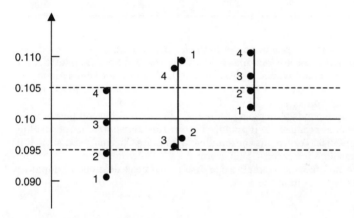

Figure 2.6 Template design 2.

the end user diagnose what went wrong and how to improve the machining operation.

Template design 2 This template is much simpler than template design 1. It still kept all the information about the thickness measurements. However, the shapes of the cross section of the plates are not shown explicitly (see Fig. 2.6). It may not communicate as well as template design 1, but it is much easier to make.

Example 2.3: Chernoff Faces (Bruckner, 1978) Chernoff (1971) introduced the idea of using faces to represent multidimensional data. This example is about using faces to represent several oil companies. Table 2.1 gives the definitions and the values of 15 variables for several companies.

How is it possible to make sense of this data set and make a comparison of these companies? Bruckner (1978) makes use of "Chernoff faces" and uses the "facial structure" variable to represent these 15 variables, as in Table 2.2.

TABLE 2.1 Definitions and Values of Variables for Several Companies

Variables	Arco	Texaco	Shell	Union	Chevron	Exxon
X_1: Net bonus ($B)	0.56	1.16	0.97	0.53	0.84	1.44
X_2: Excess$/lease	1.1	2.7	1.7	1.2	1.2	2.9
X_3: Net acreage	0.78	0.56	1.59	0.49	1.16	1.02
X_4: No. of leases won	306	176	336	203	378	250
X_5: Ave. ownership	49	66	95	47	70	84
X_6: Pct. prod. lease won	10	8	13	4	13	8
X_7: Avg. yrs. to prod.	4.5	7.8	3.6	4.2	5.8	5.2
X_8: Net gas prod.	0.38	0.31	1.9	1.22	0.70	0.99
X_9: Net liq. prod.	66	56	430	103	197	276
X_{10}: Net royalty	62	50	378	99	141	199
X_{11}: Royalty/bonus	0.11	0.04	0.39	0.19	0.17	0.14
X_{12}: Royalty/prod. yr.	174	277	656	527	355	609
X_{13}: R^2	0.84	0.91	0.38	0.98	0.50	0.58
X_{14}: Roy/PYR/PR.Ac	2.8	2.5	2.7	8.5	1.6	4.3
X_{15}: Pct lease terminated	35	34	54	38	32	36

Based on the assignment of variables to the facial features, Bruckner produced the Chernoff faces for these companies as shown in Fig. 2.7. For example, the face width corresponds to net bonus, X_1. Exxon has the highest net bonus among all the oil companies, so the Chernoff face of Exxon has the widest face width. Union has the lowest net bonus, so its Chernoff face has the narrowest face width. The ear diameter corresponds to net gas production, X_8. So the Chernoff face of Shell has the largest ears because Shell produced by far the largest amount of gasoline among oil companies. Texaco has the smallest ears in its Chernoff face, since it is the smallest gasoline producer in this group.

Clearly, Chernoff faces are templates to display business data of oil companies and try to make sense out of them.

TABLE 2.2 Multidimensional Data Represented by Facial Feature Variables

Variables	Corresponding facial features	Default value	Min	Max
X_1: Net bonus ($B)	Face width	0.6	0.2	0.7
X_2: Excess$/lease	Brow length	0.5	0.3	1
X_3: Net acreage	Half-face height	0.5	0.5	1.0
X_4: No. of leases won	Eye separation	0.7	0.3	0.8
X_5: Ave. ownership	Pupil position	0.5	0.2	0.8
X_6: Pct. prod. lease won	Nose length	0.25	0.15	0.4
X_7: Avg. yrs. to prod.	Nose width	0.1	0.1	0.2
X_8: Net gas prod.	Ear diameter	0.5	0.1	1.0
X_9: Net liq. prod.	Ear level	0.5	0.35	0.65
X_{10}: Net royalty	Mouth length	0.5	0.3	1.0
X_{11}: Royalty/bonus	Eye slant	0.5	0.2	0.6
X_{12}: Royalty/prod. yr.	Mouth curvature	0.0	0.0	4.0
X_{13}: R^2	Mouth level	0.5	0.2	0.4
X_{14}: Roy/PYR/PR.Ac	Eye level	0.1	0.0	0.3
X_{15}: Pct lease terminated	Brow height	0.8	0.6	1.0

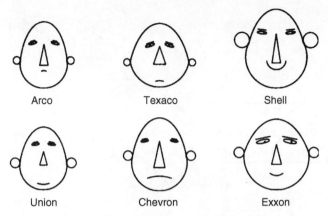

Figure 2.7 Chernoff faces corresponding to Table 2.1.

Example 2.4: Template for Different Styles of Music Microsoft Windows XP includes the ability to display a template showing certain features of music such as tempo and pitch (see Fig. 2.8). Clearly, there are many ways to design templates to visualize the meanings of data.

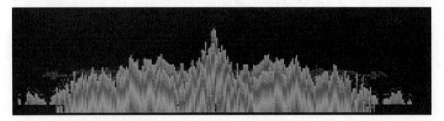

Figure 2.8 A music template in Microsoft Windows XP.

2.3 Data Visualization and Animation

2.3.1 Introduction to data visualization

A more sophisticated form of graphical data display is data visualization. Data visualization can be defined as mapping of data to a representation that can be perceived (Post et al., 2003; Rosenblum et al., 1994; Fortner, 1995). The type of mapping could be visual, audio, tactile, or a combination of these. The visual mapping can be further subdivided into sets of marks (for example, points, lines, areas, and volumes) that express position or pattern, and retinal properties (i.e., color, shape, size, orientation, and texture) that enhance the marks and may also carry additional information. Specifically, in multivariate data visualization, the multivariate data sets are transformed into two-dimensional (2D) or three-dimensional (3D) computer graphics with shape, color, etc. Usually,

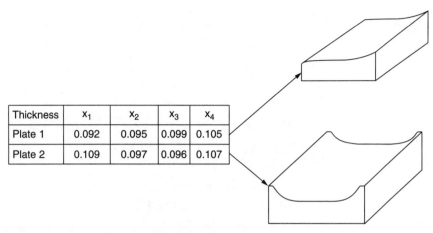

Figure 2.9 Simple data visualization for Example 2.2.

data visualization is a process of transforming multivariate data back to their original application. For example, in Example 2.2, the original multivariate data sets are the actual multiple measurements from machined parts with a certain geometrical shape. Different measurements for different parts indicate that their geometrical shapes are different. In this case, data visualization is simply transforming data into shapes, as illustrated in Fig. 2.9. Clearly, data visualization can help a great deal in communicating the data and explaining the facts to the end users. The data visualization process can be illustrated by Fig. 2.10.

The original multivariate data are often in the form of computer files such as spreadsheets. The data are processed by a data management system in which the data are preprocessed, bad pieces of data are eliminated, and good pieces of data are labeled. Based on the original application of data, appropriate 2D or 3D graphical models are built to

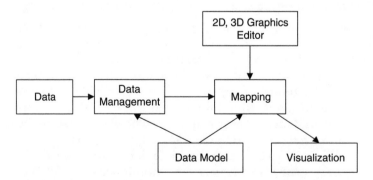

Figure 2.10 Typical data visualization process.

facilitate the visualization. Good pieces of data are processed by the data model and inputted into the mapping process, and appropriate data visualization is produced.

The most common multivariate data types involved in data visualization are 2D matrix data, 3D matrix data, and, in general, "locations and values" data.

2D matrix data. 2D matrix data usually has the following form:

x	y	Values at (x, y) $f(x, y)$ (e.g., temperature °C)
0.5	0.5	230
0.5	1.0	242
1.0	0.5	228
1.0	1.0	250

Here $f(x, y)$ usually represents a characteristic such as temperature and speed, and it is a function of a point in x-y 2D space. Clearly, 2D matrix data is the distribution of values on to a 2D plane.

The commonly used 2D visualization methods include color raster imaging, surface plot, and contour plot. Color raster imaging is to convert the data values in the 2D matrix into shades of gray or to colors to create a grayscale or color image. The surface plots, the data of the 2D matrix, are converted to a height of a rubber membrane above a plane. Let us look at the following example:

Example 2.5: Polysilicon Deposition Thickness The polysilicon deposition process is a process in semiconductor manufacturing in which a layer of polysilicon is deposited on the top of the oxide layer on the silicon wafer. A uniform thickness is expected on the layer. The following table gives the thickness measurements of a polysilicon layer at different locations on the wafer:

X Grid level	Y Grid level					
	1	2	3	4	5	6
1	2029	1975	1961	1975	1934	1949
2	2118	2109	2099	2125	2108	2149
3	2521	2499	2499	2576	2537	2552
4	2792	2752	2716	2684	2635	2606
5	2863	2835	2859	2829	2864	2839
6	3020	3008	3016	3072	3151	3139
7	3125	3119	3127	3567	2563	3520

The values in the table are the silicon film thickness (in angstroms: Å) at various x-y coordinates on the wafer.

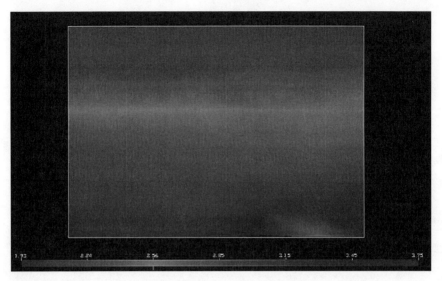

Figure 2.11 Color raster imaging of polysilicon deposition thickness distribution.

Color raster imaging of the thickness distribution is shown in Fig. 2.11. The horizontal axis of this color raster graph is in the units of 1000 Å.

Figure 2.12 is a surface plot of this thickness data set. This surface plot is also a color raster image at the same time. Different film thickness segments are assigned different colors.

3D matrix data. 3D matrix data usually has the following form:

x	y	z	Values at (x, y, z) $f(x, y, z)$, e.g., density at (x, y, z)
0.5	0.5	0.0	0.56
1.0	0.5	0.0	0.71
0.5	1.0	0.0	0.62
1.0	1.0	0.0	0.76
0.5	0.5	1.0	0.98
1.0	0.5	1.0	1.07
0.5	1.0	1.0	1.12
1.0	1.0	1.0	1.23

$f(x, y, z)$ also represents a property or characteristic, and it is a function in 3D space. Clearly, 3D matrix data are really the distribution of a characteristic onto a 3D space. They are also called volumetric data because the numbers fill a volume.

Graphical Multivariate Data Display and Data Stratification   37

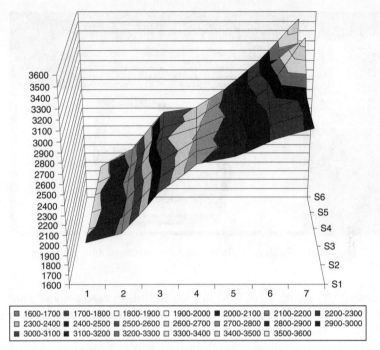

Figure 2.12   Surface plot of the thickness data set.

Visualizing 3D matrix data depends on computer power. The commonly used 3D visualization techniques include slicing and dicing, isosurfaces, and volumetric visualization.

Locations and values data.   A more general type of data has the form of $f(\mathbf{X})$, where $\mathbf{X}$ is an "identifier," or "location" that indicates a selected location of a 3D structure. $f(\mathbf{X})$ represents a characteristic, or a property, which is a function of the location $\mathbf{X}$. We may have several characteristics, or properties, and they are all functions of X. That is, we have $f_1(\mathbf{X}), f_2(\mathbf{X}), ..., f_n(\mathbf{X})$.

Example 2.6   In automobile body manufacturing, in every selected measurement location on a body panel such as a hood, the dimensional deviation of that location from the designed location is measured and recorded (see Fig. 2.13) In this case, the measurement location is $\mathbf{X}$ and the dimensional deviation is $f(\mathbf{X})$, which is a function of $\mathbf{X}$.

Example 2.7   In the automobile paint application process, paint defects are inspected and repaired at the end of the process. There are many different defective types such as crater, dirt, overspray, etc. The number of defects, locations of defects, and defect types are recorded at the inspection station. The data visualization in Fig. 2.14 can help to record all this information.

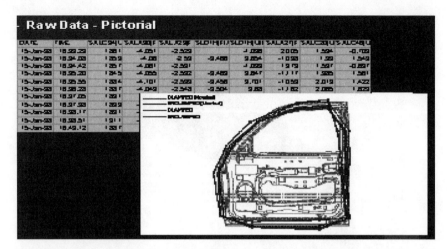

Figure 2.13 Computer visualization of dimensional measurement data for automobile manufacturing.

Data animation. The data visualization methods discussed above do not deal with the images that are changing over time. In other words, they deal with static visualization. If the visual image is changing and we want to show it, then we need animation. The way computer animation works is very much like playing movies, as here too a series of successive image frames are played on the computer screen. The difference is that the image frames in animation are created by the computer algorithm and data sets and not created by the artists.

2.4 Multivariate Data Stratification

Multivariate graphical templates and data visualization can also give visual insights into the relationships between and among multivariate variables. Through these visual exploratory analyses, we can discover

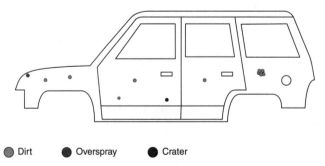

Figure 2.14 Paint defects location display.

hidden patterns in the data and divide the data into groups or strata based on some key characteristics such as batch, time, and group. This stratification can help a great deal in discovering the real causes for difficult quality problems.

The first reported multivariate data graphical stratification method in quality engineering is the multi-vari chart technique introduced by Seder (1950). We are going to discuss the multi-vari chart technique in detail.

2.4.1 Multi-vari chart technique

The multi-vari chart technique was first introduced into discrete manufacturing environment, mostly mechanical processes for quality problem diagnosis. In multi-vari chart analysis, randomly selected parts/components/products are measured. Multiple measurements are taken for each part. These measurements are usually related to key characteristics for that part. Here we illustrate that by the following example.

Example 2.8: Tube Dimensions (Seder, 1950) In a fluted tube, the dimensional consistency of the diameter of the tube is a major concern. Four measurements are taken at each tube as illustrated by Fig. 2.15, where d_1 is the minimum diameter on the left end, which is the minimum length on a diameter direction on the left end, and d_2 is the maximum diameter on the left end, which is the maximum length on a diameter direction on the left end. For a perfect tube, $d_1 = d_2$. If $d_1 < d_2$, then we call the left end of tube out of round. Similarly, d_3 is the minimum diameter on the right end and d_4 is the maximum diameter on the right end. For a perfect tube, we should have $d_1 = d_2 = d_3 = d_4$. If in practice, the diameters on one end are significantly different from those at the other end, that is, either, d_1 and/or d_2 are significantly larger than d_3 and/or d_4, or vice versa, then the tube is tapered.

The following graphical template was proposed to visually represent the information of (d_1, d_2, d_3, d_4), the tolerance limit for the diameter of the tube is (0.450 in, 0.455 in). The polygons that are drawn based on (d_1, d_2, d_3, d_4) can

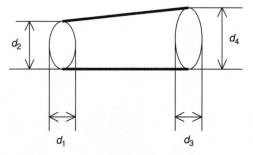

Figure 2.15 Measurements in the tube.

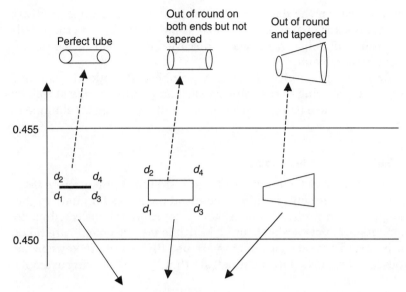

Figure 2.16 Templates for the tube example.

be used as templates to represent the different situations in machining, as illustrated in Fig. 2.16.

We can measure many units of tubes and then arrange them into the following chart (shown in Fig. 2.17), called the multi-vari chart.

Three types of variation. Actually, in most mechanical processes, it is possible to list the following three sources of variation:

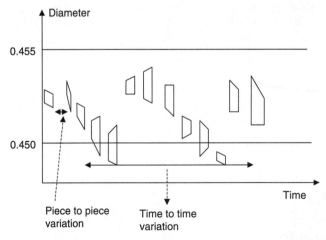

Figure 2.17 Multi-vari chart for the tube example.

Type 1. Variability on a single piece, or intrapiece variability, which is the variation among different measurements but within a single piece.

Type 2. Variability from piece to piece, or interpiece variability.

Type 3. Variability from time to time, or time variability.

Ideally, we would like to see all types of variations to become zero. However, in a practical situation, all three types of variations will exist. The task for problem solvers is to identify what type of variation is the major source of variation and identify the root causes of the major variation type.

In the discrete manufacturing process, the type 1 variation, that is, the intrapiece variability, is often associated with tool dimensions or a fault in work-tool alignment. Out-of-roundness of a turned piece may be caused by excessive distance from the cullet to tool, and out-of-parallelness on a pressed piece is caused by failure of the punch and die to be properly aligned. On the other hand, the type 2 variation, that is, the piece-to-piece variability, is often related to machine maintenance. It relates to many factors such as the clearances between bearings and spindles, between gibes and ways or other mating parts, looseness in threads, variations in cooling oil flow or temperature, and sloppiness in indexing fixtures. Type 3 variation, that is, the time-to-time variability, may be related to some factors which are often functions of time such as tool wear, chemical bath depletion, or such factors as temperature change, raw material variation from batch to batch, and operator adjustments.

The multi-vari chart, as illustrated in Fig. 2.17, is an excellent tool to break down the overall variation into the above three types of variations and identify which is the dominant type of variation. In Example 2.8, despite the fact that intrapiece variation does exist and out-of-roundness and taper are both problems, the time-to-time variation is certainly the dominant type of variation. In Fig. 2.17 it shows a gradual downward pattern over time for diameters and also sudden jumps periodically. These "signatures" in graphical templates variation can greatly help people involved in identifying root causes for the quality problem.

2.4.2 Graphical analysis of multivariate variation pattern

Besides the three types of variations that we discussed in the last subsection, there could be many other types of variation decomposition. Use of graphical templates and graphical data visualization to analyze variation types and identify major source(s) of variation can be very powerful tools for finding the root causes. This approach is quite similar to analysis of variance. Leonard Seder called it "graphical analysis of variance" (Seder, 1950).

In general, we can give the following step-by-step procedure for the graphical analysis of multivariate variation patterns:

1. Identify the single unit in which a set of multivariate data are collected and select a good data collection plan so that each data set contains sufficient information for our study.
2. Select an appropriate graphical data template as discussed in Sec. 3.2.2, or a data visualization scheme, as discussed in Sec. 3.3. The selected template or data visualization scheme should provide problem analysts sufficient information to assess the magnitude of variation and the mutual relationship among multivariate variables. The "signature" of template or visualization should provide powerful clues for root cause analysis.
3. Select the inclusive categories of variations for which all types of variations can be partitioned into these categories. The category partition of intrapiece, interpiece, and time to time is one such example.
4. Collect multivariate data samples and display the multivariate templates or data visualization graphs under each category.
5. Determine the major category(ies) that accommodate(s) most of the variations by using subject matter knowledge to unlock the root cause of variation in that category.

We will use several examples to illustrate this graphical analysis of multivariate variation patterns.

Example 2.9: Injection Molding The plastic cylindrical connectors are made by injection molding. The shape of the connector is shown in Fig. 2.18. The diameters at the left, middle, and right, that is, d_1, d_2, and d_3 are of major concern. The graphical template Fig. 2.19 is selected. Here LSL stands for lower specification limit and USL stands for upper specification limit for the diameter of the connector. The above signatures of the three-point template illustrated several situations in the molded parts.

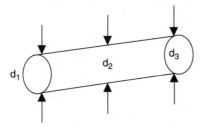

Figure 2.18 Plastic cylinder connector.

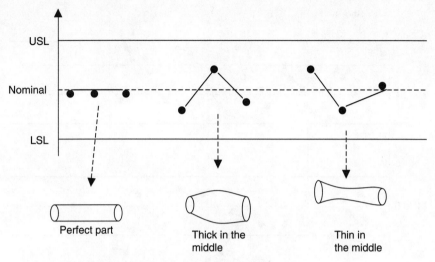

Figure 2.19 Graphical template for plastic cylinder connector.

Now we determine the categories of sources of variations. We can identify the following five sources of variations:

1. *Intrapiece.* That is the variation among (d_1, d_2, d_3).
2. *Cavity to cavity, but with mold.* As shown in Fig. 2.20, one mold has four cavities. The parts molded by different cavities may have different shapes.
3. *Mold to mold.* Each injection molding run uses several molds. There are variations from mold to mold.
4. *Shot to shot.* From each injection shot, several molds are filled. There might be variations from shot to shot.
5. *Time to time.* If we run the injection molding process for an extended duration, the time-to-time variation may be significant.

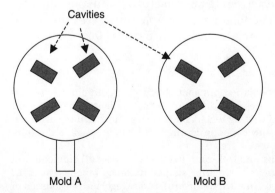

Figure 2.20 Molds and cavities.

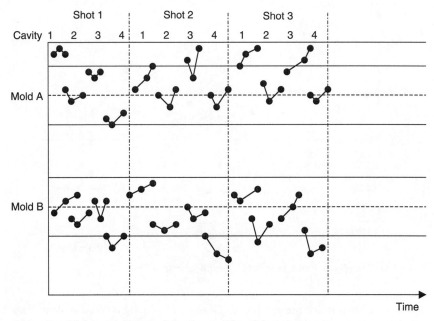

Figure 2.21 Multi-vari chart for injection molding data.

Figure 2.21 shows the graphical variation pattern for this case. We can see clearly that the variation from cavity to cavity and the mold-to-mold variation look like the leading sources of variations. Intrapiece variation shows a lot of thin-in-the-middle patterns. Mold A produces thicker parts than does mold B. We do not see a lot of shot-to-shot and time-to-time variations.

The multi-vari chart and graphical analysis of multivariate variation pattern technique that we discussed can not only be used to analyze the multivariate data in a discrete manufacturing situation, but also in many other situations. The following example illustrates the use of graphical analysis of multivariate variation pattern for elementary school performance study.

Example 2.10: Elementary Schools Performance Study In Example 2.1 we discussed the achievement test scores for elementary school students. The scores are in math, science, reading and writing tests, and they are multivariate variables. The circular profiles that we developed in Example 3.1 can be used as the template to display achievement test scores (see Fig. 2.22).

A big diamond means strong and well-rounded students, a small diamond profile means weak students, and an irregular shaped diamond means a student is only strong in some areas. Now we determine the categories of "sources of variation." We can identify the following categories:

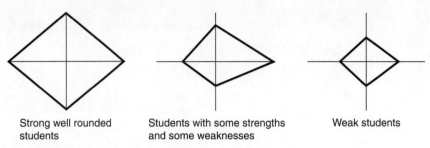

Figure 2.22 Circular profile for students.

1. Intrastudent variation: variation among the four test scores; this variation relates the discrepancies of a student in different areas.
2. Student-to-student variation but within a class
3. Class-to-class variation but within a school
4. School-to-school variation but within a school district
5. School district–to–school district variation

We can take the achievement test scores of many students from a region and plot the multi-vari chart shown in Fig. 2.23. In this multi-vari chart, the sizes of diamonds are the indicators of learning excellence. The larger the diamonds, the better the students' learning results. From the above figure, it is clear that school district 1 has larger diamond sizes than does school district 2. That is the largest source of variation. School 2 of district 1 has larger diamond sizes than does school 1. However, the school-to-school difference in school district 2 is not obvious. We can also see that within the same school, class-to-class difference is not very significant. Within each class, there are some student-to-student differences. Overall, school district–to–school district variation is the largest source of variation in achievement test scores.

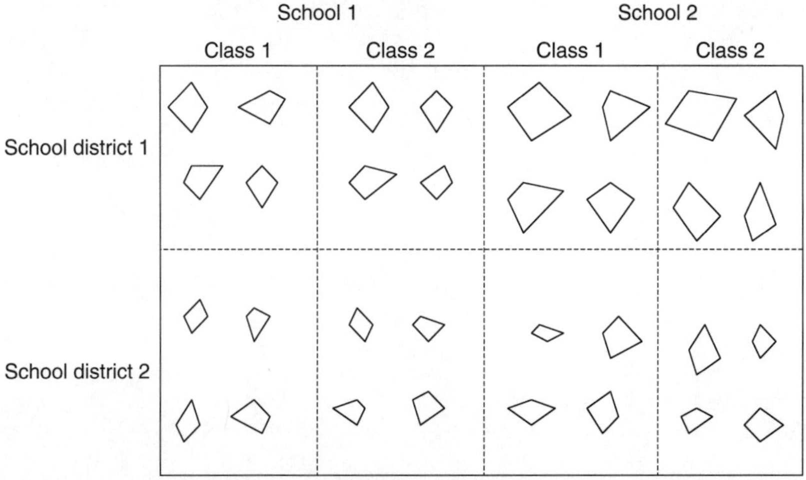

Figure 2.23 Multi-vari charts for elementary school performance study.

Chapter 3

Introduction to Multivariate Random Variables, Normal Distribution, and Sampling Properties

3.1 Overview of Multivariate Random Variables

Multivariate statistical methods deal with multivariate random variables. Unlike the univariate random variable, which is a single variable, multivariate random variables are expressed by $\mathbf{X} = (X_1, X_2, \ldots, X_p)^T$, where p represents the number of variables. So $\mathbf{X}$ is a random vector. In a multivariate random vector, the variables $X_1, X_2, \ldots, X_p$ may be correlated with each other. For example, if X_1 increases, the other random variables $X_2, \ldots, X_p$ may follow the change of X_1 in some ways.

Example 3.1: Welding Technician Training Program Let us assume that there is a welding technician training program. In this program, three courses are offered. They are basic welding technology (BWT), advanced welding technology (AWT), and practical welding workshop (PWW). The training program will teach courses to students and each student will be given grades in these three courses. The lowest possible grade is 0 and the highest is 100. For each incoming student, his/her final grade in these three courses can be represented by

$$\mathbf{X} = \begin{bmatrix} X_1 \\ X_2 \\ X_3 \end{bmatrix}$$

where X_1 = grade for BWT
X_2 = grade for AWT
X_3 = grade for PWW

In this situation, **X** is a multivariate random vector with three variables. In this example, if a student achieved a very high grade in basic welding technology (BWT), it is likely that he/she may achieve quite high grade in the other courses, that is, advanced welding technology (AWT) and practical welding workshop (PWW), and vice versa. This phenomenon is an example of correlation.

In a univariate random variable X, the mean (or expected value) $E(X) = \mu$ is a measure of central tendency (where the gravitational center of random data is for the population). For multivariate random variables, the commonly used measure of central tendency is the mean vector:

$$\boldsymbol{\mu} = \begin{bmatrix} \mu_1 \\ \vdots \\ \mu_i \\ \vdots \\ \mu_p \end{bmatrix} = \begin{bmatrix} E(X_1) \\ \vdots \\ E(X_i) \\ \vdots \\ E(X_p) \end{bmatrix} \quad (3.1)$$

which is a vector of means for all variables.

For a univariate random variable X, the variance of X, Var(X), is a scalar, Var(X) = σ^2. It is well known that σ^2 is a measure of dispersion, or variation. For the multivariate normal distribution, the measure of dispersion is a matrix, called covariance matrix, Σ. Actually the covariance matrix not only gives the measure of dispersion for all variables $X_1, X_2, ..., X_p$, but also gives the mutual variation relationship among variables. For example, if X_1 increases by one standard deviation, by how much would $X_2, ..., X_p$ change with it?

Specifically, for multivariate random variables, $\mathbf{X} = (X_1, X_2, ..., X_p)^T$. The covariance matrix Σ is defined as

$$\Sigma = \begin{bmatrix} \sigma_{11} & \sigma_{12} & \cdots & \cdots & \sigma_{1p} \\ \sigma_{21} & \cdots & \cdots & \cdots & \cdots \\ \cdots & \cdots & \sigma_{ii} & \sigma_{i,i+1} & \cdots \\ \vdots & \vdots & \vdots & \vdots & \vdots \\ \sigma_{p1} & \cdots & \sigma_{pi} & \cdots & \sigma_{pp} \end{bmatrix} \quad (3.2)$$

In the covariance matrix Σ, its diagonal elements σ_{ii}, for $i = 1, ..., p$, are variances of variables, that is,

$$\text{Var}(X_i) = \sigma_{ii}$$

The off-diagonal elements of the covariance matrix Σ, that is, σ_{ij}, for $i = 1,\ldots,p$, $j = 1,\ldots,p$, and $i \neq j$, are called covariances, specifically

$$\text{Cov}(X_i, X_j) = \sigma_{ij} = E(X_i - \mu_i)(X_j - \mu_j) \tag{3.3}$$

is called the covariance of X_i and X_j.

Some well-known properties of covariance are as follows:

1. If $\text{Cov}(X_i, X_j) > 0$, then when X_i increases, X_j also tends to increase.
2. If $\text{Cov}(X_i, X_j) < 0$, then when X_i increases, X_j tends to decrease and vice versa.
3. If two random variables X_i and X_j are independent, then $\text{Cov}(X_i, X_j) = 0$.

One problem of using $\text{Cov}(X_i, X_j)$ as a numerical measure for how closely X_i and X_j are related is that the value of $\text{Cov}(X_i, X_j)$ depends on the numerical scale and/or unit of measurements of X_i and X_j. For example, if the original units for X_i and X_j are meters and we change that to millimeters, the covariance will increase by $1000 \times 1000 = 1$ million times!

To overcome this problem, a standardized version of covariance, called the correlation coefficient, is introduced. Specifically, the correlation coefficient of X_i and X_j, for any pair of i, j, where $i \in (1,\ldots,p), j \in (1,\ldots,p)$, denoted by ρ_{ij}, is defined as

$$\text{Cor}(X_i, X_j) = \rho_{ij} = \frac{\text{Cov}(X_i, X_j)}{\sqrt{\text{Var}(X_i)\text{Var}(X_j)}} \tag{3.4}$$

Some well-known properties of correlation coefficients are as follows:

1. If $\rho_{ij} > 0$, then when X_i increases, X_j also tends to increase.
2. If $\rho_{ij} < 0$, then when X_i increases, X_j tends to decrease and vice versa.
3. If two random variables X_i and X_j are independent, then $\rho_{ij} = 0$.
4. $-1 \leq \rho_{ij} \leq +1$
5. When $\rho_{ij} = +1$, X_i and X_j have a perfect positive linear relationship, that is, there exists $\beta_1 > 0$, such that $X_j = \beta_0 + \beta_1 X_i$. When $\rho_{ij} = -1$, X_i and X_j have a perfect negative linear relationship, that is, there exists $\beta_1 < 0$, such that $X_j = \beta_0 + \beta_1 X_i$.
6. $\text{Cor}(X_i, X_i) = \rho_{ii} = +1$. This is obvious because X_i and X_i will certainly have a perfect positive linear relationship.

For the multivariate random vector $\mathbf{X} = (X_1, X_2, ..., X_p)^T$ the covariance matrix $\mathbf{P}$ is defined as

$$\mathbf{P} = \begin{bmatrix} 1 & \mathrm{Cor}(X_1, X_2) & \cdots & \mathrm{Cor}(X_1, X_p) \\ \mathrm{Cor}(X_2, X_1) & 1 & \cdots & \cdots \\ \vdots & \vdots & \vdots & \vdots \\ \mathrm{Cor}(X_p, X_1) & \mathrm{Cor}(X_p, X_2) & \cdots & 1 \end{bmatrix} = \begin{bmatrix} 1 & \rho_{12} & \cdots & \rho_{1p} \\ \rho_{21} & 1 & \cdots & \rho_{2p} \\ \vdots & \vdots & \vdots & \vdots \\ \rho_{p1} & \rho_{p2} & \cdots & 1 \end{bmatrix}$$

(3.5)

Similar to the univariate random variables, the parameters of multivariate random variables, such as μ, Σ, and $\mathbf{P}$, are population characteristics. In other words, we can know the exact values of μ, Σ, and $\mathbf{P}$ only if we can collect all the data for the whole population. That is quite impractical for real world application. However, as with that of univariate statistics, we can collect samples of data and estimate the parameters of multivariate random variables such as μ, Σ, and $\mathbf{P}$. Also, as with that of univariate statistics, the statistical estimates of as μ, Σ, and $\mathbf{P}$ are also called descriptive statistics. We will discuss multivariate data sets and descriptive statistics in the next section.

3.2 Multivariate Data Sets and Descriptive Statistics

3.2.1 Multivariate data sets

In order to study the properties of a given set of multivariate random variables $\mathbf{X} = (X_1, X_2, ..., X_p)^T$, we can collect sample data from the population and use statistics to estimate population parameters such as μ, Σ, and $\mathbf{P}$. The multivariate sample data sets are usually in the following format:

$$\mathbf{X}_{n \times p} = \begin{bmatrix} x_{11} & x_{12} & x_{13} & \cdots & x_{1p} \\ x_{21} & x_{22} & x_{23} & \cdots & x_{2p} \\ x_{31} & x_{32} & x_{33} & \cdots & x_{3p} \\ \vdots & \vdots & \vdots & \vdots & \vdots \\ x_{n1} & x_{n2} & x_{n3} & \cdots & x_{np} \end{bmatrix}$$

(3.6)

where x_{ij} is the jth variable measured on the ith sample and $i = \{1, 2, 3, ..., n\}$, $j = \{1, 2, 3, ..., p\}$.

Let us illustrate the multivariate data sets by using Example 3.2.

Example 3.2: Welding Technician Training Program For the welding technician training program described in Example 3.1, in order to estimate the training results specified by the multivariate random vector

$$\mathbf{X} = \begin{bmatrix} X_1 \\ X_2 \\ X_3 \end{bmatrix} = \begin{bmatrix} \text{BWT} - \text{Grade} \\ \text{AWT} - \text{Grade} \\ \text{PWW} - \text{Grade} \end{bmatrix}$$

we can collect a sample of scores from existing trainees as follows:

$$\mathbf{X}_{10 \times 3} = \begin{matrix} \text{Trainee 1} \\ \text{Trainee 2} \\ \text{Trainee 3} \\ \text{Trainee 4} \\ \text{Trainee 5} \\ \text{Trainee 6} \\ \text{Trainee 7} \\ \text{Trainee 8} \\ \text{Trainee 9} \\ \text{Trainee 10} \end{matrix} \begin{bmatrix} 96 & 92 & 91 \\ 85 & 91 & 80 \\ 96 & 92 & 90 \\ 90 & 86 & 97 \\ 78 & 75 & 94 \\ 98 & 87 & 92 \\ 75 & 72 & 86 \\ 97 & 93 & 95 \\ 88 & 89 & 92 \\ 79 & 73 & 80 \end{bmatrix}$$

This multivariate data set can be used to estimate μ, Σ, and $\mathbf{P}$. Clearly, this multivariate data set has the format specified by Eq. (3.6), where $p = 3$ and $n = 10$.

Multivariate data sets can be conveniently stored in a computer spreadsheet as shown in Fig. 3.1.

	C1	C2	C3
	BWT	AWT	PWW
1	95	92	91
2	85	91	80
3	96	92	90
4	90	86	97
5	78	75	94
6	98	87	92
7	75	72	86
8	97	93	95
9	88	89	92
10	79	73	86

Figure 3.1 Multivariate data set for welding trainee example.

3.2.2 Multivariate descriptive statistics

Similar to that of univariate statistics, the statistical estimates of the measure of central tendency (mean) and measure of dispersion (variance) are called descriptive statistics. In multivariate statistics, the statistical estimates of the mean vector **μ**, the covariance matrix **Σ**, and the correlation matrix **P** are corresponding multivariate descriptive statistics.

Statistical estimate of μ: Sample mean vector x̄. Similar to that of univariate statistics—where the sample mean $\bar{x}$ is used as the statistical estimate for population mean μ—in multivariate statistics, the statistical estimate of mean vector **μ**, is the sample mean vector **X̄**, as shown in Eq. (3.7):

$$\bar{\mathbf{X}} = \begin{bmatrix} \bar{x}_1 \\ \bar{x}_2 \\ \vdots \\ \bar{x}_j \\ \vdots \\ \bar{x}_p \end{bmatrix} \quad (3.7)$$

where $\bar{x}_j = (1/n)\sum_{i=1}^{n} x_{ij}$ and $i = 1, \ldots, n$ and $j = 1, \ldots, p$.

Example 3.3: Consider the data set for the welding trainee tests shown in Fig. 3.1. The sample mean vector for the welding trainee data set shown in Fig. 3.1 is

$$\bar{\mathbf{x}} = \begin{bmatrix} 88.1 \\ 85.0 \\ 90.3 \end{bmatrix}$$

Remark. We can compare the equivalent univariate descriptive statistic, the sample mean $\bar{x}$, which is defined in Eq. (3.8):

$$\bar{x} = \frac{\sum_{i=1}^{n} x_i}{n} \quad (3.8)$$

Clearly, there are some similarities.

Statistical estimate of Σ: Sample covariance matrix S. In multivariate statistics, the statistical estimate of the population covariance matrix **Σ** is the sample covariance matrix **S**, which is defined by the following Eq. (3.9):

$$\mathbf{S} = \begin{bmatrix} s_{11} & s_{12} & \cdots & s_{1j} & \cdots & s_{1p} \\ s_{21} & s_{22} & \cdots & s_{2j} & \cdots & s_{2p} \\ \vdots & \vdots & \vdots & \vdots & \vdots & \vdots \\ s_{j1} & s_{j2} & \cdots & s_{jj} & \cdots & s_{jp} \\ \vdots & \vdots & \vdots & \vdots & \vdots & \vdots \\ s_{p1} & s_{p2} & \cdots & s_{pj} & \cdots & s_{pp} \end{bmatrix} \quad (3.9)$$

where

$$s_{kj} = \frac{1}{n}\sum_{i=1}^{n}(x_{ik} - \bar{x}_k)(x_{ij} - \bar{x}_j) \qquad (3.10)$$

for all $k = 1, \ldots, p$ and $j = 1, \ldots, p$; in particular, $s_{jj} = s_j^2$ is the sample variance of the variable X_j, for all $j = 1, \ldots, p$, and

$$s_j^2 = \frac{1}{n}\sum_{i=1}^{n}(x_{ij} - \bar{x}_j)^2 \qquad (3.11)$$

where $j = 1, \ldots, p$. Clearly, the sample covariance matrix is also a symmetric matrix such that $s_{jk} = s_{kj}$ for all $j = 1, \ldots, p$ and $k = 1, \ldots, p$. In the multivariate data set specified by Eq. (3.6), if we use $\mathbf{x}_{i.} = (x_{i1}, x_{i2}, \ldots, x_{ip})^T$ to denote the ith multivariate observation, where $()^T$ represents transpose, then $\mathbf{S}$ can be represented by the following matrix equation:

$$\mathbf{S} = \frac{1}{n}\sum_{i=1}^{n}(\mathbf{x}_{i.} - \bar{\mathbf{X}})(\mathbf{x}_{i.} - \bar{\mathbf{X}})^T \qquad (3.12)$$

For small sample sizes where $n < 30$, the covariance matrix is corrected by a factor of $n - 1$ in order to get unbiased statistical estimate, that is,

$$\mathbf{S} = \frac{1}{n-1}\sum_{i=1}^{n}(\mathbf{x}_{i.} - \bar{\mathbf{X}})(\mathbf{x}_{i.} - \bar{\mathbf{X}})^T \qquad (3.13)$$

Example 3.4: Consider the data set for the welding trainee tests shown in Fig. 3.1. By using Eq. (3.13), the sample covariance matrix, corrected for small sample size for the welding trainee data set, can be calculated as follows:

$$\mathbf{S} = \begin{bmatrix} 72.99 & 62.00 & 18.19 \\ 62.00 & 70.22 & 7.33 \\ 18.19 & 7.33 & 25.57 \end{bmatrix}$$

Many statistical packages, such as MINITAB, can compute the sample covariance matrix from the raw data set. The MINITAB output for the welding trainee data set is shown below.

Covariances: BWT, AWT, PWW

	BWT	AWT	PWW
BWT	72.9889		
AWT	62.0000	70.2222	
PWW	18.1889	7.3333	25.5667

Remark. We can compare the equivalent univariate descriptive statistic, the sample variance s^2, with the multivariate counterpart $\mathbf{S}$, which is defined in Eq. (3.14).

$$s^2 = \frac{\sum_{i=1}^{n}(x_i - \bar{x})^2}{n-1} \qquad (3.14)$$

Statistical estimate of P: Sample correlation matrix R. In multivariate statistics, the statistical estimate of the population covariance matrix **P** is the sample covariance matrix **R**, which is defined by Eqs. (3.15) and (3.16):

$$\mathbf{R} = \begin{bmatrix} 1 & r_{12} & \cdots & r_{1j} & \cdots & r_{1p} \\ r_{21} & 1 & \cdots & r_{2j} & \cdots & r_{2p} \\ \vdots & \vdots & \vdots & \vdots & \vdots & \vdots \\ r_{j1} & r_{j2} & \cdots & 1 & \cdots & r_{jp} \\ \vdots & \vdots & \vdots & \vdots & \vdots & \vdots \\ r_{p1} & r_{p2} & \cdots & r_{pj} & \cdots & 1 \end{bmatrix} \qquad (3.15)$$

where $r_{jj} = 1$ are the diagonal elements in the **R** correlation matrix, for all $j = 1, \ldots, p$; $r_{jk} = r_{kj}$ is the sample correlation, the measure of association between the measurements on variables X_j and X_k, for $j = 1, \ldots, p$ and $k = 1, \ldots, p$; and

$$r_{kj} = \frac{s_{jk}}{\sqrt{s_{jj}}\sqrt{s_{kk}}} = \frac{\sum_{i=1}^{n}(x_{ik} - \bar{x}_k)(x_{ij} - \bar{x}_j)}{\sqrt{\sum_{i=1}^{n}(x_{ik} - \bar{x}_k)^2 \sum_{i=1}^{n}(x_{ij} - \bar{x}_j)^2}} \qquad (3.16)$$

Example 3.5: Consider the data set for the welding trainee tests shown in Fig. 3.1. The sample correlation matrix **R** for the welding trainee data set can be computed by using Eqs. (3.15) and (3.16):

$$\mathbf{R} = \begin{bmatrix} 1.00 & 0.87 & 0.42 \\ 0.87 & 1.00 & 0.17 \\ 0.42 & 0.17 & 1.00 \end{bmatrix}$$

MINITAB can compute the sample correlation matrix **R** conveniently from the raw data. The MINITAB computer printout of the sample correlation matrix **R** for welding trainee data is given below.

```
Correlations: BWT, AWT, PWW

          BWT           AWT
AWT     0.866
PWW     0.421         0.173

Cell Contents: Pearson correlation
```

Scatter plots of multivariate data. Scatter plots are a visual representation of the multivariate information captured in the raw multivariate data set.

Example 3.6: Visual Scatter Plot Consider the data set for the welding trainee tests shown in Fig. 3.1. We can create the following scatter plot as shown in Fig. 3.2.

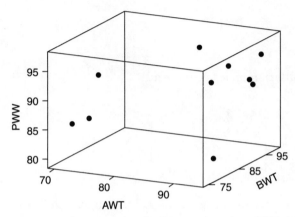

Figure 3.2 3D scatter plot of welding trainee data: BWT, AWT, PWW.

3.3 Multivariate Normal Distributions

When multivariate data are analyzed, the multivariate normal model is the most commonly used model. The multivariate normal distribution model extends the univariate normal distribution model to fit multivariate observations.

A p-dimensional vector of random variables $\mathbf{X} = (X_1, X_2, ..., X_p)$, where $-\infty < X_i < +\infty$, for $i = 1, ..., p$, is said to have a multivariate normal distribution if its density function $f(\mathbf{X})$ is of the form

$$f(\mathbf{x}) = \frac{1}{(2\pi)^{p/2} |\mathbf{\Sigma}|^{1/2}} \exp\left[-\frac{(\mathbf{x}-\boldsymbol{\mu})^T \mathbf{\Sigma}^{-1}(\mathbf{x}-\boldsymbol{\mu})}{2}\right] \quad (3.17)$$

where $\boldsymbol{\mu}$ = mean column vector
$\mathbf{\Sigma}$ = covariance matrix
$\mathbf{\Sigma}^{-1}$ = inverse of the covariance matrix
$|\mathbf{\Sigma}|$ = determinant of the covariance matrix

[Note: Bold represents a vector or matrix, lower case is a vector upper case is a matrix]

The multivariate normal distribution $\mathbf{X} = [X_1 X_2 \cdots X_p]^T$ is represented as

$$\mathbf{X} \sim N_p(\boldsymbol{\mu}, \boldsymbol{\Sigma}) \tag{3.18}$$

where

$$\boldsymbol{\mu} = \begin{bmatrix} \mu_1 \\ \mu_2 \\ \vdots \\ \mu_p \end{bmatrix}$$

is the multivariate population mean vector and

$$\boldsymbol{\Sigma} = \begin{bmatrix} \sigma_{11} & \sigma_{12} & \cdots & \sigma_{1p} \\ \sigma_{21} & \sigma_{22} & \cdots & \sigma_{2p} \\ \vdots & \vdots & \cdots & \vdots \\ \sigma_{p1} & \sigma_{p2} & \cdots & \sigma_{pp} \end{bmatrix}$$

is the multivariate population covariance matrix. The p in $N_p(\cdot)$ indicates a p-dimensional random vector.

3.3.1 Some properties of the multivariate normal distribution

1. Linear combinations of the components of a multivariate normal distribution are normally distributed. If $\mathbf{X} \sim N_p(\boldsymbol{\mu}, \boldsymbol{\Sigma})$, then $\mathbf{a}^T \mathbf{X} = a_1 X_1 + a_2 X_2 + \cdots + a_p X_p$ is univariate normally distributed and

$$\mathbf{a}^T \mathbf{X} \sim N(\mathbf{a}^T \boldsymbol{\mu}, \mathbf{a}^T \boldsymbol{\Sigma} \mathbf{a}) \tag{3.19}$$

2. All subsets of the components of $\mathbf{X}$ are multivariate normal distributions. If

$$\mathbf{X}_{p \times 1} = \begin{bmatrix} \mathbf{X}_{1_{q \times 1}} \\ --- \\ \mathbf{X}_{2_{(p-q) \times 1}} \end{bmatrix}, \quad \boldsymbol{\mu}_{p \times 1} = \begin{bmatrix} \boldsymbol{\mu}_{1_{q \times 1}} \\ --- \\ \boldsymbol{\mu}_{2_{(p-q) \times 1}} \end{bmatrix}, \quad \text{and}$$

$$\boldsymbol{\Sigma}_{p \times p} = \begin{bmatrix} \boldsymbol{\Sigma}_{11_{q \times q}} & | & \boldsymbol{\Sigma}_{12_{q \times (p-q)}} \\ --- & | & --- \\ \boldsymbol{\Sigma}_{21_{(p-q) \times q}} & | & \boldsymbol{\Sigma}_{22_{(p-q) \times (p-q)}} \end{bmatrix}$$

then

$$\mathbf{X}_1 \sim N_q(\boldsymbol{\mu}_1, \boldsymbol{\Sigma}_{11}) \quad \text{and} \quad \mathbf{X}_2 \sim N_{p-q}(\boldsymbol{\mu}_2, \boldsymbol{\Sigma}_{22}) \tag{3.20}$$

3. If $\mathbf{X}_{1_{q \times 1}}$ and $\mathbf{X}_{2_{r \times 1}}$ are independent, then $\boldsymbol{\Sigma}_{12_{q \times (p-q)}}$ and $\boldsymbol{\Sigma}_{21_{(p-q) \times q}}$ are zero matrices.

Remark. The univariate normal distribution is the special case of a multivariate normal distribution with $p = 1$. Univariate normal density is defined as

$$X \sim N(\mu_X, \sigma_X^2)$$

where

$$f(x) = \frac{1}{\sqrt{(2\pi)\sigma^2}} e^{-[(x-\mu)/\sigma]^2/2}, \quad -\infty < x < \infty \tag{3.21}$$

3.4 Multivariate Sampling Distribution

Though the multivariate normal distribution model is a very reasonable model to describe a multivariate population, samples of multivariate raw data are the only source of information for the multivariate population under study. Statistical inferences are the only viable practical means to draw conclusions about the multivariate population based on an analysis of sample data from the population. In statistical inference, the raw data from the sample is mathematically processed to become a statistic. By definition, a statistic is a function of observations in a sample that does not depend on unknown parameters. In statistical analysis, an intelligently designed statistic is the key to statistical inferences. For example, in univariate statistics, $\bar{x}$ is the sample mean. It is a function of raw sample data, so it is a statistic. It is an intelligently designed statistic because $\bar{x}$ can be used to estimate the population mean μ and $\bar{x}$ is considered the best estimate of μ because it has the smallest estimation error compared to all other alternatives.

The probability distribution of a statistic is called the sampling distribution. Obviously, knowledge regarding the sampling distribution is very important in statistical analysis. In this section, we are going to discuss the sampling distribution for multivariate statistics.

3.4.1 Sampling distribution of $\overline{\mathbf{X}}$

Assume that a multivariate random sample of n items is drawn from a multivariate normal population, $\mathbf{X} \sim N_p(\boldsymbol{\mu}, \boldsymbol{\Sigma})$. The sample data will have the following form (Srivastava, 2002):

$$\mathbf{X}_{n \times p} = \begin{bmatrix} x_{11} & x_{12} & x_{13} & \cdots & x_{1p} \\ x_{21} & x_{22} & x_{23} & \cdots & x_{2p} \\ x_{31} & x_{32} & x_{33} & \cdots & x_{3p} \\ \vdots & \vdots & \vdots & \vdots & \vdots \\ x_{n1} & x_{n2} & x_{n3} & \cdots & x_{np} \end{bmatrix}$$

Recall that

$$\overline{\mathbf{X}} = \frac{1}{n}\sum_{i=1}^{n} \mathbf{x}_{i.} \qquad (3.22)$$

Then

$$\overline{\mathbf{X}} \sim N_p\left(\boldsymbol{\mu}, \frac{1}{n}\boldsymbol{\Sigma}\right) \qquad (3.23)$$

Remark. For a sample mean $\overline{X}$ from a univariate normal population $X \sim N(\mu, \sigma^2)$, its distribution is

$$\overline{X} \sim N\left(\mu, \frac{\sigma^2}{n}\right) \qquad (3.24)$$

3.4.2 Sampling distribution of S

The sampling distribution of **S** is defined by the Wishart distribution (Johnson and Wichern, 2002). Specifically, $W_{n-1}(\mathbf{S}|\boldsymbol{\Sigma})$ is a Wishart distribution with $n-1$ degrees of freedom. It is the probability distribution of $\sum_{j=1}^{n-1} \mathbf{Z}_j \times \mathbf{Z}'_j$, where $\mathbf{Z}_j \sim N_p\left(\mathbf{0}_{p \times 1}, \boldsymbol{\Sigma}_{p \times p}\right)$.

Remark. The sample variance s^2 from a univariate normal population has the following distribution: $(n-1)s^2 \sim \sigma^2 \chi^2_{n-1}$, and χ^2_{n-1} is the distribution of $\sum_{i=1}^{n-1} Z_i^2$, where Z_i is the standard normal random variable.

3.4.3 Central limit theorem applied to multivariate samples

Let $\overline{\mathbf{X}}$ be the sample mean vector and **S** be the sample covariance matrix, based on multivariate random samples drawn from a multivariate population $\mathbf{X} \sim N_p(\boldsymbol{\mu}, \boldsymbol{\Sigma})$.

When $n \to \infty$,

$$\lim_{n\to\infty} \overline{X} = \mu, \qquad \lim_{n\to\infty} S = \Sigma$$

and

$$\sqrt{n}(\overline{X} - \mu) \sim N_p(0, \Sigma) \qquad (3.25)$$

$$n(\overline{X} - \mu)^T S^{-1}(\overline{X} - \mu) \to n(\overline{X} - \mu)^T \Sigma^{-1}(\overline{X} - \mu) \sim \chi_p^2 \qquad (3.26)$$

Equation (3.26) serves as basis for determining the control limits for many multivariate control charts, which we will discuss in Chap. 10. In practical application, usually when $n - p > 40$, Eq. (3.26) will be considered to be appropriate. When $n - p$ is significantly less than 40, the actual distribution of $n(\overline{X} - \mu)^T S^{-1}(\overline{X} - \mu)$ will be significantly different from χ_p^2. In this case, another very important multivariate sampling distribution, the T^2 distribution, is the appropriate distribution model for $n(\overline{X} - \mu)^T S^{-1}(\overline{X} - \mu)$.

3.4.4 Hotelling's T^2 distribution

A multivariate sampling distribution, that is, the multivariate counterpart of Student's t distribution is called Hotelling's T^2 distribution. The T^2 distribution, introduced by Hotelling (1947), is the basis of several multivariate control charts.

It is well known that for a random sample of size n from a univariate normal population, $X \sim N(\mu, \sigma^2)$, if we compute the sample mean $\overline{X}$ and sample variance s^2, then $t = (\overline{X} - \mu)/(s/\sqrt{n})$ will have a t distribution with $n - 1$ degrees of freedom. By using simple algebra

$$t^2 = (\overline{X} - \mu)^2/(s^2/n) = n(\overline{X} - \mu)(s^2)^{-1}(\overline{X} - \mu) \qquad (3.27)$$

When t^2 is generalized to p multivariate random variables, it becomes

$$T^2 = n(\overline{X} - \mu)^T S^{-1}(\overline{X} - \mu) \qquad (3.28)$$

T^2 is called Hotelling's T^2 statistic after its formulator, H. Hotelling. The distribution of T^2 can be determined by likelihood ratio tests (Bickel and Doksum, 1977). The sampling distribution of Hotelling's T^2 has the same shape as an F distribution. Specifically,

$$T^2 = n(\overline{X} - \mu)^T S^{-1}(\overline{X} - \mu) \sim \frac{p(n-1)}{n-p} F_{p, n-p} \qquad (3.29)$$

with $F_{p,n-p}$ representing F distribution with p numerator degrees of freedom and $n - p$ denominator degrees of freedom.

Similar to the significance of t distribution, the distribution of T^2 specified by Eq. (3.29) plays very important roles in hypothesis testing, on mean vectors, for multivariate random variables and establishing the control limits in multivariate control charts.

3.4.5 Summary

The T^2 distribution is the multivariate counterpart of the t distribution in the univariate statistical analysis:

$$T^2 = n(\overline{\mathbf{X}} - \boldsymbol{\mu})^T \mathbf{S}^{-1}(\overline{\mathbf{X}} - \boldsymbol{\mu}) \sim \frac{p(n-1)}{n-p} F_{p,n-p}$$

when $n - p$ is small (<40), and

$$T^2 = n(\overline{\mathbf{X}} - \boldsymbol{\mu})^T \mathbf{S}^{-1}(\overline{\mathbf{X}} - \boldsymbol{\mu}) \sim \chi_p^2$$

for large sample size ($n - p > 40$).

Clearly, the distributions relating T^2 are easy to deal with. On the other hand, the sampling distribution relating $\mathbf{S}$ is the Wishart distribution, which is the multivariate counter part of χ^2 distribution. However, the Wishart distribution is extremely complicated and difficult to use. As with univariate statistics, where t distribution is the basis for statistical inferences on means, T^2 is the basis for statistical inferences for mean vectors in multivariate statistics. Because T^2 is easy to use, the statistical inferences on mean vectors for multivariate random variables are easy to conduct. However, because Wishart distribution is difficult to use, there are no good approaches to statistical inferences from the covariance matrix.

3.5 Multivariate Statistical Inferences on Mean Vectors

There are two major tasks in statistical inferences. One is parameter estimation and the other is statistical hypothesis testing. In parameter estimation, we use the information in sample data to estimate the population parameters. Specifically, we design mathematical functions of sample data, called estimators, to estimate the population parameter. There are two kinds of estimators, the point estimator and interval estimator. The point estimator is the "best guess" for the population parameter to be estimated. For example, for a univariate normal population $X \sim N(\mu, \sigma^2)$, sample mean $\overline{X}$ is a point estimator for the population mean μ. However, the point estimator is based on sample data, and certainly there will be discrepancies between the point estimator and the population parameter. For example, $\overline{X}$ might be close to μ, but it will never be precisely equal to μ unless the sample size n is the entire population. The interval estimator is also called the confidence interval. It gives an interval so that we have a fair degree of confidence that the population parameter will fall in that interval. Clearly, the tighter the interval, the better.

Hypothesis testing uses information in the sample data to test a hypothesis or conjecture about a problem situation in the population. For example, a machining operation makes parts. The part dimension X has random variation and is believed to follow univariate normal distribution. That is, $X \sim N(\mu, \sigma^2)$. Certainly, it is very desirable for the population mean of part dimension μ to be equal to the designed nominal dimension or target dimension, μ_0. For example, if a machine makes a shaft with the intended diameter being 0.2 in, we would like the population mean for the produced shafts' diameter also to be 0.2 in. However, our desired result may or may not be true. There are two possibilities. In hypothesis testing, we describe these two possibilities as two hypotheses. One is called the null hypothesis H_0 and the other is called the alternative hypothesis H_1. For the machining problem, the two hypotheses are as follows:

H_0: $\mu = \mu_0$ (mean dimension is on target)

H_1: $\mu \neq \mu_0$ (mean dimension is not on target)

In statistical hypothesis testing, a random sample of data is collected. We compute a test statistic which is a function of sample data. Then, based on the values of the test statistic and relevant probability distribution theory, H_0 is either rejected or not rejected.

In the machining example, the test statistic is

$$t_0 = \frac{\overline{X} - \mu_0}{s/\sqrt{n}} \qquad (3.30)$$

Clearly, if H_0: $\mu = \mu_0$ is true, t_0 will follow a t distribution with $n - 1$ degrees of freedom, where n is the sample size. However, if H_0 is actually not true, that is, the mean part dimension is not on target, then it is quite likely that the sample mean $\overline{X}$ will show a sizable deviation from the production target. If the deviation $|\overline{X} - \mu|$ is significantly larger than $s/\sqrt{n}$, which is the sample standard deviation of $\overline{X}$, then t_0 will have a sizable value (either positive or negative) and significantly deviate from the t distribution. If that happens, then the evidence in the sample data will tend to reject H_0, that is, it will tend to indicate that the mean dimension is not on target.

Statistical inferences on univariate random variables having normal distribution are well established and described in many statistical textbooks.

Statistical inferences on multivariate random variables having multivariate normal distribution are well established on the problems regarding mean vectors because T^2 distribution is easy to use. However, statistical inferences regarding covariance matrices remain a difficult task. In this section, we will discuss the statistical inferences on mean vectors. Similar to univariate statistical inferences, the two major tasks in multivariate statistical inferences also are parameter estimation and hypothesis testing. In parameter estimation, the point estimators of multivariate population parameters $\boldsymbol{\mu}$, $\boldsymbol{\Sigma}$, and $\mathbf{P}$, are $\overline{\mathbf{X}}$, $\mathbf{S}$, and $\mathbf{R}$, respectively, and they have already been discussed in Secs. 3.3 and 3.4. Therefore, in Sec. 3.5, we will primarily discuss interval estimators for $\boldsymbol{\mu}$.

3.5.1 Small sample multivariate hypothesis testing on a mean vector

Situation. In this case, it is assumed that there is a multivariate normal population: $\mathbf{X} \sim \mathbf{N}_p(\boldsymbol{\mu}, \boldsymbol{\Sigma})$. A random sample of n multivariate observations is collected, where $n - p < 40$. Based on the sample data, $\overline{\mathbf{X}}$ and $\mathbf{S}$ are computed. The hypothesis testing problem is the following:

$$H_0: \boldsymbol{\mu} = \boldsymbol{\mu}_0, \quad H_1: \boldsymbol{\mu} \neq \boldsymbol{\mu}_0 \qquad (3.31)$$

Specifically, we test whether the population mean vector $\boldsymbol{\mu}$ is equal to a target vector $\boldsymbol{\mu}_0$.

Test statistics. If $H_0: \boldsymbol{\mu} = \boldsymbol{\mu}_0$ is true, from Eq. (3.29), a T^2-based test statistic can be derived:

$$T_0^2 = n(\overline{\mathbf{X}} - \boldsymbol{\mu}_0)^T \mathbf{S}^{-1}(\overline{\mathbf{X}} - \boldsymbol{\mu}_0) \sim \frac{p(n-1)}{n-p} F_{p,n-p} \qquad (3.32)$$

When $H_0: \boldsymbol{\mu} = \boldsymbol{\mu}_0$ is not true, $T_0^2 = n(\overline{\mathbf{X}} - \boldsymbol{\mu}_0)^T \mathbf{S}^{-1}(\overline{\mathbf{X}} - \boldsymbol{\mu}_0)$ will no longer be distributed as $[p(n-1)/n-p]F_{p,n-p}$ and its value will be significantly larger. The decision rule for this hypothesis problem is as follows:

Decision rule. Reject H_0 if

$$T_0^2 = n(\overline{\mathbf{X}} - \boldsymbol{\mu}_0)^T \mathbf{S}^{-1}(\overline{\mathbf{X}} - \boldsymbol{\mu}_0) > T_c^2 = \frac{p(n-1)}{n-p} F_{\alpha;p,n-p} \qquad (3.33)$$

where T_c^2 is called the cutoff score or critical value for T^2 statistic.

Example 3.7: To Test Whether the Welding Trainee Class Performance is Meeting the Requirements For the three training courses in the welding trainee program, the passing grade scores are given by $\boldsymbol{\mu}_0^T = [85\ 85\ 85]$. That is, it is desirable to have scores higher than 85 for all three courses. In order to evaluate the effectiveness of the training program, we test whether the population mean scores $\boldsymbol{\mu}^T = (\mu_1, \mu_2, \mu_3)$ are no less than the target specified by $\boldsymbol{\mu}_0^T = [85\ 85\ 85]$. We set up the following hypothesis:

$$H_0: \boldsymbol{\mu} \leq \boldsymbol{\mu}_0 \qquad H_1: \boldsymbol{\mu} > \boldsymbol{\mu}_0$$

In this test we select $\alpha = 0.05$. And we computed

$$\overline{\mathbf{X}} = \begin{bmatrix} 88.1 \\ 85.0 \\ 90.30 \end{bmatrix} \text{ and } \mathbf{S} = \begin{bmatrix} 72.9889 & 62 & 18.1889 \\ 62 & 70.222 & 7.333 \\ 18.1889 & 7.333 & 25.5667 \end{bmatrix}$$

The test statistic is

$$T_0^2 = n(\overline{\mathbf{X}} - \boldsymbol{\mu}_0)^T \mathbf{S}^{-1}(\overline{\mathbf{X}} - \boldsymbol{\mu}_0)$$

Test criteria: Reject the null hypothesis if $T_0^2 > T_c^2$

$$T_0^2 = n(\overline{\mathbf{X}} - \boldsymbol{\mu}_0)^T (\mathbf{S})^{-1}(\overline{\mathbf{X}} - \boldsymbol{\mu}_0)$$

$$= 10 \left(\begin{bmatrix} 88.1 \\ 85 \\ 90.32 \end{bmatrix} - \begin{bmatrix} 85 \\ 85 \\ 85 \end{bmatrix}\right)^T \begin{bmatrix} 72.9889 & 62 & 18.1889 \\ 62 & 70.222 & 7.333 \\ 18.1889 & 7.333 & 25.5667 \end{bmatrix}^{-1} \left(\begin{bmatrix} 88.1 \\ 85 \\ 90.32 \end{bmatrix} - \begin{bmatrix} 85 \\ 85 \\ 85 \end{bmatrix}\right) = 116.0611$$

where

$$\mathbf{S}^{-1} = \begin{bmatrix} 0.078647 & -0.06556 & -0.3715 \\ -0.06556 & 0.06933 & 0.026756 \\ -0.03715 & 0.026756 & 0.057868 \end{bmatrix}$$

The critical value is

$$T_c^2 = \frac{(n-1)p}{(n-p)} F_{\alpha;p,n-p} = \frac{(10-1)3}{(10-3)} F_{3,10-3}(0.05) = \frac{27}{7} F_{3,7}(0.05) = 16.7786$$

Since $T_0^2 > T_c^2$, H_0 is rejected, H_0 is a hypothesis that means the mean class performance is less than the target scores. Rejecting H_0 means that the trainee class's mean scores are better than the cutoff scores *for all three* courses.

3.5.2 Large sample multivariate hypothesis testing on a mean vector

Situation. It is assumed that there is a multivariate normal population: $\mathbf{X} \sim N_p(\boldsymbol{\mu}, \boldsymbol{\Sigma})$. A random sample of n multivariate observations is collected. However, the sample size is large such that $n - p > 40$. Based on the sample data, $\overline{\mathbf{X}}$ and $\mathbf{S}$ are computed. The hypothesis testing problem is the following:

$$H_0: \mu = \mu_0 \qquad H_1: \mu \neq \mu_0 \qquad (3.34)$$

Specifically, we are testing whether the population mean vector μ is equal to a target vector μ_0.

Test statistics. If $H_0: \mu = \mu_0$ is true, from Eq. (3.26), a T^2-based test statistic can be derived:

$$T_0^2 = n(\overline{\mathbf{X}} - \mu_0)^T \mathbf{S}^{-1}(\overline{\mathbf{X}} - \mu_0) \sim \chi_p^2 \qquad (3.35)$$

When $H_0: \mu = \mu_0$ is not true, $T_0^2 = n(\overline{\mathbf{X}} - \mu_0)^T \mathbf{S}^{-1}(\overline{\mathbf{X}} - \mu_0)$ will no longer be distributed as χ_p^2, and its value will be significantly larger. The decision rule for this hypothesis problem is as follows:

Decision rule. Reject H_0 if

$$T_0^2 = n(\overline{\mathbf{X}} - \mu_0)^T \mathbf{S}^{-1}(\overline{\mathbf{X}} - \mu_0) > T_c^2 = \chi_{\alpha;p}^2 \qquad (3.36)$$

3.5.3 Small sample multivariate hypothesis testing on the equality of two mean vectors

Situation. It is assumed that there are two multivariate normal populations: $\mathbf{X}_1 \sim N_p(\mu_1, \Sigma_1)$ and $\mathbf{X}_2 \sim N_p(\mu_2, \Sigma_2)$. Two random samples of multivariate observations, with sample size n_1 from the first population and n_2 from the second population, are collected. The sample size is small such that $n_1 + n_2 - p < 40$. Based on the sample data from both samples, two sample mean vectors $\overline{\mathbf{X}}_1$ and $\overline{\mathbf{X}}_2$ and two sample covariance matrices $\mathbf{S}_1$ and $\mathbf{S}_2$ are computed. It is further assumed that $\Sigma_1 = \Sigma_2$.

The hypothesis testing problem is the following:

$$H_0: \mu_1 - \mu_2 = \theta_0 \qquad H_1: \mu_1 - \mu_2 \neq \theta_0 \qquad (3.37)$$

When $\theta_0 = 0$, we test whether the two population mean vectors are equal, that is, if $\mu_1 = \mu_2$.

Test statistics. When $\Sigma_1 = \Sigma_2 = \Sigma$, $\sum_{j=1}^{n_1}(\mathbf{x}_{1j} - \overline{\mathbf{x}}_1)(\mathbf{x}_{1j} - \overline{\mathbf{x}}_1)^T$ is an estimate of $(n_1 - 1)\Sigma$ and $\sum_{j=1}^{n_1}(\mathbf{x}_{2j} - \overline{\mathbf{x}}_2)(\mathbf{x}_{2j} - \overline{\mathbf{x}}_2)^T$ is an estimate of $(n_2 - 1)\Sigma$. Subsequently we can pool the two variances to estimate the common variance Σ'.

$$\mathbf{S}_p = \frac{n_1 - 1}{n_1 + n_2 - 2}\mathbf{S}_1 + \frac{n_2 - 1}{n_1 + n_2 - 2}\mathbf{S}_2$$

If $H_0: \mu_1 - \mu_2 = \theta_0$ is true, a T^2-based test statistic can be derived:

$$T_0^2 = (\bar{X}_1 - \bar{X}_2 - \theta_0)^T \left[\left(\frac{1}{n_1} + \frac{1}{n_2}\right) S_p \right]^{-1} (\bar{X}_1 - \bar{X}_2 - \theta_0)$$

$$\sim \left(\frac{n_1 + n_2 - 2}{n_1 + n_2 - p - 1}\right) F_{p, n_1 + n_2 - p - 1} \qquad (3.38)$$

When $H_0: \mu_1 - \mu_2 = \theta_0$ is not true,

$$T_0^2 = (\bar{X}_1 - \bar{X}_2 - \theta_0)^T \left[\left(\frac{1}{n_1} + \frac{1}{n_2}\right) S_p \right]^{-1} (\bar{X}_1 - \bar{X}_2 - \theta_0)$$

will no longer be distributed as

$$\left(\frac{n_1 + n_2 - 2}{n_1 + n_2 - p - 1}\right) F_{p, n_1 + n_2 - p - 1}$$

and its value will be significantly larger. The decision rule for this hypothesis problem is as follows:

Decision rule. Reject H_0 if

$$T_0^2 = (\bar{X}_1 - \bar{X}_2 - \theta_0)^T \left[\left(\frac{1}{n_1} + \frac{1}{n_2}\right) S_p \right]^{-1} (\bar{X}_1 - \bar{X}_2 - \theta_0)$$

$$> T_c^2 = \left(\frac{n_1 + n_2 - 2}{n_1 + n_2 - p - 1}\right) F_{\alpha; p, n_1 + n_2 - p - 1} \qquad (3.39)$$

Example 3.8: The following data set is obtained from a study conducted to determine the impact–absorbing qualities of bicycle [see p. 81] helmets. Two measurements ($p = 2$), impact force (foot pounds) and acceleration (G units), were taken on two sides of 19 samples of helmets. The researcher wanted to determine the differences in the impact force and acceleration between the front and back of the helmets ($\alpha = 0.01$). The mean vectors are analyzed to be

$$\text{For front of helmet} \quad \bar{X}_1 = \begin{bmatrix} 140 \\ 419 \end{bmatrix}$$

$$\text{For back of helmet} \quad \bar{X}_2 = \begin{bmatrix} 155.89 \\ 412.11 \end{bmatrix}$$

The covariance matrices are analyzed to be

$$\text{For front of helmet} \quad S_1 = \begin{bmatrix} 545.78 & 213.78 \\ 213.78 & 330.88 \end{bmatrix}$$

$$\text{For back of helmet} \quad S_2 = \begin{bmatrix} 478.51 & 249.32 \\ 249.32 & 526.65 \end{bmatrix}$$

Is there any difference in impact force or acceleration between the two sides?

$$S_p = \frac{n_1 - 1}{n_1 + n_2 - 2} S_1 + \frac{n_2 - 1}{n_1 + n_2 - 2} S_2 = \begin{bmatrix} 751.4 & 356.25 \\ 356.25 & 694.09 \end{bmatrix}$$

$$\Sigma' = \left(\frac{1}{n_1} + \frac{1}{n_1} \right) S_p = \begin{bmatrix} 79.09 & 37.5 \\ 37.5 & 73.06 \end{bmatrix}$$

$$\left[\left(\frac{1}{n_1} + \frac{1}{n_2} \right) S_p \right]^{-1} = \begin{bmatrix} 0.0167 & -0.0086 \\ -0.0086 & 0.0181 \end{bmatrix}$$

Hypothesis test:

$$H_0 : \mu_1 = \mu_2 \qquad H_a : \mu_1 \neq \mu_2$$

$$T_0^2 = (\overline{X}_1 - \overline{X}_2)^T \left[\left(\frac{1}{n_1} + \frac{1}{n_2} \right) S_p \right]^{-1} (\overline{X}_1 - \overline{X}_2) = 3.89$$

$$T_c^2 = \left(\frac{n_1 + n_2 - 2}{n_1 + n_2 - p - 1} \right) F_{\alpha; p, n_1 + n_2 - p - 1} = \frac{36}{35} F_{0.01; 2, 35} = 10.86$$

Since $T_0^2 \leq T_c^2$ we cannot reject the null hypothesis and hence do not have enough evidence to support the statement that there are significant differences in impact force or acceleration between the two sides.

3.5.4 Large sample multivariate hypothesis testing on the equality of two mean vectors

Situation. It is assumed that there are two multivariate normal populations: $X_1 \sim N_p(\mu_1, \Sigma_1)$ and $X_2 \sim N_p(\mu_2, \Sigma_2)$. Two random samples of multivariate observations, with sample size n_1 from the first population and n_2 from the second population, are collected. The sample size is large such that $n_1 + n_2 - p > 40$. Based on the sample data from both samples, two sample mean vectors $\overline{X}_1$ and $\overline{X}_2$ and two sample covariance matrices S_1 and S_2 are computed. It is further assumed that $\Sigma_1 = \Sigma_2$.

The hypothesis testing problem is the following:

$$H_0 : \mu_1 - \mu_2 = \theta_0 \qquad H_1 : \mu_1 - \mu_2 \neq \theta_0 \qquad (3.40)$$

When $\theta_0 = 0$, we test whether the two population mean vectors are equal, that is, if $\mu_1 = \mu_2$.

Test statistics. When sample size is large, the test statistic is

$$T_0^2 = (\overline{\mathbf{X}}_1 - \overline{\mathbf{X}}_2 - \boldsymbol{\theta}_0)^T \left(\frac{1}{n_1} S_1 + \frac{1}{n_2} S_2 \right)^{-1} (\overline{\mathbf{X}}_1 - \overline{\mathbf{X}}_2 - \boldsymbol{\theta}_0) \sim \chi_p^2 \quad (3.41)$$

When $H_0: \boldsymbol{\mu}_1 - \boldsymbol{\mu}_2 = \boldsymbol{\theta}_0$ is not true,

$$T_0^2 = (\overline{\mathbf{X}}_1 - \overline{\mathbf{X}}_2 - \boldsymbol{\theta}_0)^T \left[\left(\frac{1}{n_1} + \frac{1}{n_2} \right) S_p \right]^{-1} (\overline{\mathbf{X}}_1 - \overline{\mathbf{X}}_2 - \boldsymbol{\theta}_0)$$

will no longer be distributed as $\sim \chi_p^2$ and its value will be significantly larger. The decision rule for this hypothesis problem is as follows:

Decision rule. Reject H_0 if

$$T_0^2 = (\overline{\mathbf{X}}_1 - \overline{\mathbf{X}}_2 - \boldsymbol{\theta}_0)^T \left[\left(\frac{1}{n_1} + \frac{1}{n_2} \right) S_p \right]^{-1} (\overline{\mathbf{X}}_1 - \overline{\mathbf{X}}_2 - \boldsymbol{\theta}_0) > T_c^2 = \chi_{\alpha;p}^2 \quad (3.42)$$

3.5.5 Overview of confidence intervals and confidence regions in multivariate statistical inferences

In univariate statistical inferences, each population parameter is a scalar. For example, both μ and σ^2 are scalars for univariate normal distribution. So the confidence interval is a bracket for which there is a high confidence level that the bracket will contain the true parameter. In multivariate statistical inferences, the population parameters $\boldsymbol{\mu}$, Σ, and $\mathbf{P}$ are vectors and matrices; the confidence intervals in turn will become confidence regions. For the mean vector $\boldsymbol{\mu}$, the confidence region is a p-dimensional region for which there is a high confidence level that the region will contain the mean vector $\boldsymbol{\mu}$. However, when p is more than 2, it becomes rather difficult to comprehend how this confidence region works. Therefore, simultaneous confidence intervals are developed. Simultaneous confidence intervals are a group of intervals where each interval contains an individual component of mean vector with a $100(1 - \alpha)\%$ confidence. Other kinds of tighter simultaneous confidence intervals are developed, called Bonferroni simultaneous confidence intervals, which work well for small p. In the following subsections, we will discuss both confidence regions and simultaneous confidence intervals for various cases. An exhaustive explanation of the multivariate statistical inferences is given by Harris (1975).

3.5.6 Confidence regions and intervals for a single mean vector with small sample size

Situation. In this case, it is assumed that there is a multivariate normal population: $\mathbf{X} \sim N_p(\mathbf{\mu}, \mathbf{\Sigma})$. A random sample of n multivariate observations is collected, where $n - p < 40$. Based on the sample data, $\overline{\mathbf{X}}$ and $\mathbf{S}$ are computed. We would like to find the confidence region or simultaneous confidence intervals for the mean vector $\mathbf{\mu}$.

Confidence region. Because

$$n(\overline{\mathbf{X}} - \mathbf{\mu}_0)^T \mathbf{S}^{-1}(\overline{\mathbf{X}} - \mathbf{\mu}_0) \sim \frac{p(n-1)}{n-p} F_{p,n-p}$$

so

$$P\left[n(\overline{\mathbf{X}} - \mathbf{\mu})^T \mathbf{S}^{-1}(\overline{\mathbf{X}} - \mathbf{\mu}) \leq \frac{(n-1)p}{(n-p)} F_{\alpha;p,n-p} \right] = 1 - \alpha \quad (3.43)$$

Thus, a $100(1 - \alpha)\%$ confidence region for $\mathbf{\mu}$ can be characterized by the following inequality:

$$n(\overline{\mathbf{X}} - \mathbf{\mu})^T \mathbf{S}^{-1}(\overline{\mathbf{X}} - \mathbf{\mu}) \leq c^2 = \frac{(n-1)p}{(n-p)} F_{\alpha;p,n-p} \quad (3.44)$$

Geometrically, the confidence region is an ellipsoid centered at $\overline{\mathbf{X}}$ with related length of axes defined by c in Eq. (3.44).

Simultaneous confidence intervals. The simultaneous confidence intervals are determined for each individual component mean and each of these confidence intervals are determined with a $100(1 - \alpha)\%$ confidence.

The simultaneous confidence intervals for this case are given by

$$\bar{x}_1 - \sqrt{\frac{(n-1)p}{(n-p)} F_{\alpha;p,n-p}} \sqrt{\frac{S_{11}}{n}} \leq \mu_1 \leq \bar{x}_1 + \sqrt{\frac{(n-1)p}{(n-p)} F_{\alpha;p,n-p}} \sqrt{\frac{S_{11}}{n}}$$

$$\bar{x}_2 - \sqrt{\frac{(n-1)p}{(n-p)} F_{\alpha;p,n-p}} \sqrt{\frac{S_{22}}{n}} \leq \mu_2 \leq \bar{x}_2 + \sqrt{\frac{(n-1)p}{(n-p)} F_{\alpha;p,n-p}} \sqrt{\frac{S_{22}}{n}} \quad (3.45)$$

$$\vdots$$

$$\bar{x}_p - \sqrt{\frac{(n-1)p}{(n-p)} F_{\alpha;p,n-p}} \sqrt{\frac{S_{pp}}{n}} \leq \mu_p \leq \bar{x}_p + \sqrt{\frac{(n-1)p}{(n-p)} F_{\alpha;p,n-p}} \sqrt{\frac{S_{pp}}{n}}$$

and all hold simultaneously with confidence coefficient $(1 - \alpha)$.

Example 3.9: Construct 95% simultaneous confidence intervals around the population means for the welding trainee data set shown in Fig. 3.1.

Using

$$\overline{X} = \begin{bmatrix} 88.1 \\ 88.5 \\ 90.30 \end{bmatrix}$$

as a sampling estimate of μ and

$$S = \begin{bmatrix} 72.9889 & 62 & 18.1889 \\ 62 & 70.222 & 7.333 \\ 18.1889 & 7.333 & 25.5667 \end{bmatrix}$$

as an estimate of $\Sigma_{\overline{x}}$, we get

$$\overline{x}_j - \sqrt{\frac{(n-1)p}{(n-p)}F_{\alpha;p,n-p}}\sqrt{\frac{S_{jj}}{n}} \leq \mu_j \leq \overline{x}_j + \sqrt{\frac{(n-1)p}{(n-p)}F_{\alpha;p,n-p}}\sqrt{\frac{S_{jj}}{n}}$$

$$\sqrt{\frac{(n-1)p}{(n-p)}F_{\alpha;p,n-p}} = \sqrt{\frac{9\times 3}{7}F_{0.05;3,7}} = 4.096$$

$$88.1 - 4.096\sqrt{\frac{72.9889}{10}} \leq \mu_1 \leq 88.1 + 4.096\sqrt{\frac{72.9889}{10}} = 77.03 \leq \mu_1 \leq 99.17$$

$$85.0 - 4.096\sqrt{\frac{70.222}{10}} \leq \mu_2 \leq 85.0 + 4.096\sqrt{\frac{70.222}{10}} = 74.15 \leq \mu_2 \leq 95.86$$

$$90.3 - 4.096\sqrt{\frac{25.667}{10}} \leq \mu_3 \leq 90.3 + 4.096\sqrt{\frac{25.667}{10}} = 83.75 \leq \mu_3 \leq 96.85$$

Bonferroni's simultaneous confidence limits. For situations where the number of simultaneous confidence limits being generated is small, the simultaneous confidence limits can be made tighter by using the Bonferroni confidence limits. These are given by

$$\overline{x}_j - t_{\alpha_j/2, n-1}\sqrt{\frac{S_{jj}}{n}} \leq \mu_j \leq \overline{x}_j + t_{\alpha_j/2, n-1}\sqrt{\frac{S_{jj}}{n}} \tag{3.46}$$

where $\sum_{j=1}^{p}\alpha_j = \alpha$. Typically the Type I error is equally split between the p dimensions.

Example 3.10: Construct 95% Bonferroni simultaneous confidence intervals around the population means for the welding trainee data set shown in Fig. 3.1.

Using

$$\overline{X} = \begin{bmatrix} 88.1 \\ 88.5 \\ 90.30 \end{bmatrix}$$

as a sampling estimate of μ and

$$S = \begin{bmatrix} 72.9889 & 62 & 18.1889 \\ 62 & 70.222 & 7.333 \\ 18.1889 & 7.333 & 25.5667 \end{bmatrix}$$

as an estimate of $\Sigma_{\overline{X}}$, we get

$$\overline{x}_j - t_{\frac{\alpha_j}{2}, n-1}\sqrt{\frac{S_{jj}}{n}} \le \mu_j \le \overline{x}_j + t_{\frac{\alpha_j}{2}, n-1}\sqrt{\frac{S_{jj}}{n}}$$

$$\alpha_j = \frac{0.05}{3} = 0.0167 \qquad t_{\frac{\alpha_j}{2}, n-1} = 2.933$$

$$88.1 - 2.933\sqrt{\frac{72.9889}{10}} \le \mu_1 \le 88.1 + 2.933\sqrt{\frac{72.9889}{10}} = 80.18 \le \mu_1 \le 96.02$$

$$85.0 - 2.933\sqrt{\frac{70.222}{10}} \le \mu_2 \le 85.0 + 2.933\sqrt{\frac{70.222}{10}} = 77.23 \le \mu_2 \le 92.77$$

$$90.3 - 2.933\sqrt{\frac{25.667}{10}} \le \mu_3 \le 90.3 + 2.933\sqrt{\frac{25.667}{10}} = 85.6 \le \mu_3 \le 94.99$$

The Bonferroni confidence limits calculated in Example 3.9 are much tighter than the simultaneous confidence limits calculated in Example 3.8.

3.5.7 Confidence regions and intervals for a single mean vector with large sample size

Situation. In this case, it is assumed that there is a multivariate normal population: $X \sim N_p(\mu, \Sigma)$. A random sample of n multivariate observations is collected, where $n - p > 40$. Based on the sample data, $\overline{X}$ and S are computed. We would like to find the confidence region or simultaneous confidence intervals for the mean vector μ.

Confidence region. Because

$$n(\overline{X} - \mu_0)^T S^{-1}(\overline{X} - \mu_0) \sim \chi_p^2$$

for large sample size, so

$$P\left[n(\overline{\mathbf{X}}-\boldsymbol{\mu})^T \mathbf{S}^{-1}(\overline{\mathbf{X}}-\boldsymbol{\mu}) \leq \chi^2_{a,p}\right] = 1-\alpha \qquad (3.47)$$

Thus, a $100(1-\alpha)\%$ confidence region for $\boldsymbol{\mu}$ can be characterized by the following inequality:

$$\left[n(\overline{\mathbf{X}}-\boldsymbol{\mu})^T \mathbf{S}^{-1}(\overline{\mathbf{X}}-\boldsymbol{\mu}) \leq c^2 = \chi^2_{a,p}\right] \qquad (3.48)$$

Geometrically, the confidence region is an ellipsoid centered at $\overline{\mathbf{X}}$ with related length of axes defined by c in Eq. (3.48).

Simultaneous confidence intervals

$$\overline{x}_j - \chi^2_{a,p}\sqrt{\frac{S_{jj}}{n}} \leq \mu_j \leq \overline{x}_j + \chi^2_{a,p}\sqrt{\frac{S_{jj}}{n}} \qquad j=1,\ldots,p \qquad (3.49)$$

Individual $100(1-\alpha)\%$ confidence intervals

$$\overline{x}_j - Z_{\alpha/2p}\sqrt{\frac{S_{jj}}{n}} \leq \mu_j \leq \overline{x}_j + Z_{\alpha/2p}\sqrt{\frac{S_{jj}}{n}} \qquad j=1,\ldots,p \qquad (3.50)$$

Bonferroni simultaneous confidence intervals

$$\overline{x}_j - Z_{\alpha/2p}\sqrt{\frac{S_{jj}}{n}} \leq \mu_j \leq \overline{x}_j + Z_{\alpha/2p}\sqrt{\frac{S_{jj}}{n}} \qquad j=1,\ldots,p \qquad (3.51)$$

3.5.8 Confidence regions and intervals for the difference in two population mean vectors for small samples

It is assumed that there are two multivariate normal populations: $\mathbf{X}_1 \sim \mathbf{N}_p(\boldsymbol{\mu}_1, \boldsymbol{\Sigma}_1)$ and $\mathbf{X}_2 \sim \mathbf{N}_p(\boldsymbol{\mu}_2, \boldsymbol{\Sigma}_2)$. Two random samples of multivariate observations, with sample size n_1 from the first population and n_2 from the second population, are collected. The sample size is small such that $n_1 + n_2 - p < 40$. Based on the sample data from both samples, two sample mean vectors $\overline{\mathbf{X}}_1$ and $\overline{\mathbf{X}}_2$ and two sample covariance matrices $\mathbf{S}_1$ and $\mathbf{S}_2$ are computed. It is further assumed that $\boldsymbol{\Sigma}_1 = \boldsymbol{\Sigma}_2$. We now find out the confidence region and intervals for $\boldsymbol{\mu}_1 - \boldsymbol{\mu}_2$.

Confidence region. The T^2 confidence region for the difference of two population means when the population covariances are equal is given by

$$P\left[\left\{(\bar{X}_1-\bar{X}_2)^T\left[\left(\frac{1}{n_1}+\frac{1}{n_2}\right)S_p\right]^{-1}(\bar{X}_1-\bar{X}_2)\right\}\le c^2\right]=1-\alpha \quad (3.52)$$

where

$$c^2=\left(\frac{n_1+n_2-2}{n_1+n_2-p-1}\right)F_{\alpha;p,n_1+n_2-p-1}$$

This region is an ellipsoid centered at $\bar{X}_1-\bar{X}_2$.

Simultaneous confidence intervals. The simultaneous confidence intervals are given by

$$\bar{x}_{1j}-\bar{x}_{2j}-c^2\sqrt{\left(\frac{1}{n_1}+\frac{1}{n_2}\right)S_{jj,pooled}}\le\bar{\mu}_{1j}-\bar{\mu}_{2j}\le\bar{x}_p+c^2\sqrt{\left(\frac{1}{n_1}+\frac{1}{n_2}\right)S_{jj,pooled}}$$

$$(3.53)$$

where

$$c^2=\left(\frac{n_1+n_2-2}{n_1+n_2-p-1}\right)F_{\alpha;p,n_1+n_2-p-1}$$

Bonferroni confidence limits. Bonferroni simultaneous confidence intervals are given by

$$\bar{x}_{1j}-\bar{x}_{2j}-c^2\sqrt{\left(\frac{1}{n_1}+\frac{1}{n_2}\right)S_{jj,pooled}}\le\bar{\mu}_{1j}-\bar{\mu}_{2j}\le\bar{x}_p+c^2\sqrt{\left(\frac{1}{n_1}+\frac{1}{n_2}\right)S_{jj,pooled}}$$

$$(3.54)$$

where $c^2=t_{\frac{\alpha}{2p},n_1+n_2-2}$.

3.5.9 Confidence regions and intervals for the difference in two population mean vectors for large samples

It is assumed that there are two multivariate normal populations: $X_1\sim N_p(\mu_1,\Sigma_1)$ and $X_2\sim N_p(\mu_2,\Sigma_2)$. Two random samples of multivariate observations, with sample size n_1 from the first population and n_2 from the second population, are collected. The sample size is large such that $n_1+n_2-p>40$. Based on the sample data from both samples, two sample mean vectors $\bar{X}_1$ and $\bar{X}_2$ and two sample covariance matrices S_1

and S_2 are computed. It is further assumed that $\Sigma_1 \neq \Sigma_2$. We now find out the confidence region and intervals for $\mu_1 - \mu_2$.

Confidence region. The confidence region should satisfy

$$P\left[\left[(\overline{X}_1 - \overline{X}_2)^T \left(\frac{1}{n_1}S_1 + \frac{1}{n_2}S_2\right)^{-1} (\overline{X}_1 - \overline{X}_2)\right] \leq c^2\right] = 1 - \alpha \qquad (3.55)$$

The $100(1 - \alpha)\%$ confidence ellipsoid for $\mu_1 - \mu_2$ is given by

$$\left[(\overline{X}_1 - \overline{X}_2)^T \left(\frac{1}{n_1}S_1 + \frac{1}{n_2}S_2\right)^{-1} (\overline{X}_1 - \overline{X}_2)\right] \leq \chi^2_{\alpha,p} \qquad (3.56)$$

3.5.10 Other Cases

Small sample sizes or samples in which population distributions significantly depart from multivariate normal distributions are not statistically comparable with certainty. Some methods can be applied but with some possible erroneous results, as suggested by Bartlett (1937) and Tika and Balakrishnan (1985). Removal of outliers or data transformations can be made to the data to correct departures from multivariate normal distributions to be able to apply Case 2. These methods can be read in Srivastava and Carter (1983).

Appendix 3A: Matrix Algebra Refresher

A.1 Introduction

Matrices provide a theoretically and practically useful way of approaching many types of problems.

A.2 Notations and basic operations

The organization of real numbers into a rectangular or square array consisting of p rows and q columns is called a *matrix* of order p by q and written as $p \times q$. Specifically

$$\mathbf{A} = \mathbf{A}_{p \times q} = \begin{bmatrix} a_{11} & a_{12} & \cdots & a_{1q} \\ a_{21} & a_{22} & \cdots & a_{2q} \\ \vdots & \vdots & \vdots & \vdots \\ a_{p1} & a_{p2} & \cdots & a_{pq} \end{bmatrix}$$

The above matrix can also be denoted as $\mathbf{A} = (a_{ij}){:}p \times q$.

When the numbers of rows and columns are equal, we call the matrix a *square matrix*. A square matrix of order p, is a ($p \times p$) matrix.

Null matrix. If all the elements of a matrix are zero, then the matrix is a null matrix or zero matrix.

$$\mathbf{0}_{3\times 2} = \begin{bmatrix} 0 & 0 \\ 0 & 0 \\ 0 & 0 \end{bmatrix}$$

Square matrix. If $p = q$, then the matrix is a square matrix of order p.

$$\mathbf{A}_{3\times 3} = \begin{bmatrix} a_{11} & a_{12} & a_{13} \\ a_{21} & a_{22} & a_{23} \\ a_{31} & a_{32} & a_{33} \end{bmatrix}$$

Column vector. If $q = 1$, then we have a p-column vector.

$$\mathbf{a}_{3\times 1} = \begin{bmatrix} 1 \\ 4 \\ 2 \end{bmatrix}$$

Row vector. If $p = 1$, we have a q-row vector.

$$\mathbf{a}^T_{1\times 3} = \begin{bmatrix} 3 & 2 & 8 \end{bmatrix}$$

Lower and upper triangular matrix. If matrix $\mathbf{A}$ is a square matrix and all elements above the main diagonal are equal to 0, then $\mathbf{A}$ is a lower triangular matrix.

$$\mathbf{A} = \begin{bmatrix} 2 & 0 & 0 \\ 6 & 3 & 0 \\ 1 & 2 & 7 \end{bmatrix}$$

If matrix $\mathbf{B}$ is a square matrix and all elements below the main diagonal are equal to 0, then $\mathbf{A}$ is an upper triangular matrix.

$$\mathbf{B} = \begin{bmatrix} 2 & 3 & 6 \\ 0 & 7 & 0 \\ 0 & 0 & 7 \end{bmatrix}$$

Diagonal matrix. A square matrix with off-diagonal elements equal to 0 is called a diagonal matrix. If the diagonal elements of $\mathbf{A}$ are $a_{11}, a_{22}, ..., a_{pp}$, then $\mathbf{A}$ is written as D_a or $\text{diag}(a_{11}, a_{22}, ..., a_{pp})$.

$$\mathbf{A} = \begin{bmatrix} 2 & 0 & 0 \\ 0 & 3 & 0 \\ 0 & 0 & 7 \end{bmatrix}$$

Identity matrix. If $\mathbf{A}$ is a diagonal $p \times p$ matrix with all diagonal entries equal to 1 (all off-diagonal entries equal to 0), then $\mathbf{A}$ is called an identity matrix, written as $I_{p \times p}$.

$$\mathbf{A} = \begin{bmatrix} 1 & 0 & 0 \\ 0 & 1 & 0 \\ 0 & 0 & 1 \end{bmatrix}$$

Transpose matrix. If the rows and colums of a matrix are interchanged, the resulting matrix is a transpose of the original matrix, denoted by $\mathbf{A}^T$ or $\mathbf{A}'$.

$$\mathbf{A} = \begin{bmatrix} 2 & 5 & 4 \\ 6 & 3 & 0 \\ 1 & 2 & 7 \end{bmatrix} \text{ and } \mathbf{A}^T = \begin{bmatrix} 2 & 6 & 1 \\ 5 & 3 & 2 \\ 4 & 0 & 7 \end{bmatrix}$$

Symmetric matrix. A square matrix is said to be symmetric if it is equal to its transpose.

$$\mathbf{A} = \begin{bmatrix} 2 & 5 & 4 \\ 5 & 3 & 0 \\ 4 & 0 & 7 \end{bmatrix} \text{ and } \mathbf{A}^T = \begin{bmatrix} 2 & 5 & 4 \\ 5 & 3 & 0 \\ 4 & 0 & 7 \end{bmatrix}$$

hence $\mathbf{A} = \mathbf{A}^T$, so $\mathbf{A}$ is a symmetric matrix.

A.3 Matrix operations

Matrix addition. Addition of two matrices $\mathbf{A}: m \times n$ and $\mathbf{B}: p \times q$ is defined only if they are of the same order. Thus if $\mathbf{A} = (a_{ij})_{p \times q}$ and $\mathbf{B} = (b_{ij})_{p \times q}$, then their sum is defined by

$$\mathbf{A} + \mathbf{B} = (a_{ij} + b_{ij})_{p \times q}$$

$$\mathbf{A} = \begin{bmatrix} 2 & 5 & 4 \\ 6 & 3 & 0 \\ 1 & 2 & 7 \end{bmatrix} \text{ and } \mathbf{B} = \begin{bmatrix} 1 & 6 & 1 \\ 3 & 3 & 2 \\ 4 & 9 & 7 \end{bmatrix}, \text{ then } \mathbf{A} + \mathbf{B} = \begin{bmatrix} 2+1 & 5+6 & 4+1 \\ 6+3 & 3+3 & 0+2 \\ 1+4 & 2+9 & 7+7 \end{bmatrix}$$

Matrix multiplication. Two matrices $\mathbf{A}$ and $\mathbf{B}$ can be multiplied only if the number of columns of $\mathbf{A}$ is equal to the number of rows of $\mathbf{B}$.

Thus if $\mathbf{A} = (a_{ij})_{p \times q}$ and $\mathbf{B} = (b_{ij})_{q \times r}$, then their product is defined by

$$\mathbf{A}_{p \times q} \times \mathbf{B}_{q \times r} = \mathbf{C}_{q \times r} = (c_{ij}), \text{ where}$$

$$c_{ij} = \sum_{k=1}^{q} a_{ik} \times b_{kj} \quad i = 1, \ldots, p \quad \text{and} \quad j = 1, \ldots, r$$

Some results that follow:

1. $(\mathbf{A} + \mathbf{B})^T = \mathbf{A}^T + \mathbf{B}^T$, $(\mathbf{A} + \mathbf{B} + \mathbf{C})^T = \mathbf{A}^T + \mathbf{B}^T + \mathbf{C}^T$, etc.
2. $(\mathbf{A} \times \mathbf{B})^T = \mathbf{B}^T \times \mathbf{A}^T$, $(\mathbf{A} \times \mathbf{B} \times \mathbf{C})^T = \mathbf{C}^T \times \mathbf{B}^T \times \mathbf{A}^T$, etc.
3. $\mathbf{A}(\mathbf{B} + \mathbf{C}) = \mathbf{A} \times \mathbf{B} + \mathbf{A} \times \mathbf{C}$

Orthogonal matrix. A square matrix $\mathbf{A}$ is said to be orthogonal if $\mathbf{A}(\mathbf{A}^T) = \mathbf{I}$.

Determinant of a matrix. The determinant of a square matrix $\mathbf{A} = (a_{ij})$: $p \times p$ is defined as $|\mathbf{A}| \equiv \sum_{\alpha}(-1)^{N(\alpha)} \prod_{j=1}^{p} a_{\alpha_j, j}$ denotes the summation over the distinct permutations α of the numbers $1, 2, \ldots, n$ and $N(\alpha)$ is the total number of inversions of a permutation. An inversion of a permutation $\alpha = (\alpha_1, \alpha_2, \ldots, \alpha_n)$ is an arrangement of two indices such that the larger index comes after the smaller index. For example, $N(2, 1, 4, 3) = 1 + N(1, 2, 4, 3) = 2 + N(1, 2, 3, 4) = 2$ since $N(1, 2, 3, 4) = 0$ $N(4, 3, 1, 2) = 1 + N(3, 4, 1, 2) = 3 + (3, 1, 2, 4) = 5$. For example,

$$\begin{vmatrix} 2 & 1 \\ 1 & 5 \end{vmatrix} = 2 \times 5 - 1 \times 1 = 9$$

Larger order (>2) matrix determinants are easily computed using Microsoft Excel, Minitab, and other mathematical/statistical/analysis software.

Submatrices. Let us represent a matrix $\mathbf{A}$ as

$$\mathbf{A} = \begin{bmatrix} \mathbf{A}_{11} & \mathbf{A}_{12} \\ \mathbf{A}_{21} & \mathbf{A}_{22} \end{bmatrix}$$

where $\mathbf{A}_{ij}$ is a $i \times j$, submatrix of $\mathbf{A}$ ($i, j = 1, 2$). A submatrix of $\mathbf{A}$ is obtained by deleting certain rows and columns.

Trace of a Matrix. The trace of a matrix tr($\mathbf{A}$) is the summation of $\mathbf{A}$'s diagonal elements, that is, tr($\mathbf{A}_{n \times n}$) = $\sum_{i=1}^{n} A_{ii}$. For example, if

$$\mathbf{A} = \begin{bmatrix} 1 & 6 & 5 & 7 \\ 6 & 9 & 10 & 12 \\ 3 & 7 & 8 & 10 \\ 2 & 5 & 9 & 11 \end{bmatrix}$$

then, tr($\mathbf{A}$) = $1 + 9 + 8 + 11$.

Property of a determinant. Let $\mathbf{A}$ be a square matrix of order n, then the determinant $|\mathbf{A} + \lambda\mathbf{I}_n|$ is a polynomial of degree n in λ and is written as

$$\left|\mathbf{A} + \lambda\mathbf{I}_n\right| = \sum_{i=1}^{n} \lambda^i \operatorname{tr}_{n-1} \mathbf{A}$$

where $\operatorname{tr}_0 \mathbf{A} = 1$. If $\lambda_1, \lambda_2, \ldots, \lambda_n$ denote the roots of this polynomial, then

$$\operatorname{tr}_1(\mathbf{A}) = \sum_{i=1}^{n} \lambda_i$$

$$\operatorname{tr}_2(\mathbf{A}) = \lambda_1 \lambda_2 + \lambda_1 \lambda_3 + \cdots + \lambda_{n-1} \lambda_n$$

$$\vdots$$

$$\operatorname{tr}_n(\mathbf{A}) = \lambda_1 \lambda_2 \cdots \lambda_{n-1} \lambda_n = |\mathbf{A}|$$

Rank of a matrix. An $m \times n$ matrix $\mathbf{A}$ is said to be of rank r, written as $\rho(\mathbf{A}) = r$ if and only if (iff) there is at least one nonzero minor of order r from $\mathbf{A}$ and all the minors of order $r + 1$ are zero.

Inverse of a nonsingular matrix. A square matrix $\mathbf{A}$ is said to be nonsingular if $|\mathbf{A}| \neq 0$. If $\mathbf{I} = \mathbf{B}^T \mathbf{A}^T = \mathbf{A}\mathbf{B}$, then $\mathbf{B}$ is an inverse of $\mathbf{A}$ and is written as $\mathbf{B} = \mathbf{A}^{-1}$. Then follows

$$(\mathbf{A}^{-1})^{-1} = \mathbf{A} \qquad (\mathbf{A}\mathbf{A}^{-1}) = \mathbf{I}$$

Eigenvalues and eigenvectors. Let $\mathbf{A}$ be an $n \times n$ square matrix, and let $\mathbf{x}$ be an $n \times 1$ nonnull vector such that $\mathbf{A}\mathbf{x} = \lambda\mathbf{I}$. Then λ is called a characteristic root or an eigenvalue of $\mathbf{A}$ and $\mathbf{x}$ is called the characteristic vector or eigenvector corresponding to characteristic root λ. The characteristic roots are solutions of

$$\left|\mathbf{A} - \lambda\mathbf{I}\right| = 0$$

Hence if $\mathbf{A}$ is a $p \times p$ matrix, then there are p characteristic roots of $\mathbf{A}$. All characteristic roots of an identity matrix are 1, and those of a diagonal matrix are the diagonal elements.

$$\mathbf{A} = \begin{bmatrix} 6 & 2 \\ 2 & 3 \end{bmatrix}$$

Then $|\mathbf{A} - \lambda\mathbf{I}| = (6 - \lambda)(3 - \lambda) - 4 = 14 - 9\lambda + \lambda^2 = (\lambda - 7)(\lambda - 2) = 0$. Hence the characteristic roots of $\mathbf{A}$ are 2 and 7. In general, when $\mathbf{A}$ is a square matrix and $\mathbf{A} = \mathbf{A}_{p \times p}$, the equation $|\mathbf{A} - \lambda\mathbf{I}| = 0$ will be an n-degree polynomial equation and will have p roots. The roots $\lambda_1, \lambda_2, \ldots, \lambda_p$ are the set of eigenvalues for $\mathbf{A}$. When we pick an eigenvalue, say λ_i, such that $i \in (1, 2, \ldots, p)$, because $|\mathbf{A} - \lambda_i \mathbf{I}| = 0$, the matrix $\mathbf{A} - \lambda_i \mathbf{I}$ is linearly dependent; so the following equation

$$(\mathbf{A} - \lambda_i \mathbf{I})\mathbf{p}_i = \mathbf{0}$$

will have a nonzero solution, where $\mathbf{p}_i$ is a $p \times 1$ vector and it is called the ith eigenvector of $\mathbf{A}$. Since there are p eigenvalues, there will be p eigenvectors. The matrix $\mathbf{P} = \lfloor \mathbf{p}_1, \mathbf{p}_2, \ldots, \mathbf{p}_p \rfloor$ is the matrix consisting of all p eigenvectors. Eigenvalues and eigenvectors play very important roles in principal component analysis.

Example: Let

$$\mathbf{A} = \begin{bmatrix} 1 & 1/2 \\ 1/2 & 1 \end{bmatrix}$$

By solving

$$|\mathbf{A} - \lambda \mathbf{I}| = \begin{vmatrix} 1 - \lambda & 1/2 \\ 1/2 & 1 - \lambda \end{vmatrix} = 0$$

We have

$$(1 - \lambda)^2 - \tfrac{1}{4} = 0 \qquad \lambda^2 - 2\lambda + \tfrac{3}{4} = 0$$

We get two roots for the above equation:

$$\lambda_1 = \tfrac{3}{2}, \quad \lambda_2 = \tfrac{1}{2}$$

These are the eigenvalues of $\mathbf{A}$. By solving $(\mathbf{A} - \lambda_1 \mathbf{I})\mathbf{p}_1 = \mathbf{0}$, we have

$$\begin{bmatrix} 1 - 3/2 & 1/2 \\ 1/2 & 1 - 3/2 \end{bmatrix} \begin{bmatrix} p_{11} \\ p_{21} \end{bmatrix} = \begin{bmatrix} 0 \\ 0 \end{bmatrix}$$

and then

$$\begin{bmatrix} p_{11} \\ p_{21} \end{bmatrix} = \begin{bmatrix} 1 \\ 1 \end{bmatrix}$$

is a possible solution. By solving $(\mathbf{A} - \lambda_2 \mathbf{I})\mathbf{p}_2 = \mathbf{0}$, we have

$$\begin{bmatrix} 1 - 1/2 & 1/2 \\ 1/2 & 1 - 1/2 \end{bmatrix} \begin{bmatrix} p_{12} \\ p_{22} \end{bmatrix} = \begin{bmatrix} 0 \\ 0 \end{bmatrix}$$

then

$$\begin{bmatrix} p_{12} \\ p_{22} \end{bmatrix} = \begin{bmatrix} 1 \\ -1 \end{bmatrix}$$

is a possible solution. So

$$\mathbf{P} = [\mathbf{p}_1, \mathbf{p}_2] = \begin{bmatrix} 1 & 1 \\ 1 & -1 \end{bmatrix}$$

However, we usually prefer normalized eigenvector. In this case

$$\mathbf{P} = [\mathbf{p}_1, \mathbf{p}_2] = \begin{bmatrix} 1/\sqrt{2} & 1/\sqrt{2} \\ 1/\sqrt{2} & -1/\sqrt{2} \end{bmatrix}$$

is the normalized eigenvector matrix.

The normalized eigenvector has several very desirable properties:

1. $\mathbf{PP}^T = \mathbf{I}$
2. $\mathbf{P}^T \mathbf{AP} = \Lambda$ and $\mathbf{PAP}^T = \Lambda$, where

$$\Lambda = \begin{bmatrix} \lambda_1 & 0 & \cdots & 0 \\ 0 & \lambda_2 & \cdots & 0 \\ \vdots & \vdots & \vdots & \vdots \\ 0 & \cdots & \cdots & \lambda_p \end{bmatrix}$$

3. $\mathbf{AP} = \mathbf{P}\Lambda$

Chapter 4

Multivariate Analysis of Variance

4.1 Introduction

The multivariate statistical inferences discussed in Chap. 3 can only conduct comparative study for up to two multivariate populations. In Example 3.7, we compared the impact-absorbing qualities for two sides of the bicycle protection helmet. If we are given *two* similar welding technician training programs (not one) as we discussed in Example 3.1, and both programs offer the same three courses, we can certainly use the multivariate statistical inferences techniques that are discussed in Chap. 3 to compare the performances of these two programs. However, what if we are asked to compare *more than two*, say, five similar welding technician training programs in a state?

This situation is called *multiple comparisons*, that is, comparison of three or more similar items. When each item has only one characteristic to compare, this multiple comparisons problem is a univariate problem, and it can be dealt with by analysis of variance (ANOVA). When each item has more than one characteristic—for example, in the welding technician training program problem, there are three courses, the grade of each course is a characteristic, and so we have three characteristics for each item—this kind of multiple characteristics and multiple comparisons problem will have to be dealt with by multivariate analysis of variance (MANOVA).

As seen in Chap. 3, there are usually similarities between each multivariate statistical method and its univariate counterpart, and comparing the similarities and differences between the univariate method and the corresponding multivariate method can greatly enhance our understanding of the more complicated multivariate method, especially for the readers who have some background in univariate statistical methods.

In this chapter, we will first briefly review the univariate analysis of variance in Sec. 4.2. Section 4.3 discusses the method of multivariate analysis of variance. Section 4.4 presents an industrial case study on the application of multivariate analysis of variance.

4.2 Univariate Analysis of Variance (ANOVA)

Analysis of variance (ANOVA) is a general technique that can be used to test the hypothesis that the means among two or more groups are equal, under the assumption that the sampled populations are normally distributed.

To begin, let us study the effect of temperature on the yield of a chemical reaction process. We select several different temperatures and observe their effect on the yield. This experiment can be conducted as follows. The process is heated to a selected temperature, then after a fixed time interval, say, 4 h, the yield is measured. In each temperature setting, we repeat the experiment for several time intervals, and measure the yield a few times. These measurements are called replicates. The temperature is called a *factor*. The different temperature settings are called *levels*. In this example there are several levels or settings of the factor, temperature. If we compare several welding technician training schools, say, five schools, then school is a factor and since we have five different schools, we have five levels for the factor, school.

In general, a factor is an independent treatment variable the settings (values) of which are controlled and varied by the experimenter. The setting of a factor is the level. In the experiment above, there is only one factor, temperature, and the analysis of variance that we will be using to analyze the effect of temperature is called a *one-way* or *one-factor ANOVA*. In this book, we are only discussing one-way analysis of variance (one-way ANOVA), and one way multivariate analysis of variance (one-way MANOVA).

The major question to be answered by the analysis of variance is "do these factor levels have a significant effect on the characteristics of the items under study?" In a chemical process problem, this question becomes "do temperature settings significantly affect the yield of the process?" In the welding technician training programs' problem, the question becomes "do different schools perform differently in terms of the grades that students are achieving?" Statistically, this question becomes a hypothesis testing problem as follows. The null hypothesis is, there is no difference in the population means of the different levels of factor. The alternative hypothesis is, there is significant difference in the population means.

Specifically, the data collection format in one-way ANOVA is illustrated in Table 4.1. In Table 4.1, k is the number of levels of a single factor. The number of replicates in each level is n. x_{ij} stands for the observation corresponding to jth replicate at ith level of the factor.

TABLE 4.1 Data Organization for One-Way ANOVA

Factor level	Observation				Total	Average
1	x_{11}	x_{12}	$\ldots$	x_{1n}	$x_{1.}$	$\overline{x}_{1.}$
2	x_{21}	x_{22}	$\ldots$	x_{2n}	$x_{2.}$	$\overline{x}_{2.}$
$\vdots$	$\vdots$	$\vdots$	$\ldots$	$\vdots$	$\vdots$	
k	x_{k1}	x_{k2}	$\ldots$	x_{kn}	$x_{k.}$	$\overline{x}_{k.}$
Total	$x_{.1}$	$x_{.2}$	$\ldots$	$x_{.n}$	$x_{..}$	$\overline{\overline{x}}_{..}$

Equation (4.1) is the statistical model for one-way ANOVA.

$$x_{ij} = \mu + \tau_i + \varepsilon_j \qquad i = 1, 2, \ldots, k, \quad j = 1, 2, \ldots, n \qquad (4.1)$$

where x_{ij} = ijth observation

μ = parameter common to all the treatments and is the overall mean

τ_i = parameter associated with the ith treatment, or the ith level, called the ith treatment effect

ε_{ij} = random error component.

It is assumed that the model errors are normally and independently distributed random variables with mean zero and variance σ^2. The variance is assumed constant for all levels of the factor.

The formal statistical hypothesis for one-way ANOVA is stated as

$$H_0: \mu_1 = \mu_2 = \cdots = \mu_k \quad \text{(all population means are equal)}$$

$$H_1: \text{at least one } \mu_1 \neq \mu_m \text{ for } i \neq m \qquad (4.2)$$

(at least some population means are not equal)

To analyze data and conduct the above hypothesis testing, ANOVA adopted a brilliant approach. This approach is featured by partitioning the total variation of data into components that correspond to different sources of variation.

The aim of this approach is to split the total variation in the data into a portion due to random error and portions due to changes in the levels of the factor.

Specifically, this approach can be expressed by the following equation:

$$SS_T = SS_B + SS_W \qquad (4.3)$$

where SS_T is called the total sum of squares and

$$SS_T = \sum_{i=1}^{k} \sum_{j=1}^{n} (X_{ij} - \overline{\overline{X}})^2 \qquad (4.4)$$

where $\bar{\bar{X}} = (1/N)\sum_{i=1}^{k}\sum_{j=1}^{n} X_{ij}$ is the grand mean of all observations, where $N = n \times k$. Clearly, SS_T is the sum of the squared deviation of each observation around the grand mean, so it is a measure of total variation in the data set.

SS_B is called the sum of squares between factor levels or between treatments:

$$SS_B = n\sum_{i=1}^{k}(\bar{X}_i - \bar{\bar{X}})^2 \qquad (4.5)$$

where $\bar{X}_i = (1/n)\sum_{j=1}^{n} X_{ij}$ is the mean of all observations at the ith level of the factor. SS_B is the summation of squared deviation of "level average" or "treatment average" minus the grand mean. So SS_B is a measure of total variation caused by different factor level settings. In other text books, SS_B is also called "total variation due to treatments," denoted by $SS_{treatment}$.

SS_W is called the sum of squares within factors and

$$SS_W = \sum_{i=1}^{k}\sum_{j=1}^{n}(X_{ij} - \bar{X}_i)^2 \qquad (4.6)$$

SS_W is the summation of the squared deviations of individual observations at the same level minus the "level average." So SS_W is a measure of the total variation that occurs when the factor level is not changing and so this variation must be caused by other unknown variation that is not related to the factor level or treatment. SS_W is also called "total variation due to error" in other text books, and is denoted by SS_E.

Therefore Eq. (4.3)

$$SS_T = SS_B + SS_W$$

is the formal mathematical expression of "partitioning the total variation of data into components that correspond to different sources of variation."

Intuitively, if SS_B is much larger in comparison with SS_W, it is an indication that most of total variation SS_T is caused by a change in factor levels. However, the magnitudes of SS_B and SS_W are also influenced by the number of levels k and the level of replicates n, and so "mean squares" are computed as a basis for statistical hypothesis testing. The mean squares are computed as follows:

MS_B is called the mean square between factor levels and

$$MS_B = \frac{n\sum_{i=1}^{k}(\bar{X}_i - \bar{\bar{X}})^2}{k-1} \qquad (4.7)$$

TABLE 4.2 ANOVA Table

Source of variation	Sum of squares	Degrees of freedom	Mean squares	F_0—computed F statistic
Between factors	SS_B	$k-1$	$MS_B = \dfrac{SS_B}{k-1}$	$F_0 = \dfrac{MS_B}{MS_W}$
Within factors	SS_W	$n(k-1)$	$MS_W = \dfrac{SS_W}{n(k-1)}$	
Total	SS_T	$nk-1$		

MS_W is called the mean square within factor levels

$$MS_W = \frac{\sum_{i=1}^{k}\sum_{j=1}^{n}(X_{ij}-\overline{X}_i)^2}{n(k-1)} \qquad (4.8)$$

After computing the mean squares, the ANOVA table will be filled and F-test will be conducted to test the hypothesis specified by Eq. (4.2).

4.2.1 The ANOVA table

The ANOVA table for the single factor completely randomized design model is given in Table 4.2.

> **Example 4.1: ANOVA** A quality engineer in a production shop is interested in determining if there are any differences in the gauge calibration ability of three quality inspectors. The response variables are the measurements taken by a gauge on a gauge calibration device. The treatment variables are the three auditors. There are three replicates of each auditor measuring the same gauge with the same gauge calibration device.
>
> **Objective** Is there any difference in measurements due to auditor differences at a maximum Type I error rate of 5% (i.e., $\alpha = 0.05$)?
> The data are shown in Table 4.3.
>
> **Analysis results** One-way ANOVA: Response versus Auditor (see Table 4.4).
>
> **Interpretation of results** The F_{test} is computed as 5.26 (Table 4.4) and the corresponding $F_{critical}$ value is determined from F statistical tables as F(numerator df, denominator df). $F_{critical} = F_{0.05,\,2,6} = 5.14$. Since $F_{test} > F_{critical}$, we reject the null

TABLE 4.3 Data for ANOVA Example

Auditor	Response	Auditor	Response	Auditor	Response
1	368.75	2	426.25	3	483.75
1	270	2	471.25	3	521.23
1	398.75	2	432.5	3	420

TABLE 4.4 Analysis of Variance for Response

Source	DF	SS	MS	F	P
Auditor	2	27191	13595	5.26	0.048
Error	6	15503	2584		
Total	8	42694			

hypothesis that there are no differences between the three auditor measurements. Rejection of H_0 suggests that at least one pair of group means (auditor means) is not equal (i.e., something other than chance variation is generating the differences among the k group means). A further analysis of specific paired comparisons leads to the identification of the source of the specific variation causing the rejection of the null hypothesis. These tests, such as Scheffe, Bonferonni, etc., can be reviewed in Montgomery and Hines (2003).

4.3 Multivariate Analysis of Variance

There are many similarities between univariate analysis of variance (ANOVA) and multivariate analysis of variance (MANOVA). For example, in the role of factor and its levels, the meaning of replicates is essentially the same for both ANOVA and MANOVA. However, in multivariate analysis of variance, we are dealing with more than one characteristic. In the last section, we discussed the example of temperature and its effect on yield in a chemical process. In that example, temperature is a factor, the different settings of temperature are levels, and yield is the characteristic under study. If we extend this problem by including other characteristics of the chemical product besides the yield, such as viscosity and density, then we have a multiple characteristics situation and this problem will have to be dealt with by multivariate analysis of variance. Another example of multiple characteristics is the quality of a painted surface. The quality could be determined by a number of characteristics such as paint hardness, paint surface finish, and paint thickness. A plant manager may be interested in testing three different painting processes to determine which process delivers a quality painted product when collectively measured on the above multiple characteristics. The analysis of these types of experiments in which a number of characteristis are measured is done by using MANOVA.

4.3.1 MANOVA model

There are some noticeable differences in the statistical model of MANOVA compared to the statistical model in ANOVA, denoted by Eq. (4.1). Because the MANOVA problem deals with multiple characteristics or a vector of characteristics, the MANOVA statistical model is a vector model specified by Eq. (4.9).

$$\mathbf{X}_{ij} = \mathbf{\mu} + \mathbf{\tau}_i + \mathbf{e}_{ij} = \mathbf{\mu}_i + \mathbf{e}_{ij} \quad \text{for all } j = 1,\ldots,n_i \quad \text{and} \quad i = 1,\ldots,k. \quad (4.9)$$

In this model, every item, such as $\mathbf{X}_{ij}$, $\boldsymbol{\mu}$, $\boldsymbol{\tau}_i$, $\boldsymbol{\mu}_i$, and $\mathbf{e}_{ij}$ are p dimensional vectors,

where p = number of characteristics
  $i = 1, \ldots, k$, i denotes level, where k is the number of levels
  $j = 1, \ldots, n_i$, j denotes jth observation, n_i is the number of observations (replicates) at the ith level
  $\mathbf{e}_{ij}$ = vector of error terms, and $\mathbf{e}_{ij} \sim N_p(\mathbf{0}, \boldsymbol{\Sigma})$
  $\boldsymbol{\mu}$ = overall mean vector
  $\boldsymbol{\tau}_i$ = vector of the ith treatment effect; it is assumed that
    $\sum_{i=1}^{k} n_i \boldsymbol{\tau}_i = \mathbf{0}$
  $\boldsymbol{\mu}_i$ = mean vector corresponding to ith treatment

The data collected in MANOVA would have the following structure:

Level 1	$\mathbf{X}_{11}$	$\mathbf{X}_{12}$	$\cdots$	$\mathbf{X}_{1n_1}$
Level 2	$\mathbf{X}_{21}$	$\mathbf{X}_{22}$	$\cdots$	$\mathbf{X}_{2n_2}$
$\vdots$	$\vdots$	$\vdots$	$\cdots$	
Level k	$\mathbf{X}_{k1}$	$\mathbf{X}_{k2}$	$\cdots$	$\mathbf{X}_{kn_k}$

where each vector $\mathbf{X}_{ij}$ is a vector of p characteristics, $\mathbf{X}_{ij}$ is the jth sample drawn on group i. The detailed MANOVA analysis data structure is as follows:

$$\text{Level 1} \quad \begin{bmatrix} \text{Sample 1} \\ X_{111} \\ X_{112} \\ \vdots \\ X_{11p} \end{bmatrix} \begin{bmatrix} \text{Sample 2} \\ X_{121} \\ X_{122} \\ \vdots \\ X_{12p} \end{bmatrix} \cdots \begin{bmatrix} \text{Sample } n_1 \\ X_{1n_1 1} \\ X_{1n_2 2} \\ \vdots \\ X_{1n_1 p} \end{bmatrix}$$

$$\text{Level 2} \quad \begin{bmatrix} \text{Sample 1} \\ X_{211} \\ X_{212} \\ \vdots \\ X_{21p} \end{bmatrix} \begin{bmatrix} \text{Sample 2} \\ X_{221} \\ X_{222} \\ \vdots \\ X_{22p} \end{bmatrix} \cdots \begin{bmatrix} \text{Sample } n_2 \\ X_{2n_2 1} \\ X_{2n_2 2} \\ \vdots \\ X_{2n_2 p} \end{bmatrix}$$

$$\vdots$$

$$\text{Level } k \quad \begin{bmatrix} \text{Sample 1} \\ X_{k11} \\ X_{k12} \\ \vdots \\ X_{k1p} \end{bmatrix} \begin{bmatrix} \text{Sample 2} \\ X_{k21} \\ X_{k22} \\ \vdots \\ X_{k2p} \end{bmatrix} \cdots \begin{bmatrix} \text{Sample } n_k \\ X_{kn_k 1} \\ X_{kn_{2k} 2} \\ \vdots \\ X_{kn_k p} \end{bmatrix}$$

where X_{ijm} is the experimental observation for sample number j ($j = 1, \ldots, n_i$) of group i ($i = 1, \ldots, k$) measured on multivariate characteristic m ($m = 1, \ldots, p$), and

$$\mathbf{X}_{ij} = \begin{bmatrix} X_{ij1} \\ X_{ij2} \\ \vdots \\ X_{ijp} \end{bmatrix}$$

is the vector of p multivariate observations from sample j of group i.

Assumptions

- $(X_{i1} \ X_{i2} \cdots X_{in_i})$ is a multivariate random sample of size n_i from a multivariate population for level i with mean $\boldsymbol{\mu}_i$ being a p dimensional multivariate mean vector. It is assumed that the random samples from different levels ($i = 1, \ldots, k$) are independent.
- All the levels have a common population covariance matrix $\boldsymbol{\Sigma}$.
- Each level corresponds to a multivariate normal population $\sim N_p(\boldsymbol{\mu}_i, \boldsymbol{\Sigma})$.
- Random error for each observation are independent random variables.

4.3.2 The decomposition of total variation under MANOVA model

Comparing the statistical model of ANOVA specified by Eq. (4.1) and the statistical model of MANOVA specified by Eq. (4.9), we can see that they are very similar except in MANOVA where every term becomes a p dimensional vector:

$$\mathbf{X}_{ij} = \boldsymbol{\mu} + \boldsymbol{\tau}_i + \mathbf{e}_{ij} = \boldsymbol{\mu}_i + \mathbf{e}_{ij}$$

By using some simple vector decomposition

$$\mathbf{x}_{ij} = \bar{\mathbf{x}} + (\bar{\mathbf{x}}_i - \bar{\mathbf{x}}) + (\mathbf{x}_{ij} - \bar{\mathbf{x}}) \quad (4.10)$$

where $\bar{\mathbf{x}}$ stands for the overall mean vector of all vector observations $\mathbf{x}_{ij}$, and $\bar{\mathbf{x}}_i$ stands for the mean vector of all observations at level i.

The parameters for Eq. (4.9) can be estimated as follows:

$\bar{\mathbf{x}} = \hat{\boldsymbol{\mu}}$ = overall sample grand mean vector

$(\bar{\mathbf{x}}_i - \bar{\mathbf{x}}) = \hat{\boldsymbol{\tau}}_i$ = estimate of the ith treatment effect

$(\mathbf{x}_{ij} - \bar{\mathbf{x}}) = \hat{\mathbf{e}}_{ij}$ = estimate of the residual of the ith treatment of the jth observation

In MANOVA, the total variation cannot be expressed by a total sum of squares, it can be actually expressed by a total covariance matrix. Specifically, the decomposition of total variation in a MANOVA context is described by Eqs. (4.11) to (4.16).

$$\sum_{i=1}^{k}\sum_{j=1}^{n_i}(\mathbf{x}_{ij}-\bar{\mathbf{x}})(\mathbf{x}_{ij}-\bar{\mathbf{x}})^T \times [(\mathbf{x}_{ij}-\bar{\mathbf{x}}_i)+(\bar{\mathbf{x}}_i-\bar{\mathbf{x}})][(\mathbf{x}_{ij}-\bar{\mathbf{x}}_i)+(\bar{\mathbf{x}}_i-\bar{\mathbf{x}})]^T$$

$$=(\mathbf{x}_{ij}-\bar{\mathbf{x}}_i)(\bar{\mathbf{x}}_{ij}-\bar{\mathbf{x}}_i)^T + (\bar{\mathbf{x}}_{ij}-\bar{\mathbf{x}}_i)(\bar{\mathbf{x}}_i-\bar{\mathbf{x}})^T$$

$$+(\bar{\mathbf{x}}_i-\bar{\mathbf{x}})(\bar{\mathbf{x}}_{ij}-\bar{\mathbf{x}}_i)^T + (\bar{\mathbf{x}}_i-\bar{\mathbf{x}})(\bar{\mathbf{x}}_i-\bar{\mathbf{x}})^T \qquad (4.11)$$

Summation of the sum of squares and cross products over subscripts i and j is the total multivariate sum of squares and cross products (SSP).

Thus, since $(\mathbf{x}_{ij}-\bar{\mathbf{x}}_i) = \bar{\mathbf{e}}_{ij}$, therefore,

$$\frac{1}{\sum_{i=1}^{k}n_i}(\mathbf{x}_{ij}-\bar{\mathbf{x}}_i) = E(\mathbf{e}_{ij}) = \mathbf{0} \qquad (4.12)$$

$$\sum_{i=1}^{k}\sum_{j=1}^{n_i}(\mathbf{x}_{ij}-\bar{\mathbf{x}})(\mathbf{x}_{ij}-\bar{\mathbf{x}})^T = \sum_{i=1}^{k}(\bar{\mathbf{x}}_i-\bar{\mathbf{x}})(\mathbf{x}_{ij}-\bar{\mathbf{x}})^T + \sum_{i=1}^{k}\sum_{j=1}^{n_i}(\mathbf{x}_{ij}-\bar{\mathbf{x}}_i)(\mathbf{x}_{ij}-\bar{\mathbf{x}}_i)^T$$

$$(4.13)$$

where $\sum_{i=1}^{k}\sum_{j=1}^{n_i}(\mathbf{x}_{ij}-\bar{\mathbf{x}})(\mathbf{x}_{ij}-\bar{\mathbf{x}})^T$ = total sum of squares and cross products = SSPTotal

$\sum_{i=1}^{k}(\bar{\mathbf{x}}_i-\bar{\mathbf{x}})(\mathbf{x}_{ij}-\bar{\mathbf{x}})^T$ = between sum of squares and cross products = SSP_B

$\sum_{i=1}^{k}\sum_{j=1}^{n_i}(\mathbf{x}_{ij}-\bar{\mathbf{x}}_i)(\mathbf{x}_{ij}-\bar{\mathbf{x}}_i)^T$ = within sum of squares and cross products = SSP_W

Expressing the SSPs as a covariance matrix:

$$SSP_W = \sum_{i=1}^{k}\sum_{j=1}^{n_i}(\mathbf{x}_{ij}-\bar{\mathbf{x}}_i)(\mathbf{x}_{ij}-\bar{\mathbf{x}}_i)^T \qquad (4.14)$$

From Chap. 3 on multivariate normal distributions and random samples, we know that, for any given characteristic $i = 1, \ldots, k$, the covariance matrix is given by

$$\mathbf{S}_i = \frac{1}{n_i-1}\sum_{j=1}^{n_i}(\mathbf{x}_{ij}-\bar{\mathbf{x}}_i)(\mathbf{x}_{ij}-\bar{\mathbf{x}}_i)^T \qquad (4.15)$$

Hence

$$\text{SSP}_W = (n_1 - 1)\mathbf{S}_1 + (n_2 - 1)\mathbf{S}_2 + \cdots + (n_k - 1)\mathbf{S}_k \qquad (4.16)$$

Manova hypothesis

$$H_0: \tau_1 = \tau_2 = \cdots = \tau_i = \mathbf{0}$$

$$H_a: \tau_i \neq \mathbf{0} \quad \text{for at least one } i = 1, \ldots, k$$

Wilks (1932) proposed the statistic Wilks' lambda (Λ)

$$\Lambda = \frac{|\text{SSP}_W|}{|\text{SSP}_B| + |\text{SSP}_W|} \qquad (4.17)$$

This statistic is based on the generalized sample variance statistics of a multivariate covariance matrix where the generalized sample variance is defined by $|\mathbf{S}|$, if $\mathbf{S}$ is a sample covariance matrix. The generalized sample variance is a multivariate statistic that captures most (but not all) information on the sample variation and covariances as a single number. For a detailed evaluation of the completeness of $|\mathbf{S}|$ as a singular statistical measure of the sample variance and covariances, see Anderson (1984). Intuitively, the magnitude of $|\text{SSP}_B|$ is proportional to the treatment effect or factor level effect. The more significant the treatment effect, the larger $|\text{SSP}_B|$ and smaller the Λ should become. Therefore, a small Wilks' lambda (Λ) is an indicator of strong treatment effect.

Λ can also be computed by

$$\Lambda = \frac{|\text{SSP}_W|}{|\text{SSP}_B + \text{SSP}_W|} = \frac{\left|\sum_{i=1}^{k}\sum_{j=1}^{n_k}(\mathbf{x}_{ij} - \bar{\mathbf{x}}_i)(\mathbf{x}_{ij} - \bar{\mathbf{x}}_i)^T\right|}{\left|\sum_{i=1}^{k}\sum_{j=1}^{n_i}(\mathbf{x}_{ij} - \bar{\mathbf{x}})(\mathbf{x}_{ij} - \bar{\mathbf{x}})^T\right|} \qquad (4.18)$$

$$\Lambda = \frac{|(n_1 - 1)\mathbf{S}_1 + (n_2 - 1)\mathbf{S}_2 + \cdots + (n_k - 1)\mathbf{S}_k|}{\left|\sum_{i=1}^{k}\sum_{j=1}^{n_i}(\mathbf{x}_{ij} - \bar{\mathbf{x}})(\mathbf{x}_{ij} - \bar{\mathbf{x}})^T\right|} \qquad (4.19)$$

Criteria for rejection of H_0. The criteria is to reject H_0 if Wilks' lambda (Λ) is small. The determination of cutoff values for Λ is studied by many researchers. Wilks (1932) and Bartlett (1954) have determined the exact distributions of the Wilks' lambda (Λ) for a number of cases that are dependent on the number of characteristics p and the number of treatments (levels) k.

Case I: $p = 1$ and $k \geq 2$

$$\left(\frac{\sum_{i=1}^{k} n_i - q}{q - 1}\right)\left(\frac{1 - \Lambda}{\Lambda}\right) \sim F\left(k - 1, \sum_{i=1}^{k} n_i\right) \quad (4.20)$$

Case II: $p = 2$ and $k \geq 2$

$$\left(\frac{\sum_{i=1}^{k} n_i - k - 1}{k - 1}\right)\left(\frac{1 - \sqrt{\Lambda}}{\sqrt{\Lambda}}\right) \sim F\left[2(k-1), 2\left(\sum_{i=1}^{k} n_i - k - 1\right)\right] \quad (4.21)$$

Case III: $p \geq 1$ and $k = 2$

$$\left(\frac{\sum_{i=1}^{k} n_i - p - 1}{p}\right)\left(\frac{1 - \Lambda}{\Lambda}\right) \sim F\left(p, \sum_{i=1}^{k} n_i - p - 1\right) \quad (4.22)$$

Case IV: $p \geq 1$ and $k = 3$

$$\left(\frac{\sum_{i=1}^{k} n_i - p - 2}{p}\right)\left(\frac{1 - \sqrt{\Lambda}}{\sqrt{\Lambda}}\right) \sim F\left[2p, 2\left(\sum_{i=1}^{k} n_i - p - 2\right)\right] \quad (4.23)$$

Case V: For all other combinations of p and k not in Cases I through IV, or when sample sizes are large, i.e., $\sum_{i=1}^{k} n_i = N$ is large,

$$\left[(n-1) - \left(\frac{p+k}{2}\right)\right] \ln_e \Lambda \sim \chi^2_{p(k-1)} \quad (4.24)$$

For example, when we have a MANOVA problem in which $p = 4$, $k = 3$, Case IV will be applicable. First we should compute

$$\Lambda = \frac{|SSP_W|}{|SSP_B| + |SSP_W|}$$

Then we compute

$$\left(\frac{\sum_{i=1}^{k} n_i - p - 2}{p}\right)\left(\frac{1 - \sqrt{\Lambda}}{\sqrt{\Lambda}}\right)$$

TABLE 4.5 MANOVA Table

Source of variation	Sum of squares and cross products
Between treatment	$\text{SSP}_B = \sum_{i=1}^{k}(\bar{\mathbf{x}}_i - \bar{\mathbf{x}})(\mathbf{x}_{ij} - \bar{\mathbf{x}})^T$
Within treatment (error)	$\text{SSP}_W = \sum_{i=1}^{k}\sum_{j=1}^{n_i}(\mathbf{x}_{ij} - \bar{\mathbf{x}}_i)(\mathbf{x}_{ij} - \bar{\mathbf{x}}_i)^T$
Total	$\text{SSP}_{\text{Total}} = \sum_{i=1}^{k}\sum_{j=1}^{n_i}(\mathbf{x}_{ij} - \bar{\mathbf{x}})(\mathbf{x}_{ij} - \bar{\mathbf{x}})^T$

and compare that with $F_{\alpha;\, 2p,\, 2(\sum_{i=1}^{k} n_i - p - 2)}$. If

$$\left(\frac{\sum_{i=1}^{k} n_i - p - 2}{p}\right)\left(\frac{1 - \sqrt{\Lambda}}{\sqrt{\Lambda}}\right) > F_{\alpha;\, 2p,\, 2(\sum_{i=1}^{k} n_i - p - 2)}$$

then H_0 should be rejected.

MANOVA table. The MANOVA table (Table 4.5) summarizes the variation decomposition according to the sources.

Besides Wilks' lambda Λ other similar statistics for MANOVA statistical hypothesis testing have also been developed by other researchers, and corresponding F distribution based test statistics have also been developed. These statistics are summarized as follows:

Other statistics to test the MANOVA model

$$\text{Pillai's statistic} = \text{tr}\lfloor(\text{SSP}_B)(\text{SSP}_B + \text{SSP}_W)^{-1}\rfloor \qquad (4.25)$$

$$\text{Lawley-Hotelling statistics} = \text{tr}\lfloor(\text{SSP}_B)(\text{SSP}_W)^{-1}\rfloor \qquad (4.26)$$

Example 4.2 A quality engineer is interested in testing the differences in the performance of welding trainees, determined by their level of education, based on the scores they receive in two written tests given to them after completion of the welding training.

Multivariate characteristics ($p = 2$) The two tests that are given are BWT (basic welding technology) and AWT (advanced welding technology).

Factor: Education background (three levels) AD ($i = 1$)—high school diploma, GED ($i = 2$)—general education diploma, ND ($i = 3$)—no degree. The data are shown in Fig. 4.1.

MANOVA calculations

$$\bar{\bar{\mathbf{X}}} = \begin{bmatrix} 88.1 \\ 85 \end{bmatrix} \quad \bar{\mathbf{X}}_1 = \begin{bmatrix} 96.0 \\ 92.3 \end{bmatrix} \quad \bar{\mathbf{X}}_2 = \begin{bmatrix} 90.25 \\ 88.25 \end{bmatrix} \quad \bar{\mathbf{X}}_3 = \begin{bmatrix} 77.33 \\ 73.33 \end{bmatrix}$$

$$\bar{\mathbf{X}}_1 - \bar{\bar{\mathbf{X}}} = \begin{bmatrix} 7.9 \\ 7.33 \end{bmatrix} \quad \bar{\mathbf{X}}_2 - \bar{\bar{\mathbf{X}}} = \begin{bmatrix} 2.15 \\ 3.25 \end{bmatrix} \quad \bar{\mathbf{X}}_3 - \bar{\bar{\mathbf{X}}} = \begin{bmatrix} -10.77 \\ -11.67 \end{bmatrix}$$

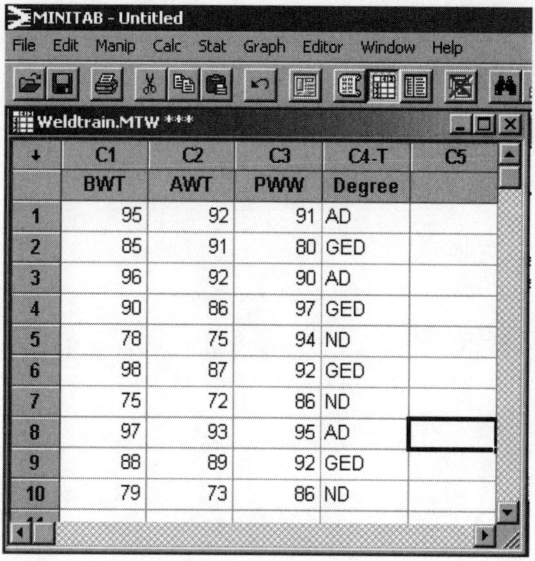

	C1	C2	C3	C4-T	C5
	BWT	AWT	PWW	Degree	
1	95	92	91	AD	
2	85	91	80	GED	
3	96	92	90	AD	
4	90	86	97	GED	
5	78	75	94	ND	
6	98	87	92	GED	
7	75	72	86	ND	
8	97	93	95	AD	
9	88	89	92	GED	
10	79	73	86	ND	

Figure 4.1 Data for Manova Example.

$$n_1[\overline{\mathbf{X}}_1 - \overline{\overline{\mathbf{X}}}] \times [\overline{\mathbf{X}}_1 - \overline{\overline{\mathbf{X}}}]^T = 3\begin{bmatrix} 62.41 & 57.93 \\ 57.93 & 53.78 \end{bmatrix} = \begin{bmatrix} 187.23 & 173.8 \\ 173.8 & 161.33 \end{bmatrix}$$

$$[\overline{\mathbf{X}}_2 - \overline{\overline{\mathbf{X}}}] \times [\overline{\mathbf{X}}_2 - \overline{\overline{\mathbf{X}}}]^T = 4\begin{bmatrix} 4.62 & 6.89 \\ 6.89 & 10.56 \end{bmatrix} = \begin{bmatrix} 18.49 & 27.95 \\ 27.95 & 42.25 \end{bmatrix}$$

$$n_3 \times [\overline{\mathbf{X}}_3 - \overline{\overline{\mathbf{X}}}] \times [\overline{\mathbf{X}}_3 - \overline{\overline{\mathbf{X}}}]^T = 3\begin{bmatrix} 115.92 & 125.61 \\ 125.61 & 136.11 \end{bmatrix} = \begin{bmatrix} 347.76 & 376.83 \\ 376.83 & 408.33 \end{bmatrix}$$

$$\mathbf{B} = \sum_{i=1}^{3} n_i \times [\overline{\mathbf{X}}_i - \overline{\overline{\mathbf{X}}}] \times [\overline{\mathbf{X}}_i - \overline{\overline{\mathbf{X}}}]^T = \begin{bmatrix} 553.48 & 578.58 \\ 578.58 & 611.92 \end{bmatrix}$$

$$\mathbf{S}_1 = \begin{bmatrix} 1 & 0.5 \\ 0.5 & 0.33 \end{bmatrix} \quad \mathbf{S}_2 = \begin{bmatrix} 30.92 & -8.42 \\ -8.42 & 4.91 \end{bmatrix} \quad \mathbf{S}_3 = \begin{bmatrix} 4.33 & 1.83 \\ 1.83 & 2.33 \end{bmatrix}$$

$$\mathbf{W} = \sum_{i=1}^{3} (n_i - 1) \times \mathbf{S}_i = \begin{bmatrix} 103.42 & -20.58 \\ -20.58 & 20.08 \end{bmatrix} \quad |\mathbf{W}| = 1652.59$$

$$\mathbf{B} + \mathbf{W} = \begin{bmatrix} 656.90 & 557.99 \\ 557.99 & 631.99 \end{bmatrix} \quad |\mathbf{B} + \mathbf{W}| = 103\,792.44$$

$$\Lambda = \frac{1652.59}{103\,792.44} = 0.0159 \quad \sqrt{\Lambda} = 0.126$$

$$F_{\text{test}} = \left(\frac{\sum_{i=1}^{k} n_i - k - 1}{k - 1}\right)\left(\frac{1 - \sqrt{\Lambda}}{\sqrt{\Lambda}}\right) = \left(\frac{10 - 3 - 1}{3 - 1}\right)\left(\frac{1 - .0126}{.0126}\right) = 20.775$$

$$F_{\text{critical}} = F_{0.05;\,2(k-1),\,2(\sum_{i=1}^{k} n_i - k - 1)} = F_{0.05,\,4,\,12} = 3.26$$

Conclusion The F_{test} is computed as 20.775 and the corresponding $F_{critical}$ value is 3.26. Since $F_{test} > F_{critical}$, we reject the null hypothesis that there are no differences between the three trainee groups.

MANOVA Minitab output for welding training example

```
General linear model: BWT, AWT versus degree

Factor     Type  Levels  Values
Degree     fixed      3  AD  GED  ND

Analysis of Variance for BWT, using Adjusted SS for Tests

Source    DF     Seq SS    Adj SS    Adj MS       F      P
Degree     2     553.48    553.48    276.74   18.73  0.002
Error      7     103.42    103.42     14.77
Total      9     656.90

Analysis of Variance for AWT, using Adjusted SS for Tests

Source    DF     Seq SS    Adj SS    Adj MS       F      P
Degree     2     611.92    611.92    305.96  106.64  0.000
Error      7      20.08     20.08      2.87
Total      9     632.00

Least Squares Means

                 ....  BWT  .....    ....  AWT  .....
Degree        Mean       SE Mean      Mean       SE Mean
AD           96.00        2.2191     92.33        0.9779
GED          90.25        1.9218     88.25        0.8469
ND           77.33        2.2191     73.33        0.9779

MANOVA for Degree           s =    2   m = -0.5   n =    2.0

Criterion          Test Statistic         F         DF        P
Wilk's                    0.01593    20.771  (  4,   12)  0.000
Lawley-Hotelling         59.40715    74.259  (  4,   10)  0.000
Pillai's                  1.02191     3.657  (  4,   14)  0.031
Roy's                    59.36714

SSCP Matrix (adjusted) for Degree

             BWT        AWT
BWT        553.5      578.6
AWT        578.6      611.9

SSCP Matrix (adjusted) for Error

             BWT        AWT
BWT       103.42     -20.58
AWT       -20.58      20.08

Partial Correlations for the Error SSCP Matrix

             BWT        AWT
BWT      1.00000   -0.45165
AWT     -0.45165    1.00000

EIGEN Analysis for Degree
```

```
Eigenvalue    59.3671     0.0400
Proportion     0.9993     0.0007
Cumulative     0.9993     1.0000

Eigenvector         1           2
BWT           0.07692     0.07893
AWT           0.23865    -0.07484
```

4.4 MANOVA Case Study

Quality management practice is a well-researched area in the global scope. Park et al. (2001) have studied the quality management practices among suppliers in Korea and have found that there are differences among suppliers whose performance is rated high, medium, or low by a common buying company, and have identified what specific practices contribute to the differences. The entire population of first-tier suppliers to a Korean autoassembler was surveyed to measure the use of quality management practices. Usable returns were received from 25% of the suppliers surveyed. To measure conformance to quality and overall rating, suppliers were categorized into high, medium, and low performing groups based on the buying company's data. Multivariate analysis of Variance (MANOVA) was done using general linear model (GLM-MANOVA) to explore differences in the high, medium, and low performing groups based on their quality practices. The multivariate measures were derived from the 1995 MBNQAC for the production systems category. Four dimensions included in the survey were Information and Analysis (10 items), Strategic Planning (11 items), Human Resources Development and Management (20 items), and Process Management (9 items). Factor analysis was used to compress the 50 item scores into seven multivariate characteristics and they were the following:

Information Management (INFO)

Strategic Planning (STRA)

Employee Satisfaction (HSAT)

Empowerment (HEMP)

Employee education (HEDUC)

Process Management (PROS)

Supplier Management (SMGT)

The MANOVA results using the supplier rating grouping criteria are as follows:

MANOVA

Test name	Value	Approx. F	d.f. Between	d.f. Within	Significance of statistic, p
Multivariate effects					
Pillai's criterion	0.311	2.527	14	192	0.002 *
Wilks' λ	0.709	2.551	14	190	0.002 *

Between Subjects Effects					
Dependent variables	Sum of squares	d.f.	Mean square	F	p
INFO	4.265	2	2.132	2.218	0.114
STRA	2.417	2	1.209	1.241	0.294
HSAT	8.788	2	4.394	4.615	0.012 *
HEMP	3.508	2	1.754	1.861	0.161
HEDU	3.699	2	1.835	1.935	0.15
PROS	10.809	2	5.404	6.271	0.003 *
SMGT	3.515	2	1.757	1.765	0.176

The results show that quality management practices do differentiate suppliers with different overall ratings. Of the seven quality management practices, the emphasis placed on process management (PROS) by the high performers was significantly greater than the emphasis placed by low performers. Thus, the authors state, process management can have a multifaceted effect by improving quality, delivery, and cost performance.

Chapter 5

Principal Component Analysis and Factor Analysis

5.1 Introduction

Many data from industrial applications are multivariate data. For example, in automobile assembly operation, the dimensional accuracies of body panels are very important in "fit and finish" for automobiles. Many sensors, such as optical coordinate measurement machines (OCMM), are used to measure dozens of actual dimensional locations, for every automobile, at selected key points (see Fig. 5.1). These measurements are definitely multivariate data and they are correlated with each other. Because the body structure is physically connected, the dimensional variation of one location often affects the dimensional variation of other locations. In many industries, such as chemical industries (Mason and Young, 2002), there are plenty of cases where data are multivariate in nature and are correlated. With the rapid development of computer information technology and sensor technology, we are swamped with this kind of multivariate data.

However, most of the analysis for this kind of data is still performed by univariate statistical methods which deal with these multivariate variables on a one-variable-at-a-time basis (see Fig. 5.2). In the above automobile assembly example, the common practice is that if there are 50 measurement points, then 50 statistical process control charts will be maintained. These 50 control charts could easily overwhelm process operators.

Also, each control chart can only tell you how large the variation is at each individual point, and, if the variation is out of control, this kind of information provides little help in searching the causes of variation. On the other hand, the variations of these kinds of multivariate variables are usually highly correlated. The pattern of variation, that is, how

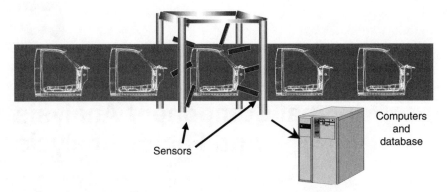

Figure 5.1 Automobile body panel dimensional measurement process.

these variations at different points are related, can actually reveal what the underlying causes could be. Clearly, univariate statistical methods are not able to serve this purpose.

Principal component analysis (PCA) and factor analysis (FCA) are multivariate statistical methods that can be used for data reduction. By data reduction we mean that these methods are very powerful in extracting a small number of hidden factors in a massive amount of multivariate data, and these factors account for most of the variation in the data. The structures of these hidden factors can help a great deal in searching for the root causes of the variation.

In this chapter, we will discuss principal component analysis and factor analysis in detail. There are two kinds of PCA. They are PCA on covariance matrices (Sec. 5.2) and PCA on correlation matrices (Sec. 5.3). These two PCA methods have substantially different properties and the choice of which PCA should be used depends on the nature of the application. In this chapter, we will discuss them in detail. Section 5.4 discusses principal component analysis for dimensional measurement data analysis. We illustrate how to explain the results of principal component analysis

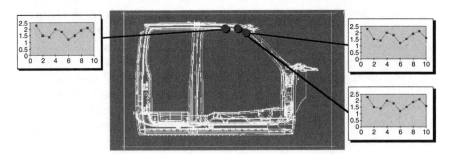

Figure 5.2 SPC chart for individual point.

by variation mode chart, visual display, and animation. Actually, many other types of multivariate data also can be analyzed and explained by the methods discussed in Sec. 5.4. Several case studies using PCA are presented in Sec. 5.5. Factor analysis is introduced in Sec. 5.6. One special feature of factor analysis is factor rotation. Factor rotation is a tool that is designed to improve the interpretation of the factor analysis, and it is presented in Sec. 5.7. Several industrial case studies in applying factor analysis are presented in Sec. 5.8.

5.2. Principal Component Analysis Based on Covariance Matrices

Dillon and Goldstein (1984) provide the following formal definition of principal components analysis (PCA):

> Principal components analysis transforms the original set of variables into a smaller set of linear combinations that account for most of the variance in the original set. The purpose of PCA is to determine factors (i.e., principal components) in order to explain as much of the total variation in the data as possible.

Principal component analysis was developed by Hotelling (1933) after its origination by Pearson (1901). Other significant contributers to the development PCA include Rao (1964), Jolliffe (1986), and Jackson (1991).

Specifically, suppose we have $\mathbf{X} = (X_1, X_2, ..., X_p)^T$ being a vector of p multivariate variables distributed as a multivariate normal distribution, that is, $\mathbf{X} \sim N(\boldsymbol{\mu}, \boldsymbol{\Sigma})$, where

$$\boldsymbol{\mu} = \begin{bmatrix} \mu_1 \\ \mu_2 \\ \vdots \\ \mu_p \end{bmatrix}$$

is the vector of population means, and

$$\boldsymbol{\Sigma} = \begin{bmatrix} \sigma_{11} & \sigma_{12} & \cdots & \sigma_{1p} \\ \sigma_{21} & \sigma_{22} & \cdots & \sigma_{2p} \\ \vdots & \vdots & \vdots & \vdots \\ \sigma_{n1} & \sigma_{n2} & \cdots & \sigma_{pp} \end{bmatrix}$$

is the population covariance matrix.

Because $\boldsymbol{\Sigma}$ is a covariance matrix, it is at least a positive semidefinite matrix. All eigenvalues of $\boldsymbol{\Sigma}$ should be greater or equal to zero. That is, if $\boldsymbol{\lambda} = (\lambda_1, \lambda_2, ..., \lambda_k, ..., \lambda_p)^T$ is the vector of eigenvalues of $\boldsymbol{\Sigma}$, where λ_1 is the largest eigenvalue, λ_2 is the second largest eigenvalue, etc., then $\lambda_1 \geq \lambda_2 \geq \cdots \geq$

$\lambda_k \geq \cdots \geq \lambda_p > 0$. When Σ has the full rank, then $\lambda_1 > \lambda_2 > \cdots > \lambda_k > \cdots > \lambda_p > 0$. We further assume that

$$A = A_{p \times p} = (A_{.1} \quad A_{.2} \quad \cdots \quad A_{.p}) = \begin{bmatrix} a_{11} & a_{12} & \cdots & a_{1p} \\ a_{21} & a_{22} & \cdots & a_{2p} \\ \vdots & \vdots & \vdots & \vdots \\ a_{p1} & a_{p2} & \cdots & a_{pp} \end{bmatrix}$$

and $A_{.j}$ is the jth eigenvector of Σ. It is a well-known result that A will also be an orthogonal matrix so that $A^T A = I$ and

$$A^T \Sigma A = D_\lambda = \begin{bmatrix} \lambda_1 & 0 & \cdots & 0 \\ 0 & \lambda_2 & \cdots & 0 \\ \vdots & \vdots & \vdots & \vdots \\ 0 & 0 & \cdots & \lambda_p \end{bmatrix} \qquad (5.1)$$

5.2.1 Two mathematical representations of principal component analysis

Principal component analysis consists of the following linear transformation:

$$Y = A^T(x - \mu) \qquad (5.2)$$

where $Y^T = (Y_1, \ldots, Y_p)$, $A = (a_{ij})_{p \times p}$, and $A_{.j}$ is the jth eigenvector of Σ. A is also an orthogonal matrix so that $A^T A = I$. Specifically, we have

$$Y_1 = a_{11}(X_1 - \mu_1) + a_{21}(X_2 - \mu_2) + \cdots + a_{p1}(X_p - \mu_p)$$
$$Y_2 = a_{12}(X_1 - \mu_1) + a_{22}(X_2 - \mu_2) + \cdots + a_{p2}(X_p - \mu_p)$$
$$\cdots\cdots\cdots\cdots\cdots\cdots\cdots\cdots\cdots\cdots\cdots\cdots\cdots\cdots\cdots\cdots$$
$$Y_i = a_{1i}(X_1 - \mu_1) + a_{2i}(X_2 - \mu_2) + \cdots + a_{pi}(X_p - \mu_p) \qquad (5.3)$$
$$\cdots\cdots\cdots\cdots\cdots\cdots\cdots\cdots\cdots\cdots\cdots\cdots\cdots\cdots\cdots\cdots$$
$$Y_\mu = a_{1p}(X_1 - \mu_1) + a_{2p}(X_2 - \mu_2) + \cdots + a_{pp}(X_p - \mu_p)$$

Equation (5.3) expresses Y_i as a linear combination of $X_1, X_2, \ldots, X_p$, by using the eigenvectors of the covariance matrix of X, for $i = 1, \ldots, p$. This is the *first type of mathematical representation* of principal component analysis, in which Y_i is the ith principal component, for $i = 1, \ldots, p$.

Principal component analysis can also be expressed in another form. By premultiplying both sides of Eq. (5.2) by A, we have

$$X - \mu = AY = \sum_{j=1}^{p} A_j Y_j \qquad (5.4)$$

Specifically,

$$\begin{bmatrix} X_1 - \mu_1 \\ X_2 - \mu_2 \\ \vdots \\ X_p - \mu_p \end{bmatrix} = \begin{bmatrix} a_{11} \\ a_{21} \\ \vdots \\ a_{p1} \end{bmatrix} Y_1 + \begin{bmatrix} a_{12} \\ a_{22} \\ \vdots \\ a_{p2} \end{bmatrix} Y_2 + \cdots + \begin{bmatrix} a_{1i} \\ a_{2i} \\ \vdots \\ a_{pi} \end{bmatrix} Y_i + \cdots + \begin{bmatrix} a_{1p} \\ a_{2p} \\ \vdots \\ a_{pp} \end{bmatrix} Y_p \quad (5.5)$$

Equation (5.5) represents the vector of $\mathbf{X} - \boldsymbol{\mu}$ by using a vector combination of principal components, that is, Y_i's, $i = 1,\ldots,p$. Equation (5.5) is the *second type of mathematical representation of principal component analysis*.

Both Eqs. (5.3) and (5.5) are valid forms to represent principal component analysis. The choice between these two mathematical representations depends on the types of applications, which we will discuss in subsequent sections.

5.2.2 Properties of principal component analysis

Theorem 5.1 *For the principal component analysis defined by Eq. (5.3) or (5.5), we have*

1. $Var(Y_i) = \lambda_i$, where Y_i is the ith principal component, and λ_i is the ith largest eigenvalue for the covariance matrix Σ, for all $i = 1,\ldots, p$.
2. $Cov(Y_i, Y_j) = 0$, for $i \neq j$, for all $i, j \in (1, 2,\ldots, p)$

Proof. Obviously

$$\mathbf{E}(\mathbf{Y}) = \mathbf{E}\mathbf{A}^T(\mathbf{x} - \boldsymbol{\mu}) = \mathbf{A}^T \mathbf{E}(\mathbf{x} - \boldsymbol{\mu}) = \mathbf{A}^T(\boldsymbol{\mu} - \boldsymbol{\mu}) = 0$$

By Eq. (5.3), it is clear that the covariance matrix of $\mathbf{Y}$,

$$Cov(\mathbf{Y}) = \begin{bmatrix} Var(Y_1) & Cov(Y_1, Y_2) & \cdots & Cov(Y_1, Y_p) \\ Cov(Y_2, Y_1) & Var(Y_2) & \cdots & \cdots \\ \vdots & \vdots & \vdots & \vdots \\ Cov(Y_p, Y_1) & Cov(Y_p, Y_2) & Cov(Y_p, Y_{p-1}) & Var(Y_p) \end{bmatrix}$$

$$= E(\mathbf{Y}\mathbf{Y}^T) = E[\mathbf{A}^T(\mathbf{x}-\boldsymbol{\mu})(\mathbf{x}-\boldsymbol{\mu})^T \mathbf{A}]$$

$$= \mathbf{A}^T E[(\mathbf{x}-\boldsymbol{\mu})(\mathbf{x}-\boldsymbol{\mu})^T]\mathbf{A}$$

$$= \mathbf{A}^T \Sigma \mathbf{A} = \mathbf{D}_\lambda = \begin{bmatrix} \lambda_1 & 0 & \cdots & 0 \\ 0 & \lambda_2 & \cdots & 0 \\ \vdots & \vdots & \vdots & \vdots \\ 0 & 0 & \cdots & \lambda_p \end{bmatrix} \quad (5.6)$$

From Eq. (5.4), clearly $\mathrm{Var}(Y_i) = \lambda_i$, for all $i = 1,\ldots, p$, and $\mathrm{Cov}(Y_i, Y_j) = 0$, for $i \neq j$.

Theorem 5.2

$$\mathrm{Var}(X_1) + \cdots + \mathrm{Var}(X_p) = \sum_{i=1}^{p} \sigma_i^2 = \sum_{i=1}^{p} \mathrm{Var}(Y_i) = \lambda_1 + \cdots + \lambda_p \quad (5.7)$$

Proof. This is because λ_j's are obtained by letting $\det(\lambda \mathbf{I} - \mathbf{\Sigma}) = 0$, where $\det(.)$ refers to the determinant, and $\lambda_1 + \cdots + \lambda_p = \mathrm{trace}(\mathbf{\Sigma}) = \mathrm{Var}(x_1) + \cdots + \mathrm{Var}(x_p) = \sum_{j=1}^{p} \sigma_j^2$.

From Theorems 5.1 and 5.2, we can easily derive the following properties for principal component analysis:

1. The principal components $\mathbf{Y} = (Y_1, Y_2, \ldots, Y_p)$ of $\mathbf{X}$ are mutually independent random variables, and their variances are $\lambda_1, \lambda_2, \ldots, \lambda_p$, respectively.

2. The sum of total variances of $\mathbf{X}$ is equal to the sum of total variance of the principal components, that is,

$$\mathrm{Var}(X_1) + \cdots + \mathrm{Var}(X_p) = \sum_{i=1}^{p} \sigma_i^2 = \sum_{i=1}^{p} \mathrm{Var}(Y_i) = \lambda_1 + \cdots + \lambda_p$$

3. In many cases, the eigenvalues of $\mathbf{\Sigma}$, that is, $\lambda_1, \lambda_2, \ldots, \lambda_p$, are not equal in magnitude. Many small eigenvalues are much smaller than the top ones such as λ_1, λ_2. That is,

$$\lambda_1 + \lambda_2 + \cdots + \lambda_s \approx \lambda_1 + \lambda_2 + \lambda_3 + \cdots + \lambda_p$$

where $s \ll p$. Therefore

$$\lambda_1 + \lambda_2 + \cdots + \lambda_s \approx \lambda_1 + \lambda_2 + \lambda_3 + \cdots + \lambda_p = \mathrm{Var}(X_1) + \cdots + \mathrm{Var}(X_p)$$

That is, the summation of several large eigenvalues will be very close to the summation of total variance on the original multivariate random variables $\mathbf{X}$.

4. Therefore, we can use the following vector combination of the first few principal components:

$$\mathbf{X} - \boldsymbol{\mu} = \begin{pmatrix} X_1 - \mu_1 \\ X_2 - \mu_2 \\ \ldots \\ X_p - \mu_p \end{pmatrix} \approx \mathbf{A}_{.1} Y_1 + \mathbf{A}_{.2} Y_2 + \cdots + \mathbf{A}_{.s} Y_s \quad (5.8)$$

as an approximation of the original multivariate random variable $\mathbf{X}$. And these first few principal components will account for a big percentage of total variance in $\mathbf{X}$.

5.2.3 Covariance and correlation between X and principal components Y

For any $i, j \in (1, 2, ..., p)$, and from Eq. (5.2), we have

$$\text{Cov}(x_i, y_j) = \text{Cov}(x_i, (\mathbf{x} - \boldsymbol{\mu})\mathbf{A}_{\cdot j})$$
$$= [\text{Cov}(x_i, x_1), \text{Cov}(x_i, x_2), ..., \text{Cov}(x_i, x_p)]\mathbf{A}_{\cdot j}$$
$$= [\sigma_{i1}, \sigma_{i2}, ..., \sigma_{ip}]\mathbf{A}_{\cdot j}$$
$$= \sum_{k=1}^{p} a_{kj}\sigma_{ik} \tag{5.9}$$

But

$$\Sigma \mathbf{A} = \mathbf{A}\mathbf{D}_\lambda$$

or

$$\begin{bmatrix} \sigma_{11} & \sigma_{12} & \cdots & \sigma_{1p} \\ \sigma_{21} & \sigma_{22} & \cdots & \sigma_{2p} \\ \vdots & \vdots & & \vdots \\ \sigma_{n1} & \sigma_{n2} & \cdots & \sigma_{pp} \end{bmatrix} \begin{bmatrix} a_{11} & a_{12} & \cdots & a_{1p} \\ a_{21} & a_{22} & \cdots & a_{2p} \\ \vdots & \vdots & & \vdots \\ a_{p1} & a_{p2} & \cdots & a_{pp} \end{bmatrix} = \begin{bmatrix} a_{11} & a_{12} & \cdots & a_{1p} \\ a_{21} & a_{22} & \cdots & a_{2p} \\ \vdots & \vdots & & \vdots \\ a_{p1} & a_{p2} & \cdots & a_{pp} \end{bmatrix} \begin{bmatrix} \lambda_1 & 0 & \cdots & 0 \\ 0 & \lambda_2 & \cdots & 0 \\ \vdots & \vdots & \cdots & \vdots \\ 0 & 0 & \cdots & \lambda_p \end{bmatrix}$$

So,

$$\text{Cov}(x_i, y_j) = \lfloor \sigma_{i1}, \sigma_{i2}, ..., \sigma_{ip} \rfloor \mathbf{A}_{\cdot j} = [a_{i1}, a_{i2}, ..., a_{ip}] \begin{bmatrix} 0 \\ 0 \\ \vdots \\ \lambda_j \\ \vdots \\ 0 \end{bmatrix} = a_{ij}\lambda_j \tag{5.10}$$

Therefore,

$$\text{Cor}(x_i, y_j) = \frac{\text{Cov}(x_i, y_j)}{\sqrt{[\text{Var}(x_i)\text{Var}(y_j)]}} = \frac{a_{ij}\lambda_j}{\sigma_{ii}^{1/2}\lambda_j^{1/2}} = a_{ij}\left(\frac{\lambda_j}{\sigma_{ii}}\right)^{1/2} \tag{5.11}$$

5.2.4 Principal component analysis on sample covariance matrix

In real world data analysis, the theoretical population covariance matrix Σ and population mean vector $\boldsymbol{\mu}$ are not available. In such a case, the sample covariance matrix **S** and sample mean vector $\overline{\mathbf{X}}$ will be used as substitutes.

In this case, principal component analysis consists of the following linear transformation:

$$\mathbf{Y} = \mathbf{B}^T(\mathbf{X} - \overline{\mathbf{X}}) \tag{5.12}$$

where $\mathbf{Y}^T = (Y_1, \ldots, Y_p)$, $\mathbf{B} = (b_{ij})_{p \times p}$, and $\mathbf{B}_{.j}$ is the jth eigenvector of **S**. **B** is also an orthogonal matrix so $\mathbf{B}^T\mathbf{B} = \mathbf{I}$. Equation (5.12) is the first type of mathematical representation of principal component analysis on the sample covariance matrix.

Principal component analysis can also be expressed as

$$\mathbf{X} - \overline{\mathbf{X}} = \mathbf{BY} = \sum_{j=1}^{p} \mathbf{B}_{.j} Y_j \tag{5.13}$$

Equation (5.13) is the second type of mathematical representation of principal component analysis on sample covariance matrix. In this case, we will use $l = (l_1, l_2, \ldots, l_p)$ to represent the eigenvalues for **S**.

Most of the properties for PCA will hold true if we have sufficient sample size in computing **S** and $\overline{\mathbf{X}}$. However, because **S** and $\overline{\mathbf{X}}$ are statistical estimates of Σ and μ, l and **B** are statistical estimates of λ and **A**. Here we list the following statistical properties of l and **B**.

1. By Girshick (1936), the maximum likelihood estimate of jth population eigenvector of Σ, $\mathbf{A}_{.j}$, is $\mathbf{B}_{.j}$. By Anderson (1984), the maximum likelihood estimate of the ith eigenvalue λ_i is $[(n-1)/n]l_i$. By Girshick (1939), the eigenvalues l_i, $i = 1, \ldots, p$, are distributed independently of their associated eigen vectors $\mathbf{B}_{.j}$, for $i = 1, \ldots, p$.

2. The variance of l_i is asymptotically $2\lambda_i^2/n$ (Girshick, 1939). Therefore, the quantity $\sqrt{n}(l_i - \lambda_i)$ is normally distributed with mean zero and standard deviation $2\lambda_i^2$. The asymptotic confidence interval for the true eigenvalues of A, λ_i, for $i = 1, \ldots, p$, is

$$\frac{l_i}{1 + z_{\alpha/2}\sqrt{2/n}} \leq \lambda_i \leq \frac{l_i}{1 - z_{\alpha/2}\sqrt{2/n}} \tag{5.14}$$

3. By Girshick (1939), the elements $(\mathbf{B}_{.i} - \mathbf{A}_{.i})$ are multivariate normally distributed with mean vector **0** and covariance matrix

$$\frac{\lambda_i}{n} \sum_{k \neq i}^{p} \frac{\lambda_k}{(\lambda_k - \lambda_i)^2} \mathbf{A}_{.k} \mathbf{A}_{.k}^T \tag{5.15}$$

The asymptotic standard error for b_{ki}, that is, the kth element of the ith eigenvector of **S**, $\mathbf{B}_{\cdot j}$, is

$$\sqrt{\frac{l_i}{n}\sum_{h \neq i}^{p} \frac{l_h}{(l_h - l_i)^2} b_{kh}^2} \tag{5.16}$$

Example 5.1 (Yang, 1996) Figure 5.3 presents the measurement data and run charts of four measuring points for each of the 10 flat panels. The deviations of the check points from their nominal location are clearly the observations of $\mathbf{x} = (x_1, x_2, x_3, x_4)$.

The actual measurement scheme and data are given in Fig. 5.4. The deviations of actual measurements and nominal dimensions are of our concern. Table 5.1 gives the deviation data (x_1, x_2, x_3, x_4).

The sample covariance matrix for this data set is

$$\mathbf{S} = \begin{bmatrix} s_{11} & s_{12} & s_{13} & s_{14} \\ s_{21} & s_{22} & s_{23} & s_{24} \\ s_{31} & s_{32} & s_{33} & s_{34} \\ s_{41} & s_{42} & s_{43} & s_{44} \end{bmatrix} = \begin{bmatrix} 9.88 & 5.11 & -7.40 & -10.66 \\ 5.11 & 6.00 & -5.78 & -7.11 \\ -7.40 & -5.78 & 8.27 & 9.18 \\ -10.66 & -7.11 & 9.18 & 15.66 \end{bmatrix}$$

The eigenvalues for this sample covariance matrix are

$$l_1 = 34.048, \quad l_2 = 2.898, \quad l_3 = 1.921, \quad l_4 = 0.934$$

The eigenvectors of the sample covariance matrix are

$$B = \begin{bmatrix} b_{11} & b_{12} & b_{13} & b_{14} \\ b_{21} & b_{22} & b_{23} & b_{24} \\ b_{31} & b_{32} & b_{33} & b_{34} \\ b_{41} & b_{42} & b_{43} & b_{44} \end{bmatrix} = [\mathbf{B}_{.1} \ \mathbf{B}_{.2} \ \mathbf{B}_{.3} \ \mathbf{B}_{.4}] = \begin{bmatrix} 0.499 & 0.365 & 0.655 & 0.434 \\ 0.349 & -0.700 & -0.253 & 0.569 \\ -0.453 & 0.454 & -0.357 & 0.679 \\ -0.651 & -0.412 & 0.616 & 0.166 \end{bmatrix}$$

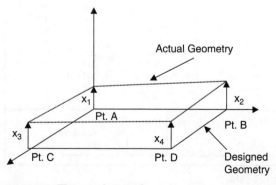

Figure 5.3 Flat panel example.

Chapter Five

Point A Panel #	Measurement	Nominal	Deviation
1	99	100	−1
2	105	100	5
3	102	100	2
4	95	100	−5
5	99	100	−1
6	104	100	4
7	97	100	−3
8	100	100	0
9	98	100	−2
10	102	100	2
Average	100.1		0.1
		Standard Dev =	3.1

Point B Panel #	Measurement	Nominal	Deviation
1	96	100	−4
2	101	100	1
3	102	100	2
4	98	100	−2
5	97	100	−3
6	100	100	0
7	97	100	−3
8	102	100	2
9	96	100	−4
10	101	100	1
Average	100.1		−0.2
		Standard Dev =	2.4

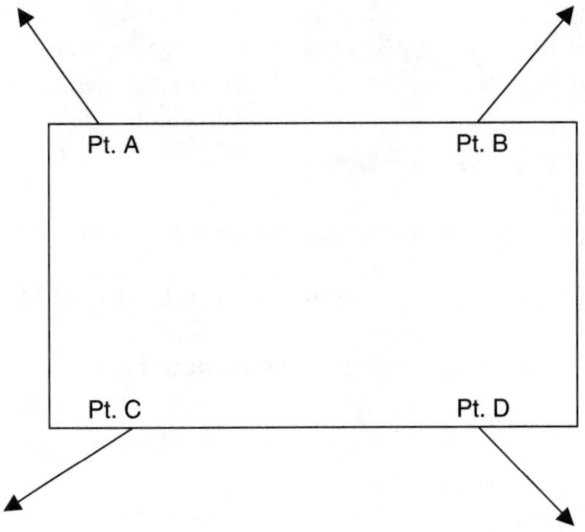

Pt. A   Pt. B

Pt. C   Pt. D

Point C Panel #	Measurement	Nominal	Deviation
1	104	100	4
2	99	100	−1
3	100	100	0
4	106	100	6
5	101	100	1
6	100	100	0
7	105	100	5
8	98	100	−2
9	104	100	4
10	99	100	−1
Average	101.4		1.6
		Standard Dev =	2.9

Point D Panel #	Measurement	Nominal	Deviation
1	102	100	2
2	99	100	−1
3	96	100	−4
4	106	100	6
5	102	100	2
6	94	100	−6
7	103	100	3
8	99	100	−1
9	103	100	3
10	95	100	−5
Average	100.1		−0.1
		Standard Dev =	4.0

Figure 5.4 Flat panel measurement data.

TABLE 5.1 Measurement Deviation Data

x_1	x_2	x_3	x_4
−1	−4	4	2
5	1	−1	−1
2	2	0	−4
−5	−2	6	6
−1	−3	1	2
4	0	0	−6
−3	−3	5	3
0	2	−2	−1
−2	−4	4	3
2	1	−1	−5

Many statistical software, such as MINITAB, can perform principal component analysis. For this example, MINITAB gives the following results for PCA:

Principal Component Analysis: A, B, C, D

```
Eigenanalysis of the Covariance Matrix

Eigenvalue    34.048     2.898     1.921     0.934
Proportion     0.855     0.073     0.048     0.023
Cumulative     0.855     0.928     0.977     1.000


Variable         PC1       PC2       PC3       PC4
A              0.499     0.365     0.655     0.434
B              0.349    -0.700    -0.253     0.569
C             -0.453     0.454    -0.357     0.679
D             -0.651    -0.412     0.616     0.166
```

In this MINITAB output, we can see that the percentage of variation explained by the first principal component is 85.5%, which is actually

$$\frac{l_1}{l_1+l_2+l_3+l_4} = \frac{34.048}{34.048+2.898+1.921+0.934} = 0.855 = 85.5\%$$

Similarly, the second component explains 7.3% of total variation. The interpretation of principal components for Example 5.1 will be discussed in detail in Sec. 5.4.

By using Eq. (5.14), we can find the 95% confidence interval for λ_1, by using $\alpha = 0.05$,

$$\frac{34.048}{1+1.96\sqrt{2/10}} \le \lambda_1 \le \frac{34.048}{1-1.96\sqrt{2/10}}$$

Therefore, $18.14 \le \lambda_1 \le 275.78$.

Also by using Eq. (5.16), we can find the standard error for the first eigenvector is

$$s_{b_{ki}} = \sqrt{\frac{l_i}{n} \sum_{h \ne i}^{p} \frac{l_h}{(l_h - l_i)^2} b_{kh}^2}$$

For example, for b_{11},

$$s_{b_{11}} = \sqrt{\frac{l_1}{n}\left(\frac{l_2}{(l_2-l_1)^2}b_{12}^2 + \frac{l_3}{(l_3-l_1)^2}b_{13}^2 + \frac{l_4}{(l_4-l_1)^2}b_{14}^2\right)}$$

$$= \sqrt{\frac{34.048}{10}\left(\frac{2.898}{(2.898-34.048)^2}0.365^2 + \frac{1.921}{(1.921-34.048)^2}0.655^2 + \frac{0.934}{(0.934-34.048)^2}0.434^2\right)}$$

$$= 0.0046$$

Similarly, we can get the standard error for the first column of eigenvector:

$$s_{B_1} = (0.0046, 0.0063, 0.0042, 0.0042)^T$$

5.3 Principal Component Analysis Based on Correlation Matrices

There are many situations where conducting principal component analysis based on covariance matrices would not be appropriate, especially for the following two cases:

1. The original variables are in different units. For example, if a variable is expressed in ounces, its variance will be $16 \times 16 = 256$ times of that expressed in pounds. Then this variable will exert substantially more influence on the types of principal components we would get because PCA is concerned with explaining total variances.

2. The original variables have different meanings and have vastly different numerical magnitudes. For example, one variable could be clearance in inches. It could range in the units of 0.001 in. The other variable might be pressure; it could range between 35 and 45 psi. Clearly, the second variable will exert much more influence on the PCA analysis.

In these cases, we standardize multivariate variables and use correlation matrix to perform principal component analysis.

Specifically, we would first standardize the random vector $\mathbf{X} = (X_1, X_2, \ldots, X_p)$ by

$$Z_i = \frac{X_i - \overline{X}_i}{\sqrt{s_{ii}}} \tag{5.17}$$

for $i = 1, \ldots, p$, where $\overline{X}_i$ is the sample mean for X_i and s_{ii} is the sample variance for X_i. Let

$$\mathbf{R} = \begin{bmatrix} 1 & r_{12} & \cdots & r_{1p} \\ r_{21} & 1 & \cdots & r_{2p} \\ \vdots & \vdots & \vdots & \vdots \\ r_{n1} & r_{n2} & \cdots & 1 \end{bmatrix}$$

be the sample correlation matrix for $\mathbf{X}$, and $\mathbf{l} = (l_1, l_2, \ldots, l_p)$ be the vector of eigenvalues for $\mathbf{R}$. In this case, principal component analysis consists of the following linear transformation:

$$\mathbf{Y} = \mathbf{C}^T \mathbf{Z} \quad (5.18)$$

where $\mathbf{Z} = (Z_1, Z_2, \ldots, Z_p)^T$
$\mathbf{Y} = (Y_1, \ldots, Y_p)^T$
$\mathbf{C} = (c_{ij})_{p \times p}$

and $\mathbf{C}_j$ is the jth eigenvector of $\mathbf{R}$. $\mathbf{C}$ is also an orthogonal matrix so that $\mathbf{C}^T \mathbf{C} = \mathbf{I}$. Equation (5.18) is the first type of mathematical representation of principal component analysis for a correlation matrix.

The principal component analysis can also be expressed as

$$\mathbf{Z} = \mathbf{CY} = \sum_{j=1}^{p} \mathbf{C}_j Y_j \quad (5.19)$$

Equation (5.19) is the second type of mathematical representation for principal component analysis for a correlation matrix.

In most literature, especially the multivariate statistical books in which most of the PCA examples are from social science, psychology, physiology, or education, the principal component analysis on correlation matrix is preferred because the multivariate random variables involved in these examples are often different in meaning and measurement units. Also, often the first type of mathematical representation is used, because in this representation, each principal component Y_i is represented by a linear combination of standardized variables, specifically,

$$Y_i = c_{1i} Z_1 + c_{2i} Z_2 + \cdots + c_{ki} Z_k + \cdots + c_{pi} Z_p \quad (5.20)$$

For $i = 1, \ldots, p$, $k = 1, \ldots, p$, the magnitude of c_{ki}, that is, $|c_{ki}|$, is often used to represent the relative importance of variable Z_k in the ith principal component. The sign of c_{ki}'s is often used to represent the variation direction of Z_k in the ith component. For example, if c_{1i} and c_{2i} have different signs—one is positive, another is negative—then we say that in the ith principal component, Z_1 and Z_2 vary in different directions. When Z_1 increases, Z_2 must decrease and vice versa.

Example 5.2 (Young and Sarle, 1983) Table 5.2 shows the weight (lbs), waist (inches), and pulse of 20 middle-aged men in a fitness club. We use principal component analysis to find hidden components that can explain most of the variations.

Since, in this data set, all three variables are in different physical units and numerical scales, we use principal component analysis with correlation matrix.

Chapter Five

TABLE 5.2 Data for 20 Middle-Aged Men

X_1 Weight (lbs)	X_2 Waist (inches)	X_3 Pulse
191	36	50
189	37	52
193	38	58
162	35	62
189	35	46
182	36	56
211	38	56
167	34	60
176	31	74
154	33	56
169	34	50
166	33	52
154	34	64
247	46	50
193	36	46
202	37	62
176	37	54
157	32	52
156	33	54
138	33	58

The correlation matrix for this data set is

$$\mathbf{R} = \begin{bmatrix} 1 & 0.87 & -0.266 \\ 0.87 & 1 & -0.322 \\ -0.266 & -0.322 & 1 \end{bmatrix}$$

By using MINITAB, we get the following principal component analysis results:

Principal Component Analysis: Weight, Waist, Pulse

```
Eigenanalysis of the Correlation Matrix

Eigenvalue     2.0371      0.8352      0.1277
Proportion     0.679       0.278       0.043
Cumulative     0.679       0.957       1.000


Variable          PC1         PC2         PC3
weight         -0.651      -0.301       0.697
waist          -0.662      -0.226      -0.715
pulse           0.372      -0.927      -0.052
```

We can see clearly that $l_1 = 2.0371$, $l_2 = 0.8352$, and $l_3 = 0.1277$. The proportion of variance explained by the first component is

$$\frac{l_1}{l_1 + l_2 + l_3} = \frac{2.0371}{2.0371 + 0.8352 + 0.1277} = 0.679 = 67.9\%$$

The proportion of variance explained by the second component is

$$\frac{l_2}{l_1+l_2+l_3} = \frac{0.8352}{2.0371+0.8352+0.1277} = 0.278 = 27.8\%$$

So the first two principal components represent 67.9% + 27.8% = 95.7% of total variation. The first component can be expressed as

$$y_1 = -0.651 \text{ weight} - 0.662 \text{ waist} + 0.372 \text{ pulse}$$

We can explain this component "weight and waist move in the same direction, the pulse moves in a different direction." In other words, when people gain weight and their waistline increases, the pulse gets slower. When people's weight and waistline decrease, the pulse is faster. This kind of variation accounts for 67.9% of variation.

The second component can be expressed as

$$y_2 = -0.301 \text{ weight} - 0.226 \text{ waist} - 0.927 \text{ pulse}$$

This component is featured by the large coefficient magnitude on pulse, and relative small coefficient magnitudes on weight and waist. This component is primarily the variation in pulse, which accounts for 27.8% of the variation. The following plot shows the comparison of magnitudes in eigenvalues.

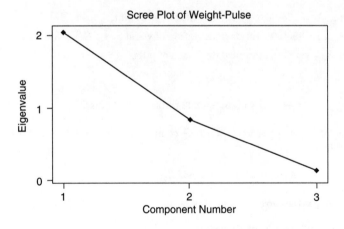

5.3.1 Principal component scores and score plots

In principal component analysis, after we identify the principal components, we may plug in the variable values of the original objects in principal component equations to get principal component scores.

Specifically, if we have an object with the normalized variable values $\mathbf{Z}_0 = (Z_{01}, Z_{02}, \ldots, Z_{0p})^T$, the corresponding ith principal component coefficient vector is $\mathbf{c}_i = (c_{1i}, c_{2i}, \ldots, c_{pi})$. Then the corresponding principal component score is

$$\mathbf{c}_i \mathbf{Z}_0 = c_{1i} Z_{01} + c_{2i} Z_{02} + \cdots + c_{pi} Z_{0p} \tag{5.21}$$

If we divide the principal component score specified by Eq. (5.21) by $\sqrt{\lambda_i}$, specifically,

$$\frac{\mathbf{c}_i \mathbf{Z}_0}{\sqrt{\lambda_i}} = \frac{c_{1i} Z_{01} + c_{2i} Z_{02} + \cdots + c_{pi} Z_{0p}}{\sqrt{\lambda_i}} \tag{5.22}$$

Then the principal component score specified by Eq. (5.22) is called the standardized principal component score. Clearly, because Var(Y_i) = λ_i and Var($Y_i/\sqrt{\lambda_i}$) = 1, the standardized principal component score should follow a standard normal distribution.

Example 5.3: Principal Component Plot for the Data in Example 5.2 First, we need to compute the means and standard deviations for our variables—weight X_1, waist X_2, and pulse X_3. We get

Mean of weight, $\bar{X}_1 = 178.60$; standard deviation of weight, $\sqrt{s_{11}} = 24.691$
Mean of waist, $\bar{X}_2 = 35.400$; standard deviation of waist, $\sqrt{s_{22}} = 3.2020$
Mean of pulse $\bar{X}_3 = 55.600$; standard deviation of pulse, $\sqrt{s_{33}} = 6.6681$

The first component can be expressed as

$$y_1 = -0.651 \text{ weight} - 0.662 \text{ waist} + 0.372 \text{ pulse}$$

So the principal component score of the first component for any (weight, waist, pulse) data set is

$$Y_{01} = -0.651 Z_{01} - 0.662 Z_{02} + 0.372 Z_{03}$$

where (Z_{01}, Z_{02}, Z_{03}) is the corresponding standardized (weight, waist, pulse) data set.

The second component can be expressed as

$$y_2 = -0.301 \text{ weight} - 0.226 \text{ waist} - 0.927 \text{ pulse}$$

So the principal component score for the second component is

$$y_{02} = -0.301 Z_{01} - 0.226 Z_{02} + 0.927 X_{03}$$

For the first data record in Example 5.2, weight = 191, waist = 36, and pulse = 50.

TABLE 5.3 Principal Component Scores

Sample no.	Weight	Waist	Pulse	Score 1	Score 2
1	191	36	50	−0.76356	0.58492
2	189	37	52	−0.80582	0.26075
3	193	38	58	−0.78278	−0.6923
4	162	35	62	0.87759	−0.65909
5	189	35	46	−0.7276	1.2357
6	182	36	56	−0.19126	−0.13932
7	211	38	56	−1.36885	−0.63345
8	167	34	60	0.84079	−0.37143
9	**176**	**31**	**74**	**2.00556**	**−2.21469**
10	154	33	56	1.16664	0.41323
11	169	34	50	0.22955	0.99385
12	166	33	52	0.62699	0.82303
13	154	34	64	1.40679	−0.76906
14	**247**	**46**	**50**	**−4.30604**	**−0.80252**
15	193	36	46	−1.03968	1.11643
16	202	37	62	−0.58988	−1.2871
17	176	37	54	−0.35152	0.14105
18	157	32	52	1.07084	1.00315
19	156	33	54	1.00223	0.66681
20	138	33	58	1.7	0.33004

Its principal component score for the first component is

$$y_{01} = -0.651 Z_{01} - 0.662 Z_{02} + 0.372 Z_{03}$$

$$= -0.651 \frac{191 - 178.60}{24.691} - 0.662 \frac{36 - 35.4}{3.202} + 0.372 \frac{50 - 55.6}{6.6681} = -0.76356$$

Its principal component score for the second component is

$$y_{02} = -0.301 Z_{01} - 0.226 Z_{02} + 0.927 Z_{03}$$

$$= -0.301 \frac{191 - 178.60}{24.691} - 0.226 \frac{36 - 35.4}{3.202} + 0.927 \frac{50 - 55.6}{6.6681} = 0.58492$$

Similarly, we can compute the principal component scores for the first two components for all samples, as listed in Table 5.3.

The samples that have high scores on a particular component are featured items of that component. For example, sample 9 and sample 14 have high (positive or negative) scores for the first component.

For sample 9, weight = 176, waist = 31, pulse = 74, score = 2.0056; this is a "low weight, thin waist, high pulse" guy.

For sample 14, weight = 247, waist = 46, pulse = 50, score = −4.303; this is a "fat and low pulse" guy.

The items that have high principal component scores are also influential items. Adding or deleting these items will usually drastically influence the principal component analysis results.

MINITAB can also plot the principal component scores as follows.

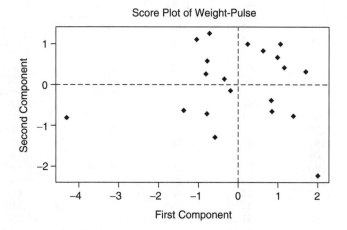

5.4 Principal Component Analysis of Dimensional Measurement Data

The data that we discussed in Example 5.1 are also called dimensional measurement data because each variable is a measurement of length or distance. Every variable has the same practical or physical meaning as well as the same unit. In this section, we are going to discuss the special principal component analysis procedure developed by Yang (1996) to analyze this kind of data. Later we will show that this special procedure cannot only be used to analyze dimensional measurement data, but can also be used to analyze many multivariate data in which every variable has the same physical or practical meaning, for example, temperature measurement data at different surface locations of objects.

A well-known example of dimensional measurement data is the painted turtle example, which is discussed in almost every book that discusses principal component analysis.

Example 5.4: The Painted Turtle Example Jolicoeur and Mosimann (1960) studied the relationship of size and shape for painted turtles. Table 5.4 contains the measurements on carapaces of 24 female turtles.

Figure 5.5 shows the meaning of each variable. Clearly, this is also a dimensional measurement data set.

The objective of principal component analysis on dimensional measurement data is mainly to analyze the major principal components in dimensional variation. In many industrial applications, dimensions are key quality characteristics and it is desirable to reduce the dimensional variation. In this analysis, it is desirable to analyze original variables,

TABLE 5.4 Measurements of Turtle Dimensions

Sample no.	Length (mm) x_1	Width (mm) x_2	Height (mm) x_3
1	98	81	38
2	103	84	38
3	103	86	42
4	105	86	42
5	109	88	44
6	123	92	50
7	123	95	46
8	133	99	51
9	133	102	51
10	133	102	51
11	134	100	48
12	136	102	49
13	138	98	51
14	138	99	51
15	141	105	53
16	147	108	57
17	149	107	55
18	153	107	56
19	155	115	63
20	155	117	60
21	158	115	62
22	159	118	63
23	162	124	61
24	177	132	67

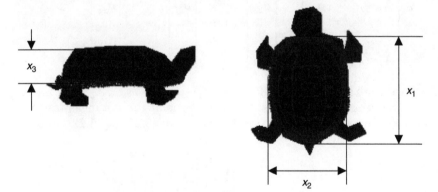

Figure 5.5 An engineer's view of the turtle problem.

$\mathbf{X} = (X_1, X_2, \ldots, X_p)$ and not the normalized variables $\mathbf{Z} = (Z_1, Z_2, \ldots, Z_p)$ for the following reasons:

1. In most industrial applications, the goal is to reduce dimensional variation. Usually, the larger the absolute dimensional deviation, the greater the harm to the product quality. For example, 4-mm dimensional

variation is usually worse than 2-mm dimensional variation. Therefore, we analyze the data in their original scale and not the normalized data.

2. Multivariate dimensional measurement data are snapshots of the "shape," or "geometry" of the objects that we are studying. Original measurement data are snapshots of the actual shape. For example, in the turtle example, this real length, width, and height data set is a snapshot for the real shape of the turtle shell. Normalized length, width, and height do not really provide the true information about the shape. The purpose of principal component analysis on dimensional measurement data is to identify major components that account for most of the variation in real geometrical shape, not the "normalized" shape. Normalized shape does not make much sense in this kind of study.

Since principal component analysis (PCA) on a covariance matrix deals with original variables, it is more appropriate to use PCA on a covariance matrix and not PCA on a correlation matrix.

Also, it is more appropriate to use the second type of mathematical expression of PCA, that is,

$$\mathbf{X} - \boldsymbol{\mu} = \mathbf{A}^T \mathbf{Y} \qquad (5.23)$$

or

$$\begin{bmatrix} X_1 - \mu_1 \\ X_2 - \mu_2 \\ \dots \\ \dots \\ X_p - \mu_p \end{bmatrix} = \begin{bmatrix} a_{11} \\ a_{21} \\ \dots \\ \dots \\ a_{p1} \end{bmatrix} Y_1 + \begin{bmatrix} a_{12} \\ a_{22} \\ \dots \\ \dots \\ a_{p2} \end{bmatrix} Y_2 + \dots + \begin{bmatrix} a_{1i} \\ a_{2i} \\ \dots \\ \dots \\ a_{pi} \end{bmatrix} Y_i + \dots + \begin{bmatrix} a_{1p} \\ a_{2p} \\ \dots \\ \dots \\ a_{pp} \end{bmatrix} Y_p \qquad (5.24)$$

Because the random vector $\mathbf{X} - \boldsymbol{\mu} = (X_1 - \mu_1, X_2 - \mu_2, \dots, X_p - \mu_p)$ represents the geometrical dimensional variations at these p measurement locations, by using the equation forms of Eq. (5.21) or (5.22), and by defining

$$\mathbf{u}_j = \mathbf{A}_{.j} Y_j = \begin{bmatrix} a_{1j} \\ a_{2j} \\ \dots \\ \dots \\ a_{pj} \end{bmatrix} Y_j = \begin{bmatrix} u_{1j} \\ u_{2j} \\ \dots \\ \dots \\ u_{pj} \end{bmatrix} \qquad (5.25)$$

we have

$$\mathbf{X} - \boldsymbol{\mu} = \mathbf{u}_1 + \mathbf{u}_2 + \dots + \mathbf{u}_p \qquad (5.26)$$

or

$$\begin{bmatrix} X_1 - \mu_1 \\ X_2 - \mu_2 \\ \vdots \\ \vdots \\ X_p - \mu_p \end{bmatrix} = \begin{bmatrix} u_{11} \\ u_{21} \\ \vdots \\ \vdots \\ u_{p1} \end{bmatrix} + \begin{bmatrix} u_{12} \\ u_{22} \\ \vdots \\ \vdots \\ u_{p2} \end{bmatrix} + \cdots + \begin{bmatrix} u_{1i} \\ u_{2i} \\ \vdots \\ \vdots \\ u_{pi} \end{bmatrix} + \cdots + \begin{bmatrix} u_{1p} \\ u_{2p} \\ \vdots \\ \vdots \\ u_{pp} \end{bmatrix} \qquad (5.27)$$

Therefore, the vector representing the geometrical dimensional variation $\mathbf{x} - \boldsymbol{\mu}$ is represented as the summation of random vectors, that is, $\mathbf{u}_1 + \mathbf{u}_2 + \cdots + \mathbf{u}_p$. Each $\mathbf{u}_j$ is a product of a random scalar Y_j and a deterministic vector $\mathbf{A}_{.j}$. Therefore, each $\mathbf{u}_j$ characterizes a totally dependent and distinguishable geometrical variation pattern. Each $\mathbf{u}_j$ can be defined as a *geometrical variation mode*.

If we use the first type of mathematical expression of PCA, that is,

$$Y_i = a_{1i}(X_1 - \mu_1) + a_{2i}(X_2 - \mu_2) + \cdots + a_{pi}(X_p - \mu_p)$$

for $i = 1,\ldots, p$, then, Y_i is expressed as a linear combination of the dimensional variations at different measurement locations. It is really difficult to explain the exact meaning of such linear combination.

5.4.1 Properties of the geometrical variation mode

1. Total variance of jth mode $\mathbf{u}_j$ is equal to λ_j because

$$\mathrm{Var}(u_{1j}) + \cdots + \mathrm{Var}(u_{pj}) = \left(a_{1j}^2 + \cdots + a_{pj}^2\right)\mathrm{Var}(Y_j) = \left(a_{1j}^2 + \cdots + a_{pj}^2\right)\lambda_j = \lambda_j$$

   where $\lambda = (\lambda_1,\ldots,\lambda_p)$ are the eigen values of Σ and where it is assumed that $\lambda_1 > \cdots > \lambda_p$.

2. Variation mode decomposition:

$$\mathrm{Var}(X_1) + \cdots + \mathrm{Var}(X_p) = \sum_{i=1}^{p} \sigma_{ii}^2 = \sum_{i=1}^{p}\sum_{j=1}^{p} \mathrm{Var}(u_{ij}) = \lambda_1 + \cdots + \lambda_p$$

   In summary, the total dimensional variation of $\mathbf{X}$ is equal to the summation of the total variances of the geometrical variation modes. The total variance for the ith mode is equal to λ_j and the ratio $\lambda_j/(\lambda_1 + \lambda_2 + \cdots + \lambda_p)$ represents the percentage of total variation explained by the jth geometrical variation mode.

3. Variance and standard deviation of u_{ij}.

$$\mathrm{Var}(Y_j) = \lambda_j \qquad \mathrm{Var}(u_{ij}) = a_{ij}^2 \mathrm{Var}(Y_j) = a_{ij}^2 \lambda_j \qquad \sigma(u_{ij}) = |a_{ij}|\sqrt{\lambda_j}$$

5.4.2 Variation mode chart

The variation mode chart (Yang, 1996) is an effective tool for interpreting geometrical variation modes. It uses the fact that the interval represented by $\mu \pm 3\sigma$ contains 99.73%, or virtually nearly 100%, of the observations in a normal distribution. The variation mode chart tries to establish a ±3-Sigma zone that describes the pattern and magnitude for each geometrical variation mode. Specifically, for each variation mode, $\mathbf{u}_j = (u_{1j}, ..., u_{pj})^T$, from property 3 of the preceding subsection, $\text{Var}(u_{ij}) = a_{ij}^2 \lambda_j$, we can construct the following variation mode chart, which consists of

1. Centerline

$$E(\mathbf{u}_j) = (0, ..., 0) \quad (5.28)$$

In actual dimensional variation, the mean vector of $\mathbf{X}$, that is, $\boldsymbol{\mu} = (\mu_1, \mu_2, ..., \mu_p)$ is the centerline for geometrical dimensional variation.

2. The variation extent limits for $\mathbf{u}_j$ are as follows:

Variation extent limit 1 (3-Sigma line 1)

$$\text{VEL}_1(\mathbf{u}_j) = \left(3a_{1j}\sqrt{\lambda_j}, ..., 3a_{pj}\sqrt{\lambda_j}\right) \quad (5.29)$$

Variation extent limit 2 (3-Sigma line 2)

$$\text{VEL}_2(\mathbf{u}_j) = \left(-3a_{1j}\sqrt{\lambda_j}, ..., -3a_{pj}\sqrt{\lambda_j}\right) \quad (5.30)$$

Since the variation extent limits are based on 3-Sigma, we can also call them 3-Sigma lines.

The following example illustrates how to use a variation mode chart in order to characterize the exact pattern and magnitude of a variation mode.

Example 5.5: Variation Mode Chart of Flat Panel Example We will use the previous flat panel, as shown in Fig. 5.4, as an example of the application of PCA in the analysis of dimensional quality. Figure 5.4 presents the measurement data and run charts of four measuring points for each of the 10 such flat panels. The deviations of the check points from their nominal location are clearly the observations of $\mathbf{X} = (X_1, X_2, X_3, X_4)$ and are listed in Table 5.1.

Example 5.1 presented the result of principal component analysis for this flat panel data. By using Eq. (5.27) or (5.28), the variation mode chart for the first principal component can be established. In this example, $\lambda_1 = 34.048$, $\mathbf{A}_{.1} = (0.499, 0.349, -0.453, -0.651)$, so

$$\text{VEL}_1(\mathbf{u}_1) = 3\sqrt{\lambda_1}\mathbf{A}_{.1}^T = 3\sqrt{34.048} \times (0.499, 0.349, -0.453, -0.653)$$
$$= (8.73, 6.11, -7.93, -11.4) \quad (5.31)$$

Similarly,

$$\text{VEL}_2(\mathbf{u}_1) = (-8.73, -6.11, 7.93, 11.4) \quad (5.32)$$

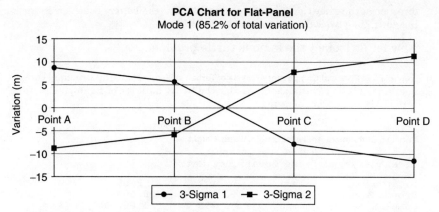

Figure 5.6 Variation mode chart for mode 1.

The variation mode chart for mode 1 of this example is presented in Fig. 5.6. Since $\lambda_1/(\lambda_1 + \lambda_2 + \lambda_3 + \lambda_4) = 0.85$, this first geometrical variation mode accounts for 85 percent of the total variation in the flat panel.

In this variation mode chart, the range of variation (in a ±3-Sigma sense) is the range between the two variation extent limit (VEL) lines or "3-Sigma" lines [Eqs. (5.29) and (5.30)]. Each VEL line indicates the direction (either + or −) along which the points vary together, and the magnitude of movements for each measuring point (proportional to the corresponding magnitude in 3-Sigma lines). For the flat panel example, by following VEL_1 line (3-Sigma line 1)[Eq. (5.29)], it can be inferred that when points A and B vary in the *positive* direction, Points C and D vary in the *negative* direction. Point A may move in the positive direction up to 8.73 mm, and at the same time, B will move in the positive direction up to 6.11 mm, point C will move in the negative direction up to 7.93 mm, and point D will move in the negative direction up to 11.4 mm. This is because VEL_1 line are +8.73 and +6.11 for points A and B and are −7.93 and −11.4 for points C and D and vice versa. When read together, the variation mode chart indicates that the panel is rotating, and it is illustrated in Fig. 5.7.

Figure 5.7 provides the exact geometrical interpretation for variation mode 1. If we deal with an industrial dimensional quality problem, the goal is to reduce the dimensional variation. After this kind of PCA analysis, the dimensional variation problem is much easier to resolve, since the pattern of variation is identified. Analysis of variation mode charts can help in understanding the different types of motion or variation of parts, such as *translations* (all points moving in the same

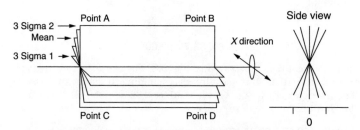

Figure 5.7 Graphical illustration of variation mode 1.

direction at the same time), *rotations*, or *twists*. This can be very beneficial in tracking down the physical root causes of variation.

Now let us look at the famous turtle problem.

Example 5.6: Principal Component Analysis of Turtle Problem By using principal component analysis on covariance matrix for the turtle data set of Example 5.3, from MINITAB we get the following.

Principal Component Analysis: Length, Width, Height

```
Eigenanalysis of the Covariance Matrix
Eigenvalue      678.37          6.77          2.85
Proportion        0.986         0.010         0.004
Cumulative        0.986         0.996         1.000

Variable          PC1           PC2           PC3
Length           -0.814         0.555         0.172
Width            -0.496        -0.818         0.291
Height           -0.302        -0.151        -0.941
```

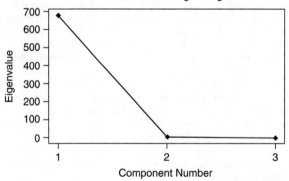

Scree Plot of Length-Height

Clearly, the first component is predominant; it explains 98.6 percent of the variation. The variation extent limits of the first mode of the turtle problem are as follows:

$$\text{VEL}_1(\mathbf{z}_1) = (3a_{11}\sqrt{\lambda_1}, \ldots, 3a_{p1}\sqrt{\lambda_1})$$

$$= (3(-0.814)\sqrt{678.37},\ 3(-0.496)\sqrt{678.37},\ 3(-0.302)\sqrt{678.37})$$

$$= (-63.60,\ -38.75,\ -23.60)$$

Similarly,

$$\text{VEL}_2(\mathbf{z}_1) = (63.60,\ 38.75,\ 23.60)$$

The variation mode chart is illustrated in Fig. 5.8. The centerline in the variation mode chart $E(\mathbf{u}_1) = (0, 0, 0)$ is actually corresponding to the mean vector of (X_1, X_2, X_3). The averages of (X_1, X_2, X_3) are (136.04, 102.58, 52.042). Therefore, the 3-Sigma low limit of the size of turtle is

$$(136.04 - 63.60,\ 102.58 - 38.75,\ 52.042 - 23.60) = (72.44,\ 63.83,\ 28.44)$$

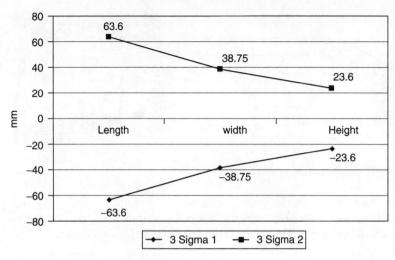

Figure 5.8 Variation mode chart for turtle problem.

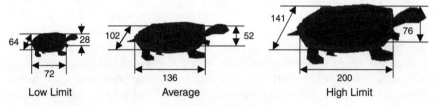

Figure 5.9 Illustration of the first principal component of turtle data.

The 3-Sigma high limit of the size of turtle is

(136.04 + 63.60, 102.58 + 38.75, 52.042 + 23.60) = (199.54, 141.33, 75.64)

It can be illustrated in Fig. 5.9. Therefore, the dominant mode of variation for turtle data is proportional growth with the ratio of growth featured by (0.814, 0.496, 0.302), which is the coefficient of the first principal component, or approximately a 8:5:3 ratio. That is, for every 8 mm growth in length, there is going to be 5 mm growth in width, and 3 mm growth in height.

5.4.3 Visual display and animation of principal component analysis

The PCA method discussed in this section can show and explain the exact meaning of principal components for dimensional measurements data by using variation mode charts and figures such as Figs. 5.7 and 5.9. Computer graphics can be used to enhance the visual display. Figure 5.10 shows a computer visual display of the PCA variation mode for automobile dimensional quality analysis. Animation (that is, with graphic wire frames

122 Chapter Five

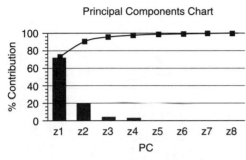 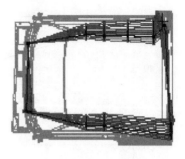

Figure 5.10 Computer graphical display of PCA.

moving) can show the variation modes even better. Graphical display and animation can greatly help explain the complicated analysis results to people who have very little background in multivariate statistics.

5.4.4 Applications for other multivariate data

The special principal component analysis method that we discussed in this section may also be used to analyze many other types of multivariate data. For example, in a paint cure oven in an automobile assembly plant, there could be several temperature sensors measuring temperatures at different locations inside the oven. Every variable is a temperature reading that has the same physical meaning and measurement unit. The temperature data collected by these sensors can also be analyzed by the PCA method that we developed in this section in order to study the temperature variation profile inside the oven.

We use Example 5.6 to show how the method developed in this section can be used to analyze other types of problems.

Example 5.7: Boiler Temperature Data (Mason and Young, 2002) Table 5.5 listed a set of temperature readings from four burners on a boiler. Each data set corresponds to a reading of measurement at a certain time.

TABLE 5.5 Temperature Readings for Four Burners in a Boiler

Temp 1	Temp 2	Temp 3	Temp 4
507	516	527	516
512	513	533	518
520	512	537	518
520	514	538	516
530	515	542	525
528	516	541	524
522	513	537	518
527	509	537	521
530	512	538	524
530	512	541	523

By using principal component analysis on a covariance matrix, we get the following MINITAB output:

Principal Component Analysis: Temp 1, Temp 2, Temp 3, Temp 4

```
Eigenanalysis of the Covariance Matrix
Eigenvalue     90.790      5.331      4.006      0.873
Proportion      0.899      0.053      0.040      0.009
Cumulative      0.899      0.952      0.991      1.000

Variable          PC1        PC2        PC3        PC4
Temp1          -0.841      0.183      0.110     -0.496
Temp2           0.061     -0.858     -0.207     -0.465
Temp3          -0.433     -0.139     -0.722      0.522
Temp4          -0.318     -0.458      0.651      0.514
```

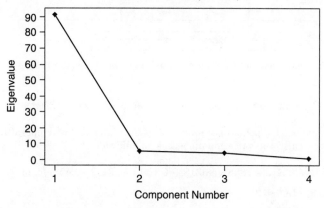

The mean temperatures of these four burners are 522.6, 513.2, 537.1, and 520.3. The first principal component is very dominant. It accounts for 90% of temperature variation, whereas the second component accounts for 5.3% of temperature variation.

The first two eigenvalues are $l_1 = 90.79$ and $l_2 = 5.331$. Therefore for variation mode 1

$$\text{VEL}_1 = (3(-0.841)\sqrt{90.79}, 3(0.061)\sqrt{90.79}, 3(-0.433)\sqrt{90.79}, 3(-0.318)\sqrt{90.79})$$

$$= (-24.0, 1.7, -12.4, -9.1)$$

$$\text{VEL}_2 = (24.0, -1.7, 12.4, 9.1)$$

The variation mode chart for the first principal component is illustrated in Fig. 5.11 From this variation mode chart, we can clearly see that the first principal component is dominated by the synchronized temperature changes for burners 1, 3, and 4; that is, if the temperature in burner 1 is high, then so is it in burner 3 and burner 4 (burner 2 does not follow the trend of other burners in this principal component). Also, the ratio of change for temperatures in burners 1, 3, and 4 is approximately 24:12:9, or 8:4:3. If burner 1 temperature fluctuates (increase or decrease) by 8°, then burner 3 temperature will fluctuate by 4°, and that of burner 4 will

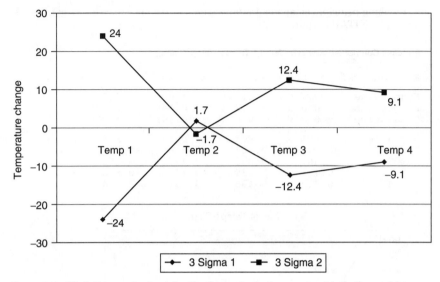

Figure 5.11 Variation mode chart for the first principal component in boiler problem.

fluctuate by 3°. The maximum variation extents in temperatures for this mode are illustrated in Fig. 5.11.

The second principal component is dominated by the high value of temperature in burner 2. Specifically, the variation mode 2 is as follows:

$$VEL_1 = (3(0.183)\sqrt{5.331}, 3(-0.858)\sqrt{5.331}, 3(-0.139)\sqrt{5.331}, 3(-0.458)\sqrt{5.331})$$

$$= (1.3, -5.9, -1.0, -3.2)$$

$$VEL_2 = (-1.3, 5.9, 1.0, 3.2)$$

Clearly, this variation mode is mainly about temperature of burner 2. In this variation mode, burner 1 and burner 3 temperatures will not be affected too much by the fluctuation of temperature in burner 2, but temperatures in burner 4 will be affected.

5.5 Principal Component Analysis Case Studies

5.5.1 Improving automotive dimensional quality by using principal component analysis

Dimensional quality is a measure of conformance of the actual geometry of manufactured products with their designed geometry. Many products are required to maintain dimensional consistency. In the real world, manufacturing and assembly processes are inherently imprecise, resulting in dimensional inconsistencies. It is very important to control the dimensional quality as it will affect both the function and appearance of the product. Some products, such as automotive body structures, have elaborate surface profiles and very complex manufacturing processes. Maintaining good dimensional quality is very difficult and critical to the success of the end product.

Dimensional variability must initially be accommodated for in design and then fully controlled in production. Dimensional variations at different measuring points are usually correlated with each other. The method presented in this case study is based on the use of principal component analysis (PCA) (Yang, 1996). PCA is able to extract major variation patterns in multivariate random variables. The pattern of variation may lead to clues for the causes of variation.

Automotive body assembly process. An automotive body structure consists of many substructures such as the underbody, doors, and body-sides. Each substructure consists of numerous metal panels fabricated at stamping process. Figure 5.12 provides the substructure breakup of the

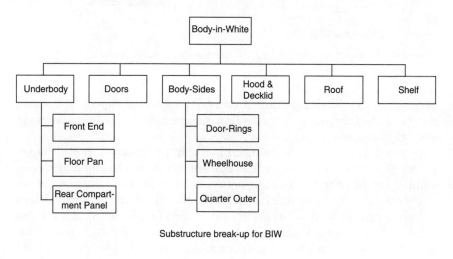

Substructure break-up for BIW

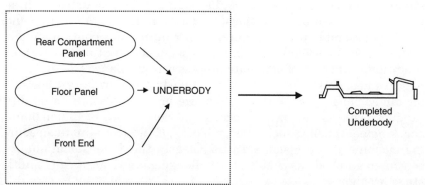

Figure 5.12 Substructures breakup for automotive body structure.

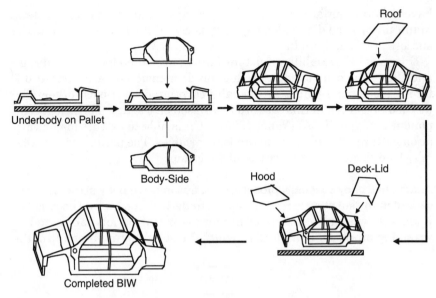

Figure 5.13 Process flow diagram for automotive body assembly process.

automotive body-in-white (unpainted full autobody). Figure 5.13 shows a typical manufacturing process flow diagram for an automotive body assembly process.

In most automotive body manufacturing processes, metal panel pieces that are fabricated in the stamping process are matched and welded together to form substructures. Usually these substructures are not made on the final assembly line. The substructures are matched and welded sequentially (as shown in Fig. 5.13) to form the final body-in-white. Clearly, dimensional variation could come from any of the steps in the manufacturing process. In the metal panel stamping process, variation could originate from sources such as metal thickness or die conditions. In the substructure fabrication and the final assembly process, sources contributing to variation include tooling conditions, clamping and weld sequences, fixture and locator conditions, and lack of adequate process control plans. Therefore, all major components, substructures, and final bodies-in-white are checked for dimensional quality throughout the process to assure a quality product. This is performed through various types of dimensional checking equipment, such as gage blocks, CMM (coordinate measuring machines), and vision stations. Clearly, the dimensional quality control in autobody assembly is a challenging multivariate and multistage problem.

Dimensional quality measurement. The actual surface profiles of the products are usually of concern in dimensional quality. In reality, continuously measuring the whole surface is not possible. Thus, the dimensional quality is monitored by the measurements of the actual geometry at a selected set of measuring points.

In general, one can assume that there are p measurement points. X_i denotes the deviation of the actual location of measuring point i from its designed geometry (such as that of Fig. 5.3). Clearly, for a population of a product, $\mathbf{X} = (X_1, X_2,\ldots, X_p)$ will be random variables characterizing the dimensional variation of products. $\mu_i = E(X_i)$ is the mean deviation at ith measuring point, and $\sigma_i^2 = \text{Var}(X_i)$ can be used to measure the piece-to-piece variation at the ith measuring point. We can define

$$L = \sum_{i=1}^{p} \mu_i^2 + \sum_{i=1}^{p} \sigma_i^2 \qquad (5.33)$$

as the total squared deviation due to dimensional variation, where $\sum_{i=1}^{p} \mu_i^2$ stands for the squared deviations due to mean deviation, and $\sum_{i=1}^{p} \sigma_i^2$ stands for the squared deviations due to piece-to-piece variation. Usually, piece-to-piece variation is of major concern since the mean deviation is relatively easy to adjust. Piece-to-piece variations at different measuring points are likely to be dependent on each other since each product belongs to the same structure. It is reasonable to assume that $\mathbf{X}$ is a vector of the multivariate normal random variable with a covariance matrix Σ. PCA can effectively deal with these kinds of dimensional measurement data and extract variation patterns occuring in the manufacturing process. These variation patterns can help in determining the root causes of variation and thus, with process corrections, can reduce the manufacturing variation.

Case study: Root-cause analysis of an automobile hinge variation by using PCA. The aim of this case study is to determine the root cause of high cross-car (the distance between the sides of the car) variation at measuring point UPHIN (Upper Hinge) on the right side of the full automobile body-in-white as presented in Fig. 5.14.

Since the underbody is built before the full body and its variation could be the cause of variation of the full body, 10 sets of underbody measurement data, along with 10 sets of full-body assembly measurement data, are used in the analysis. These full-body assemblies were built from the 10 underbodies for which data were obtained. Underbody and full body data are merged, resulting in 10 combined sets of data. Measuring points on the right side of full body and underbody are illustrated in Fig. 5.15.

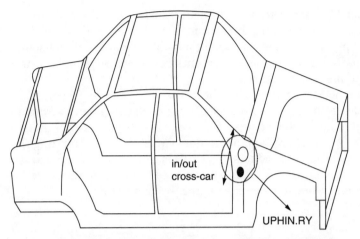

Figure 5.14 High cross-car variation at right UPHIN measuring point.

The combined data set is analyzed by PCA in order to check for any correlations between in/out variation (variation perpendicular to side body surface), at the hinge/rocker region of the underbody, and the in/out variation at check point UPHIN at the right side of the full body.

PCA identified two major modes of variation which are summarized in Table 5.6. The variation mode charts are shown in Figs. 5.16 and 5.18, and graphically illustrated, respectively, in Figs. 5.17 and 5.19.

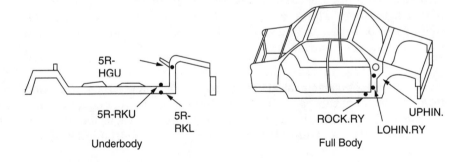

All measuring points are measured in cross car direction

Underbody measuring points:
5R-HGU: Upper hinge, right side
5R-RKL: Lower rocker, right side
5R-RKU: Upper rocker, right side

Full body measuring points:
UPHIN.RY: Upper hinge, right side
LOHIN.RY: Lower hinge, right side
ROCK.RY: Rocker, right side
UPHIN.LY: Upper hinge, left side
LOHIN.LY: Lower hinge, left side
ROCK.LY: Rocker, left side

Figure 5.15 Selected measuring points in underbody and full body.

TABLE 5.6 Principal Component Analysis for Combined Data Set

Eigenvalue	1.0752	0.8089
Proportion	0.456	0.343
Cumulative	0.456	0.799
Point	PC1	PC2
LOHIN.LY	−0.224	0.163
UPHIN.LY	−0.209	0.123
ROCK.LY	−0.174	0.178
LOHIN.RY	0.376	−0.295
UPHIN.RY	0.115	−0.496
ROCK.RY	0.356	−0.262
5R-HGU I	0.110	−0.084
5R-RKU I	0.742	0.338
5R-RKL I	0.178	0.635

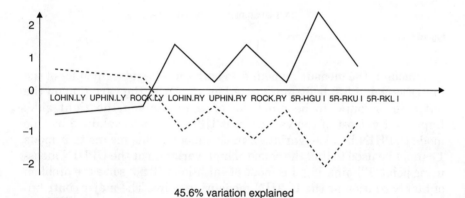

45.6% variation explained

Figure 5.16 Variation mode chart 1.

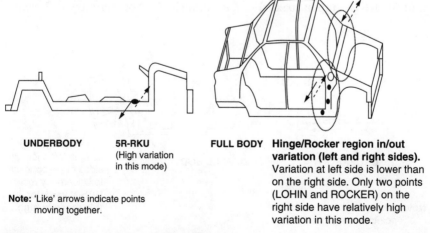

UNDERBODY   5R-RKU
(High variation in this mode)

Note: 'Like' arrows indicate points moving together.

FULL BODY   **Hinge/Rocker region in/out variation (left and right sides).** Variation at left side is lower than on the right side. Only two points (LOHIN and ROCKER) on the right side have relatively high variation in this mode.

Figure 5.17 Mode 1 variation of hinge region.

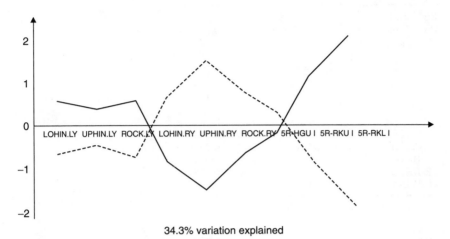

Figure 5.18 Variation mode chart 2.

In mode 1, the measuring points on both the left and right sides of the underbody and full body move together. However, the variation on the left side is found to be much lower than that on the right side. Importantly, most of the variation at UPHIN is not explained by this mode (UPHIN has low variation in this mode). This means that mode 1 cannot be used to find the cause of high variation at the UPHIN measuring point. Eliminating this mode of variation will not solve the problem of high variation at the UPHIN measuring point. The major contributor to this mode is the in/out variation at the upper rocker (5R-RKU) of the underbody.

Most of the variation at UPHIN is explained by mode 2. This variation is strongly correlated to the variation at points 5R-RKL (lower rocker) and 5R-RKU (upper rocker) of the underbody. The primary difference

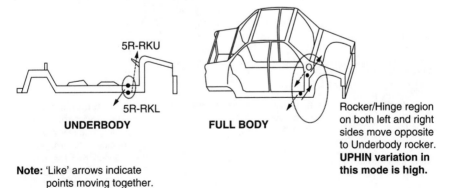

Figure 5.19 Mode 2 of hinge variation.

between modes 1 and 2 is the direction of variation. In this mode, the cross-car measuring points on the full body move opposite to the underbody rocker. In other words, the variation at UPHIN is due to the full-body varying in a direction opposite to that of the underbody. This results in a twisting motion that maximizes the variation at the UPHIN measuring point.

Based on the clue generated by mode 2 of PCA study, it was found that the twisting motion is due to the interference of substructures in the hinge region. By eliminating this interference, the variation at the upper hinge area is greatly reduced and production quality is enhanced.

Summary. In this case study, the features of dimensional quality measurement and analysis have been presented with practical examples. The quality loss due to dimensional variation can be partitioned into mean deviation and piece-to-piece variation. By using PCA, the piece-to-piece variation can be further decomposed into a set of independent geometrical variation modes. Each geometrical variation mode represents a distinguishable variation pattern. These patterns can provide valuable clues about the causes of dimensional variation, thereby developing procedures for in-process variation reduction. The geometrical interpretation can be explicitly represented by a variation mode chart or three-dimensional animation, which can make the principal component analysis easily understood. The case study of an automotive body structure dimensional quality analysis has illustrated the value and power of PCA in solving real world engineering problems with the use of an available statistical tool.

5.5.2 Performance degradation analysis for IRLEDs (Yang and Yang, 2000)

Infrared light emitting diodes (IRLEDs) are optical devices in which spontaneous photon emissions originate from an injection excitation. The p-n junction is needed to emit light by recombination of injected electrons and holes. For recombination to occur, electrons and holes must exit at the same time and place. This requirement is met by bonding a p-type and an n-type semiconductor together and injecting carriers under electrical bias. In the diodes, two types of recombinations occur. One is radiative recombination, which results in the emission of photons. The other is nonradiative recombination, which releases a phonon of energy. The amount of each is determined by the material properties of the semiconductor. The radiative recombination rate over time is strongly influenced by lattice imperfections such as dislocations.

IRLEDs are widely used and play a key role in satellite orientation systems. Such systems require high reliability of the components within them. However, the performance of IRLEDs degrades over time in field applications. The performance of the devices is usually characterized by the following three performance characteristics: reverse current, forward voltage, and a variation ratio of luminous power. The life of an IRLED is often defined in terms of some specified critical values. Failure occurs when one of the performance characteristics exceeds its critical value. For example, if the reverse current becomes too large, the IRLED will be considered failed. Therefore, the reliability of the devices depends on the degradation rates of the performance characteristics.

The performance characteristics contain a large amount of credible and useful information on the reliability of the devices because the degradation of reliability is usually exhibited in the degradation of performance. Therefore, the reliability of some products can be assessed using performance degradation measurements. On the other hand, performance characteristics are usually signals of degradation mechanisms. Examining the degradation process of performance characteristics provides a way to trace the evolution of degradation mechanisms. Alleviating the evolution of the degradation mechanisms can greatly improve the reliability of the devices. Therefore, analyzing performance characteristics is an effective approach to better understanding the reliability and degradation mechanisms of the devices.

Experimental description. The devices used in this work are GaAs/GaAs IRLEDs, the wavelength of which is 880 nm and the design forward current is 50 mA. The devices are highly reliable and their performance characteristics degrade slowly over time in field applications. To observe informative performance degradation in a reasonable amount of time, two samples of the devices were tested using elevated currents. The test currents were selected to be 220 and 280 mA, respectively. The devices were inspected periodically during the test and the three performance characteristics were measured at each inspection time. Table 5.7 summarizes the test plans.

Specifically, at stress level 220 mA, for each measurement time, such as at 0 hour, 96th hour, and so on, the three characteristics for each of the 15 sample units were measured. At stress level 280 mA, again for each measurement time, such as 0 hour, 48th hour, and so on, the three characteristics for each of 20 sample units were measured.

TABLE 5.7 Accelerated Degradation Test Plans for the IRLEDs

Stress level (mA)	Sample size	Censoring time (h)	Measurement time (h)
220	15	1368	0, 96, 168, 240, 312, 408, 504, 648, 792, 984, 1176, 1368
280	20	890	0, 48, 96, 168, 240, 312, 432, 552, 720, 890

Principal component analysis. In degradation analysis of the multiperformance characteristics, it is important to analyze the performance measurement data and relate the data to the reliability and the degradation mechanisms of the devices. One technique used to accomplish this is PCA. The objectives of PCA are data condensations and interpretations.

In this case study, since the three characteristics—reverse current X_1, forward voltage X_2, and variation ratio of luminous power X_3—have very different meanings and measurement units, the PCA on correlation matrix will be used.

For each stress level and measurement time, a sample of N ($N = 15$, or 20 in this study) observation vectors are collected as follows:

$$\mathbf{X} = \begin{bmatrix} x_{11} & x_{12} & x_{13} \\ \vdots & \vdots & \vdots \\ x_{N1} & x_{N2} & x_{N3} \end{bmatrix} \quad (5.34)$$

The correlation matrix $\mathbf{R}$ will be computed on each data set specified by Eq. (5.32).

The measurement data are standardized:

$$z_{ij} = \frac{x_{ij} - \bar{x}_j}{s_j} \quad i = 1, 2, \ldots, N, \; j = 1, 2, 3 \quad (5.35)$$

where

$$\bar{x}_j = \frac{1}{N} \sum_{i=1}^{N} x_{ij} \quad s_j^2 = \frac{1}{N-1} \sum_{i=1}^{N} (x_{ij} - \bar{x}_i)^2$$

Then the standardized observations can be written as

$$\mathbf{Z} = \begin{bmatrix} z_{11} & z_{12} & z_{13} \\ \vdots & \vdots & \vdots \\ z_{N1} & z_{N2} & z_{N3} \end{bmatrix} = [Z_1 \; Z_2 \; Z_3] \quad (5.36)$$

The jth principal component is a linear combination of the standardized variables Z_1, Z_2, and Z_3:

$$Y_j = c_{1j} Z_1 + c_{2j} Z_2 + c_{3j} Z_3 \quad j = 1, 2, 3 \quad (5.37)$$

where c_{1j}, c_{2j}, and c_{3j} are called loadings, they are the elements of $\mathbf{a}_j$, the jth eigen vector of $\mathbf{S}$. The scores of the ith IRLED for jth principal component are as follows:

$$y_{ij} = a_{1j} z_{i1} + a_{2j} z_{i2} + a_{3j} z_{i3} \quad i = 1, 2, \ldots, N, \; j = 1, 2, 3 \quad (5.38)$$

The variances of the three principal components are λ_1, λ_2, and λ_3, respectively. $\lambda_i / \mathrm{tr}\mathbf{R}$, or $\lambda_i/3$ in this case, represent the relative importance

TABLE 5.8 Summary of the First Principal Component for IRLEDs Tested at 220 mA

Measurement time (h)	Loading X_1	X_2	X_3^*	Percent variance
0	0.422	0.555	0.717	55.6
96	−0.388	−0.660	0.643	64.2
168	−0.482	−0.625	0.614	73.9
240	0.307	0.668	−0.678	66.0
312	−0.318	0.672	−0.669	67.2
408	0.118	−0.702	0.702	62.7
504	0.115	−0.703	0.703	62.9
648	0.477	−0.638	0.605	76.1
792	0.460	−0.635	0.621	73.7
984	0.475	−0.634	0.611	73.7
1176	0.462	−0.630	0.625	74.2
1368	0.476	−0.624	0.620	73.8

*For $t = 0$, X_3 standards for the original luminous power.

of the ith principal component. The algebraic sign and magnitude of c_{ij} indicates the direction and importance of the contribution of the ith performance characteristic to the jth principal component. The first principal component usually explains a large percentage of total sample variance and is most important in data interpretations. Consequently, the scores of this principal component contain useful information on reliability and degradation mechanisms of the products.

PCA for the experimental data. The three performance characteristics of the IRLEDs were measured before testing and periodically after testing. The principal component method was used to analyze the measurement data at each inspection time, based on the correlation matrix. Tables 5.8 and 5.9 summarize the loadings and the percentages of total

TABLE 5.9 Summary of the First Principal Component for IRLEDs Tested at 280 mA

Measurement time (h)	Loading X_1	X_2	X_3^*	Percent variance
0	0.462	0.503	0.732	46.7
48	0.042	0.713	−0.699	53.3
96	0.547	−0.613	0.570	78.6
168	0.535	−0.614	0.580	78.2
240	0.523	−0.618	0.587	77.6
312	0.506	−0.625	0.595	77.2
432	0.516	−0.617	0.594	75.1
552	0.480	−0.617	0.623	72.0
720	0.454	−0.624	0.636	71.1
890	0.438	−0.633	0.638	71.1

*For $t = 0$, X_3 standards for the original luminous power.

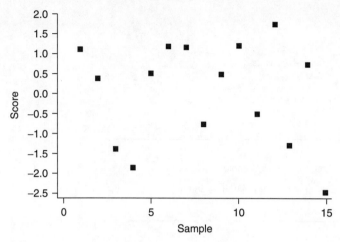

Figure 5.20 Score vs. sample of units tested at 220 mA.

variance of the first principal component at various measurement times. The results are the basis of the analysis.

Relationship between performance and reliability

Relative reliability metric. PCA shows that before testing, the first principal component, as shown in Tables 5.8 and 5.9, is a weighted average of the standardized performance characteristics. It reflects the general performance of the devices. Therefore, the first principal component can be considered a quality factor. A particular device's score represents the quality of the device.

For the first principal component, the score of each IRLED was calculated in the samples before testing. The scores are plotted versus samples in Fig. 5.20 and 5.21. We find that, before testing, the scores of the sample units follow the standard normal distribution, as shown in Figs. 5.22 and 5.23. The normality indicates that the quality of the sample is homogeneous. If a score appears to be an outlier, the corresponding unit then has exceptionally good/poor quality. For example, when $t = 240$ h, the scores of units 1, 7, and 13, tested at 280 mA, significantly digress from normality, as shown in Fig. 5.24. The measurement data at this time is listed in Table 5.10. Actually, these three IRLEDs have failed by the measurement time, while the others in the sample still survive.

In this test, the IRLEDs, which have the extreme scores before testing, are found to be most reliable or unreliable. For example, as depicted in Fig. 5.20, units 4 and 15 tested at 220 mA have extremely low scores. Measurement data show that the luminous power of these units degrades more rapidly than others. As a result, the lives of the two units are much shorter. On the other hand, unit 12, tested at 220 mA, has an extremely

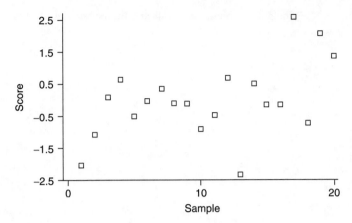

Figure 5.21 Score vs. sample of units tested at 280 mA.

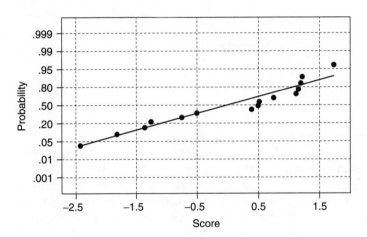

Figure 5.22 Score distribution of units tested at 220 mA.

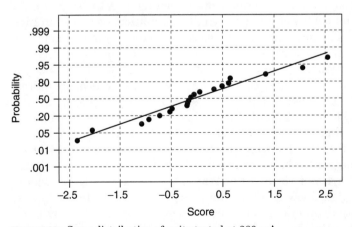

Figure 5.23 Score distribution of units tested at 280 mA.

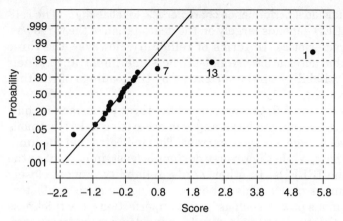

Figure 5.24 Score distribution of units tested at 280 mA ($t = 240$ h).

high score. The variation ratio of the luminous power of this unit is only 6.4% at the censoring time, while the variation ratios of units 4 and 15 are 58% and 60%, respectively, at the same time. In fact, this is the most reliable device in the sample. Analysis for the units tested at 280 mA arrives at the same conclusions. This is not surprising, because the relative reliability of units in a sample is closely related to the difference in performance between units with the passage of time. Therefore, the first principal component can be regarded as a metric of relative reliability.

The relative reliability metric, however, cannot be used to compare the reliability of units having close scores. The metric may not account for some minor performance characteristics, and the reliability of the units is affected by a number of unpredictable factors during operation. Conservatively, the metrics on extreme values indicate the significant difference in reliability of the units. This provides an efficient way of reliability screening. In some cases, especially in military applications, it is desirable to select the most reliable unit from a sample to assure high reliability of a critical system. On the other hand, the most unreliable unit may need to be screened out from a sample. For example, in the sample tested at 220 mA, unit 12 can be selected to serve the purpose of high reliability, while units 4 and 15 may be considered as unreliable devices compared to others.

TABLE 5.10 Some Measurement Data of IRLEDs Tested at 280 mA ($t = 240$ h)

Unit # no.	Reverse current (nA)	Forward voltage (V)	Variation ratio of luminous power (%)
1	12.9×10^6	1.118	50.7
7	38.3	1.134	37.2
13	23.6	1.125	56.2

Importance of individual performance characteristic to reliability. The reliability of an IRLED depends largely on the degradation of the three performance characteristics. In terms of reliability, however, the three performance characteristics are not of the same importance. Their importance to reliability can be measured by the loadings of the first principal component.

From Tables 5.8 and 5.9, the loadings of the three performance characteristics, before testing, can be arranged in decreasing order: luminous power > forward voltage > reverse current. This indicates that the luminous power is most, while the reverse current is least, important to the reliability of the IRLEDs. In other words, reliability is most closely related to luminous power. This is not surprising, because the factors that affect luminous power, such as lattice imperfections, the thickness of active layer, dielectrics on surface and sawing damage, are more complicated than those that affect reverse current. These factors are more susceptible and difficult to control. The importance of order suggests that, if it is desirable to improve the reliability of the devices, most of the efforts should be made to improve the luminous power.

After testing, the loadings of the first principal component of the forward voltage and the variation ratios of luminous power are very close in magnitude, but opposite in sign. It seems that these two characteristics have almost the same importance for the reliability of the devices. Actually, the degradation rate of forward voltage is much smaller than that of luminous power and its degradation is less detrimental. These two characteristics are highly negatively correlated. The correlation between the two characteristics rapidly increases over a short period of aging time and then remains approximately constant, as shown in Fig. 5.25. The correlation indicates that the degradation of the two characteristics is governed basically by the same mechanisms. Luminous

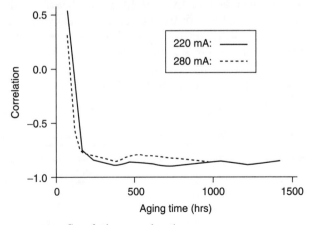

Figure 5.25 Correlation vs. aging time.

power, however, is more sensitive to the mechanisms. Moreover, the loadings of reverse current are relatively small and fluctuate greatly at some inspection times. The importance of reverse current can sometimes be negligible. The degradation of this characteristic is not observed in most devices during testing. This characteristic is less susceptible to test stress and is less important to reliability. Consequently, the luminous power is a dominant characteristic over aging time.

Degradation mechanism analysis. The degradation of IRLEDs exhibits the deterioration of luminous power, forward voltage, reverse current, or any combination of these characteristics. The first two characteristics usually degrade gradually, while the reverse current may deteriorate rapidly. In this test, the decrease of luminous power was found to be highly correlated to the decrease of forward voltage. In addition, the sudden increase of reverse current can only occur in the IRLED, of which the other two characteristics have significantly degraded, although the correlation between luminous power and reverse current is not significant.

The degradation modes of the IRLEDs can be differentiated using the first and second principal components because they explain more than 95% of the total variance. Figures 5.26 and 5.27 show the score plots of the first versus second principal component (PC) at the censoring times. It is clear from the two figures that the test units can be divided into three groups for each test sample. The first group contains the units that degraded fatally in the three characteristics, i.e., No. 15 tested at 220 mA and No. 1 tested at 280 mA. The second group contains the units that degraded tremendously in both luminous power and forward voltage, i.e., Nos. 3, 4, 13 tested at 220 mA and Nos. 5, 7, 11, 12, 13, 14, 15, 18 tested at 280 mA. The third group contains the units

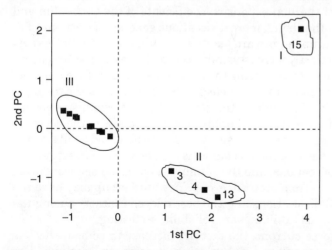

Figure 5.26 Score plot of first vs. second principal components for units tested at 220 mA.

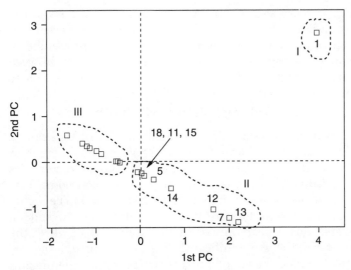

Figure 5.27 Score plot of first vs. second principal components for units tested at 280 mA.

which were excluded in the first and second groups. These units degraded significantly in luminous power and forward voltage. Among the three groups the severity of degradation can be arranged in ascending order: Group III < Group II < Group I.

The degradation of the devices in luminous power and forward voltage is mainly due to dislocations of crystals and contamination of gold. The speed of the dislocation growth depends on the rate of nonradiative recombination, and increases with injection current and mechanical stress. Mechanical stress is created during chip bonding and device operation due to the mismatch of thermal expansion coefficients of the GaAs chip and heat sink. On the other hand, more dislocations generate more nonradiative recombinations under forward bias. As a result, the nonradiative current component distinctly increases, indicating an increase in space charge recombination, and thus decreasing the injection efficiency of the p-n junction. Moreover, contaminants from electrodes are able to enter the crystal under forward bias and impair the devices. These mechanisms govern the degradation of all devices in the three groups because of the decrease in luminous power and forward voltage, which were observed in all devices. However, the activeness of the mechanisms depends on the injection current, the initial dislocations, and the stability of electrode-semiconductor interface, which vary from group to group. The third group may have the least severe initial dislocations and the best alloyed contacts among the three groups, and thus exhibits more reliability in testing.

In terms of reverse current, the distinct increase in some devices is mainly due to the impurity accumulating at some dislocations. The accumulation forms spots of high concentration of doping material and contaminants. The spots create high electrical fields and cause an avalanche breakthrough. This mechanism also explains why the tremendous increase of reverse current can occur only in the devices that have already degraded significantly in luminous power and forward voltage. This mechanism was active in the first group, which may have the most severe initial dislocations and the poorest contacts between the electrode and the semiconductor.

5.6 Factor Analysis

Factor analysis is a multivariate tool that is very similar to PCA. Factor analysis is also used to condense a set of observed variables into a smaller number of transformed variables called components or factors. Both factor analysis and PCA are used to identify an underlying structure or pattern beneath a set of multivariate data. However, the emphasis in factor analysis is on the identification of underlying "factors" that might explain the mutual correlative relationships. Figure 5.28 illustrates the flow chart of factor analysis.

The data set for factor analysis is essentially the same as that for principal component analysis. Factor analysis can be represented by the following equations:

$$\mathbf{X} - \boldsymbol{\mu} = \mathbf{L}_{(p \times m)} \mathbf{Y}_{(m \times 1)} + \boldsymbol{\varepsilon}_{(p \times 1)} \qquad (5.39)$$

where

$$\mathbf{L}_{p \times m} = \begin{bmatrix} \lambda_{11} & \lambda_{12} & \cdots & \lambda_{1m} \\ \lambda_{21} & \lambda_{22} & \cdots & \lambda_{2m} \\ \vdots & \vdots & \vdots & \vdots \\ \lambda_{p1} & \lambda_{p2} & \cdots & \lambda_{pm} \end{bmatrix}$$

is called the factor loading matrix.

$$\mathbf{Y}_{m \times 1} = \begin{bmatrix} Y_1 \\ Y_2 \\ \vdots \\ Y_m \end{bmatrix}$$

is the vector representing m factors, where Y_j is the jth factor, for $j = 1, \ldots, m$, $m < p$, usually m is much smaller than p.

In Eq. (5.37),

$$\boldsymbol{\varepsilon}_{p \times 1} = \begin{bmatrix} \varepsilon_1 \\ \varepsilon_2 \\ \vdots \\ \varepsilon_m \end{bmatrix}$$

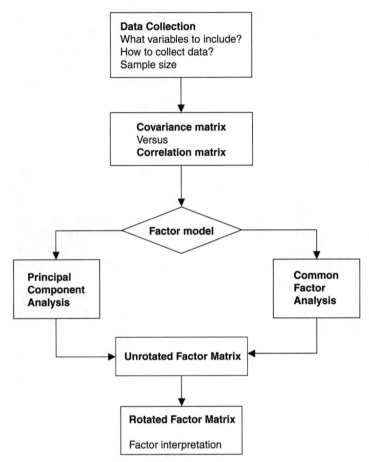

Figure 5.28 Factor analysis flow chart.

represents the error vector that is the difference between original variables and factors, where ε_i is called ith specific factor error, for $i = 1,..., p$.

If we examine Eq. (5.37), we can find that it is very similar to Eq. (5.3), except that in Eq. (5.37), we want to use the vector combination of m factors to approximately represent the original multivariate variables. While in principal component analysis, there are p components; however, we are usually interested only in the first few principal components that explain most of the variations.

Actually in factor analysis (see Fig. 5.28) there are two kinds of factor models: one is the principal component analysis, the other is common factor analysis.

If we want to find a small number of factors that explain most of the variations, then we select principal component analysis as the factor model. In such a case, we usually first run a regular PCA, and find out how many principal components are needed to explain a big portion of variation, say 80%. Then we choose that number of components as our choice of m.

5.6.1 Common factor analysis

The common factor analysis model is selected if we want to explain the structure of correlation and not the total amount of variation. Specifically, common factor analysis has the following model:

$$\mathbf{X} - \boldsymbol{\mu} = \mathbf{L}_{(p \times m)} \mathbf{Y}_{(m \times 1)} + \boldsymbol{\varepsilon}_{(p \times 1)} \tag{5.40}$$

or

$$\mathbf{X} - \boldsymbol{\mu} = \mathbf{L}\mathbf{Y} + \boldsymbol{\varepsilon}$$

In common factor analysis, ε_i is required to be mutually independent such that

$$\text{Cov}(\boldsymbol{\varepsilon}) = E[\boldsymbol{\varepsilon}'\boldsymbol{\varepsilon}] \begin{bmatrix} \psi_1 & 0 & \cdots & 0 \\ 0 & \psi_2 & \cdots & 0 \\ \vdots & \vdots & \vdots & \vdots \\ 0 & 0 & \cdots & \psi_p \end{bmatrix}$$

5.6.2 Properties of common factor analysis

In common factor analysis (factor analysis), Y_i's are assumed to be identically and independently distributed normal random variables, that is, $Y_i \sim N(0, 1)$, for all $i = 1, \ldots, m$, and $\text{Cov}(\mathbf{Y}) = \mathbf{I}_{m \times m}$. Therefore, from $\mathbf{X} - \boldsymbol{\mu} = \mathbf{L}\mathbf{Y} + \boldsymbol{\varepsilon}$, or specifically,

$$\begin{bmatrix} X_1 \\ X_2 \\ \vdots \\ X_p \end{bmatrix} - \begin{bmatrix} \mu_1 \\ \mu_2 \\ \vdots \\ \mu_p \end{bmatrix} = \begin{bmatrix} l_{11} \\ l_{21} \\ \vdots \\ l_{p1} \end{bmatrix} Y_1 + \begin{bmatrix} l_{12} \\ l_{22} \\ \vdots \\ l_{p2} \end{bmatrix} Y_2 + \cdots + \begin{bmatrix} l_{1m} \\ l_{2m} \\ \vdots \\ l_{pm} \end{bmatrix} Y_m + \begin{bmatrix} \varepsilon_1 \\ \varepsilon_2 \\ \vdots \\ \varepsilon_p \end{bmatrix}$$

$$= \mathbf{L}_{.1} Y_1 + \cdots + \mathbf{L}_{.m} Y_m + \boldsymbol{\varepsilon} \tag{5.41}$$

$$\text{Cov}(\mathbf{X}) = \boldsymbol{\Sigma} = \mathbf{L}\,\text{Cov}(\mathbf{Y})\mathbf{L}^T + \text{Cov}(\boldsymbol{\varepsilon})$$

$$= \mathbf{L}\mathbf{L}^T + \boldsymbol{\Psi}$$

Communality and total variance. The diagonal elements of the matrix $\mathbf{L}\mathbf{L}^T$ are called communalities. In almost all common factor analysis applications, standardized $\mathbf{X}$ variables and correlation matrices are used.

Therefore

$$\text{Var}(X_i) = 1 \quad \text{for all } i = 1, \ldots, p.$$

For standardized **X** variables

$$\Sigma = \mathbf{LL}^T + \Psi$$

implies

$$\Sigma = \begin{bmatrix} 1 & \rho_{12} & \cdots & \rho_{1p} \\ \rho_{21} & 1 & \cdots & \rho_{2p} \\ \vdots & \vdots & \vdots & \vdots \\ \rho_{p1} & \rho_{p2} & \cdots & 1 \end{bmatrix} = \begin{bmatrix} l_{11} & \cdots & l_{1m} \\ l_{21} & \cdots & l_{2m} \\ \vdots & \cdots & \vdots \\ l_{p1} & \cdots & l_{pm} \end{bmatrix} \begin{bmatrix} l_{11} & l_{21} & \cdots & \cdots & l_{p1} \\ \vdots & \vdots & \cdots & \cdots & \vdots \\ l_{1m} & l_{2m} & \cdots & \cdots & l_{pm} \end{bmatrix} + \begin{bmatrix} \Psi_1 & 0 & \cdots & 0 \\ 0 & \Psi_2 & \cdots & 0 \\ \vdots & \vdots & \cdots & \vdots \\ 0 & 0 & \cdots & \Psi_p \end{bmatrix}$$

If we define the diagonal elements of $\mathbf{LL}^T$

$$h = \begin{bmatrix} h_1 & 0 & \cdots & 0 \\ 0 & h_2 & \cdots & \\ \vdots & \vdots & \vdots & \vdots \\ 0 & 0 & \cdots & h_p \end{bmatrix} = \text{Diag}(\mathbf{LL}^T)$$

as the vector of communalities, from Eq. (5.39)

$$h_1 = \sum_{j=1}^{m} l_{1j}^2, \ldots, h_i = \sum_{j=1}^{m} l_{ij}^2, \ldots, h_p = \sum_{j=1}^{m} l_{pj}^2 \tag{5.42}$$

Clearly

$$\text{Diag}(\Sigma) = \text{Diag}(\mathbf{LL}^T) + \text{Diag}(\Psi) \tag{5.43}$$

and

$$\text{Var}(X_i) = 1 = h_i + \Psi_i \quad \text{for all } i = 1, \ldots, p. \tag{5.44}$$

It is also clear that

$$\Sigma - \text{Diag}(\Sigma) = \mathbf{LL}^T - \text{Diag}(\mathbf{LL}^T) \tag{5.45}$$

The total variance of X is

$$\text{Var}(X_1) + \text{Var}(X_2) + \cdots + \text{Var}(X_p) = p = \sum_{i=1}^{p} h_i + \sum_{i=1}^{p} \Psi_i \tag{5.46}$$

Remarks.

1. From Eq. (5.39), the off-diagonal elements of the matrix $= \mathbf{LL}^T$ should be equal to the off-diagonal elements of the covariance (correlation matrix), which means that the common factor analysis is trying to

use a small number of factors to represent the correlative structures in the original data.

2. In common factor analysis, the original **X** variable is decomposed into two parts. One is a vector combination of common factors, Y_i, $i = 1,..., m$, which is $\mathbf{L}_{.1}Y_1 + \cdots + \mathbf{L}_{.m}Y_m$, where $\mathbf{L}_{.1}, ..., \mathbf{L}_{.m}$ are called factor loadings, in which l_{ij} is the weight of common factor Y_j in representing X_i. The other part is the special random component, $\varepsilon = (\varepsilon_1, \varepsilon_2, ..., \varepsilon_p)$ where ε_i, for $i = 1,..., p$, are independent of each other and represent the X_i variable component that cannot be represented by common factors.

3. From Eq. (5.42), that is, $\text{Var}(X_i) = 1 = h_i + \Psi_i$, it is clear that the variance of each original variable X_i is decomposed into two parts. One is communality, which is the variance component from common factors, and the other is Ψ_i, which is the special variance component that cannot be represented by the common factors.

Comparison with PCA. If we use principal component analysis based on correlation analysis and take the first m components with largest eigenvalues as factors, from Eq. (5.19), we get

$$\mathbf{X} - \mu = \mathbf{CY} = \sum_{j=1}^{p} \mathbf{C}_{.j}Y_j = \sum_{i=1}^{m} \mathbf{C}_{.i}Y_i + \sum_{j=m+1}^{p} \mathbf{C}_{.j}Y_j \tag{5.47}$$

We can rewrite the above equation as

$$\mathbf{X} - \mu = \sum_{i=1}^{m} \mathbf{C}_{.i}Y_i + \varepsilon \tag{5.48}$$

where

$$\varepsilon = (\varepsilon_1, \varepsilon_2, ..., \varepsilon_p)^T = \sum_{j=m+1}^{p} \mathbf{C}_{.j}Y_j$$

$$\varepsilon_i = \sum_{j=m+1}^{p} c_{ij}y_j \quad \text{for } i = 1,..., p$$

Clearly, every ε_i is a linear combination of $\mathbf{Y}_j$'s, for $j = m+1,..., p$. In most of the cases, ε_i will not be independent of each other. But

$$\sum_{i=1}^{p} \text{Var}(\varepsilon_i) = \sum_{j=m+1}^{p} \text{Var}(\mathbf{Y}_j) = \sum_{j=m+1}^{p} \lambda_j \tag{5.49}$$

which is the summation of the last $p - m$ smallest eigenvalues. Clearly, the total variance of residuals for principal component analysis is often less than that of the common factor analysis.

$$\sum_{j=m+1}^{p} \lambda_j \leq \sum_{i=1}^{p} \Psi_i \tag{5.50}$$

Therefore, the objective of principal component analysis is to use a small number of principal components to represent the biggest share of total variance. On the other hand, the objective of common factor analysis is to use a small number of factors to represent most of the correlation structure for the original variable.

If we are interested in extracting the hidden factors that contribute to most of the variation, we should choose principal component analysis. If we want to extract the hidden factors that explain most of the interrelationship among variables, we should choose factor analysis method.

Example 5.8: Factor Analysis on the Weight, Waist, and Pulse Data Set We can use the common factor analysis to analyze the same data set as that of Example 5.2. By using MINITAB, we get the following results:

Factor Analysis: Weight, Waist, Pulse

```
Maximum Likelihood Factor Analysis of the Correlation Matrix

* NOTE * Heywood case

Unrotated Factor Loadings and Communalities

Variable      Factor1       Factor2       Communality
weight         0.821        -0.572            1.000
waist          0.996        -0.093            1.000
pulse         -0.324         0.000            0.105

Variance       1.7694        0.3354           2.1049
% Var          0.590         0.112            0.702

Factor Score Coefficients

Variable      Factor1       Factor2
weight        -0.189        -2.021
waist          1.160         1.666
pulse         -0.000        -0.000
```

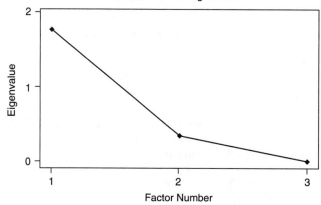

From the MINITAB output, we can find that the communalities for weight and waist are equal to one, which means that all the variations on weight and waist can be fully explained by the first two factors. However, the communality for pulse is only 0.105, which means that Var(ε_3) = ψ_3 = 0.895. That is, most of the variance of pulse cannot be represented by the two common factors. Most of the variance of pulse is independent of the two factors and can be explained by Var(ε_3) = ψ_3.

If we use PCA based factor analysis, we will get the following MINITAB results:

Factor Analysis: Weight, Waist, Pulse

```
Principal Component Factor Analysis of the Correlation Matrix

Unrotated Factor Loadings and Communalities

Variable     Factor1      Factor2      Communality
weight       -0.929       -0.275         0.938
waist        -0.945       -0.207         0.935
pulse         0.532       -0.847         1.000

Variance      2.0371       0.8352        2.8723
% Var         0.679        0.278         0.957

Factor Score Coefficients

Variable     Factor1      Factor2
weight       -0.456       -0.329
waist        -0.464       -0.247
pulse         0.261       -1.014
```

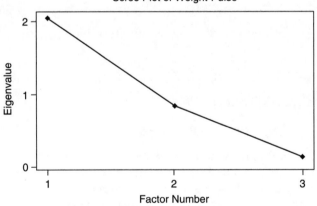

Scree Plot of Weight-Pulse

5.6.3 Parameter estimation in common factor analysis

Parameter estimation in PCA is quite straightforward. Given a sample covariance matrix **S** or sample correlation matrix **R**, we only need to extract eigenvalues and eigenvectors to estimate the principal components. In common factor analysis, the most popular approach is to use the maximum likelihood estimation. Here we briefly describe the procedure developed by Joreskog (1977). Since

$$\sum_{j=1}^{m} l_{ij}^2 + \Psi_i = 1 \quad \text{for all } i = 1, \ldots, p.$$

$$\Sigma = LL^T + \Psi$$

and S is the statistical estimate of Σ, so we estimate the matrix entries and in both L and Ψ such that

$$\begin{aligned} &\text{Minimize } \| S - LL^T - \Psi \|^2 \\ &\text{Subject to } \Sigma = LL^T + \Psi \end{aligned} \quad (5.51)$$

Assume that $\hat{L}$ and $\hat{\Psi}$ are statistical estimates for L and Ψ, then the optimal solution for the above mathematical program should follow the following conditions:

$$(S - \hat{L}\hat{L}^T - \hat{\Psi})\hat{L} = 0 \qquad \text{Diag}(S - \hat{L}\hat{L}^T - \hat{\Psi}) = 0 \quad (5.52)$$

However, these equations may not be solved directly. Joreskog (1977) proposed to solve iteratively starting from

$$(S - \Psi_0)\hat{L} = \hat{L}\hat{L}^T\hat{L} \qquad \Psi_0 = \text{Diag}(S - \hat{L}\hat{L}^T) \quad (5.53)$$

Assume that $\hat{L}^T\hat{L}$ is a diagonal matrix and let

$$S_r = S - \Psi_0 \quad (5.54)$$

We need an initial estimate of Ψ_0 to start our iterative algorithm. So from Eq. (5.51)

$$S_r = \hat{L}\hat{L}^T \quad (5.55)$$

By using singular value decomposition,

$$S_r = P\Delta\Delta^T = (P\Delta^{1/2})(P\Delta^{1/2})^T \quad (5.56)$$

where Δ is the diagonal matrix containing the eigenvalues of S_r.

Comparing Eqs. (5.51) and (5.54) we can use $\hat{L} = P\Delta^{1/2}$ as our initial estimate of L, and by using $\Psi_0 = \text{Diag}(S - \hat{L}\hat{L}^T)$ we get the new estimate of Ψ_0 and we continue this iterative process until convergence.

5.7 Factor Rotation

One of the most important issues in using factor analysis is its interpretation. Factor analysis often uses factor rotation to enhance its interpretation. Factor rotation can be expressed by the following:

$$\tilde{L} = LT \quad (5.57)$$

where $T = T_{m \times m}$ is called a *rotation matrix*. T is also an orthogonal matrix, that is, $\mathbf{TT}^T = \mathbf{I}$. Therefore,

$$\begin{aligned}
\Sigma &= \mathbf{LL}^T + \Psi \\
&= \mathbf{LTT}^T\mathbf{L} + \Psi \\
&= \mathbf{LT(LT)}^T + \Psi \\
&= \tilde{\mathbf{L}}\tilde{\mathbf{L}}^T + \Psi
\end{aligned} \tag{5.58}$$

So if we use $\tilde{\mathbf{L}} = \mathbf{LT}$ to replace $\mathbf{L}$ as the factor loading matrix, it will have the same ability to represent the correlation matrix Σ.

After rotation

$$\mathbf{X} = \mathbf{LY} + \varepsilon = \mathbf{LTT}^T\mathbf{Y} + \varepsilon = \tilde{\mathbf{L}}\tilde{\mathbf{Y}} + \varepsilon \tag{5.59}$$

where $\tilde{\mathbf{L}}$ is the new rotated factor loading matrix and $\tilde{\mathbf{Y}} = \mathbf{T}^T\mathbf{Y}$ is the new rotated factor. The new factor $\tilde{\mathbf{Y}}$ will also be a standard normal vector such that $\tilde{\mathbf{Y}} \sim N(\mathbf{0}, \mathbf{I})$. This is because that if $Y \sim N(\mathbf{0}, \mathbf{I})$, $\tilde{\mathbf{Y}} = \mathbf{T}^T\mathbf{Y}$ will also be multivariate normal random variable, $E(\tilde{\mathbf{Y}}) = E(\mathbf{T}^T\mathbf{Y}) = 0$, $\text{Cov}(\tilde{\mathbf{Y}}) = \text{Cov}(\mathbf{T}^T\mathbf{Y}) = \mathbf{T}^T\text{Cov}(\mathbf{Y})\mathbf{T} = \mathbf{T}^T\mathbf{IT} = \mathbf{I}$, due to the fact that $\mathbf{T}$ is an orthogonal matrix.

The purpose of rotation is to make the rotated factor loading matrix $\tilde{\mathbf{L}}$ have some desirable properties. There are two kinds of rotations. One is to rotate the factor loading matrix such that the rotated matrix $\tilde{\mathbf{L}}$ will have a simple structure. The other is called *procrustes* rotation, which is trying to rotate the factor matrix to a target matrix. We will discuss these two rotations in subsequent subsections.

5.7.1 Factor rotation for simple structure

Thurstone (1947) suggested the following criteria for simple structure:

1. Each row of $\tilde{\mathbf{L}}$ should contain at least one zero.
2. Each column of $\tilde{\mathbf{L}}$ should contain several responses whose loadings vanish in one column but not in the other.
3. Every pair of columns of $\tilde{\mathbf{L}}$ should contain several responses whose loadings vanish in one column but not in the other.
4. If the number of factors m is four or more, every pair of columns of $\tilde{\mathbf{L}}$ should contain a large number of responses with zero loading in both columns.
5. Conversely, for every pair of columns of $\tilde{\mathbf{L}}$, only a small number of responses should have nonzero loadings in both columns.

The rotation for simple structures is to find an orthogonal matrix **T**, such that (Kaiser, 1958; Magnus and Neudecker, 1999)

$$\text{Maximize} \quad Q = \sum_{j=1}^{m}\left[\sum_{i=1}^{p} \tilde{l}_{ij}^4 - \frac{c}{p}\left(\sum_{j=1}^{m} \tilde{l}_{ij}^2\right)^2\right] \quad (5.60)$$

$$\text{Subject to} \quad LT = \tilde{L}$$

There are four frequently used rotation techniques (Kaiser, 1958):

1. *Equimax.* This corresponds to $c = m/2$ in Eq. (5.58). Equimax rotates the loadings so that a variable loads high on one factor but low on others.
2. *Varimax.* This corresponds to $c = 1$ in Eq. (5.58). Varimax maximizes the variance of the squared loadings.
3. *Quartimax.* This corresponds to $c = 0$ in Eq. (5.58). Quartimax achieves simple loadings.
4. *Orthomax.* In Orthomax rotation, the user will determine c value, where $0 < c < 1$.

Example 5.9: Factor Rotation for Example 5.7 By using the same data as that of Example 5.7, we tried four rotation methods in MINITAB, and the results are as follows:

```
Maximum Likelihood Factor Analysis of the Correlation Matrix

* NOTE * Heywood case

Unrotated Factor Loadings and Communalities

Variable    Factor1    Factor2    Communality
weight       0.821     -0.572       1.000
waist        0.996     -0.093       1.000
pulse       -0.324      0.000       0.105

Variance     1.7694    0.3354       2.1049
% Var        0.590     0.112        0.702
```

Equimax rotation

```
Rotated Factor Loadings and Communalities
Equimax Rotation

Variable    Factor1    Factor2    Communality
weight       0.879     -0.477       1.000
waist        0.530     -0.848       1.000
pulse       -0.145      0.290       0.105

Variance     1.0736    1.0313       2.1049
% Var        0.358     0.344        0.702
```

Factor Score Coefficients

```
Variable    Factor1    Factor2
weight       1.722      1.075
waist       -0.969     -1.784
pulse        0.000      0.000
```

Varimax rotation

Rotated Factor Loadings and Communalities
Varimax Rotation

```
Variable    Factor1    Factor2    Communality
weight       0.879     -0.477      1.000
waist        0.530     -0.848      1.000
pulse       -0.145      0.290      0.105

Variance     1.0736     1.0313     2.1049
% Var        0.358      0.344      0.702
```

Factor Score Coefficients

```
Variable    Factor1    Factor2
weight       1.722      1.075
waist       -0.969     -1.784
pulse        0.000      0.000
```

Quartimax rotation

Rotated Factor Loadings and Communalities
Quartimax Rotation

```
Variable    Factor1    Factor2    Communality
weight       0.909     -0.417      1.000
waist        0.996      0.085      1.000
pulse       -0.319     -0.058      0.105

Variance     1.9206     0.1843     2.1049
% Var        0.640      0.061      0.702
```

Factor Score Coefficients

```
Variable    Factor1    Factor2
weight       0.173     -2.023
waist        0.846      1.845
pulse       -0.000     -0.000
```

Orthomax

Rotated Factor Loadings and Communalities
Orthomax Rotation with Gamma = 0.50

```
Variable    Factor1    Factor2    Communality
weight       0.860     -0.511      1.000
waist        1.000     -0.021      1.000
pulse       -0.323     -0.023      0.105

Variance     1.8431     0.2618     2.1049
% Var        0.614      0.087      0.702
```

Factor Score Coefficients

```
Variable    Factor1    Factor2
weight      -0.042     -2.030
waist        1.037      1.745
pulse       -0.000     -0.000
```

Every rotation method produces different results, and it seems that the rotated factor loadings on Orthomax rotation are the neatest ones to explain. The first factor has the loading factor $(0.86, 1.0, -0.323)^T$, which is very close to $(1, 1, -0.3)^T$. This means that the first factor is mostly weight-waist synchronized variation, and the pulse is somewhat varying in the opposite direction of weight-waist. The second factor is very close to $(-0.5, 0, 0)^T$, which is only related to weight. This is a minor factor indicating the independent variation of weight.

5.7.2 Procrustes rotation

Procrustes rotation is used to find rotation matrix $\mathbf{T}$ such that the original factor loading matrix $\mathbf{L}$ can be rotated into a target matrix $\mathbf{O}$ (Yates, 1988). Specifically, we want to find $\mathbf{T}$ such that

$$\begin{aligned} &\text{Minimize} \quad \text{Tr}(\mathbf{E}^T\mathbf{E}) \\ &\text{Subject to} \quad \mathbf{E} = \mathbf{LT} - \mathbf{O} \\ &\qquad\qquad\quad \mathbf{T}^T\mathbf{T} = \mathbf{I} \end{aligned} \qquad (5.61)$$

In the next section, we will present a case study that uses procrustes rotation in factor analysis.

5.8 Factor Analysis Case Studies

5.8.1. Characterization of texture and mechanical properties of heat-induced soy protein gels (Kang, Matsumura, and Mori, 1991)

Soy proteins play important roles in many foodstuffs because of their nutritional values and their contribution to food texture. Soy proteins' texture can be influenced by their processing conditions.

The texture of soy protein can be measured by a sequence of tests and corresponding mechanical properties.

Measurement of mechanical properties of gels. A compression-decompression test was carried out by using a compression tester. The tester had a cylindrical plunger with a cross-sectional area of 0.25 cm^2. The tests were performed at 20°C. The plunger descended at a rate of 1.2 mm per minute. The following four measures were recorded:

1. *Compression work.* It is the mechanical work in each compression-decompression cycle.
2. *Resiliency (RS).* It is the ratio of decompression work versus compression work.

3. *Compressibility (CM)* It is measured by the ratio of deformation versus the gel sample thickness.
4. *Force.* It is determined as the force at yield.

The compression-decompression tests are conducted at different levels until rupture. Specifically, four levels of compression load are used. They are small (S), medium (M), large (L), and rupture (R). They correspond to 15%, 30%, 60%, and 100% of rupture force.

Since there are four measures and each measure is recorded at four levels of compression loads, each protein gel sample test produces 16 observations of data.

Soy gel processing condition. There are three major process variables in processing soy protein gel. They are as follows:

1. *Heating temperature.* Ranging from 80 to 100°C
2. *Protein concentration.* Ranging from 18 to 20 percent
3. *Type of protein.* They are acid precipitated soy protein (APP) and low glycinin APP

Test data set. There are 24 soy protein gel samples. Each sample will go through the compression and decompression test, yielding 16 pieces of data. So in this case study, $p = 16$, $n = 24$ in the data matrix:

$$\mathbf{X}_{n \times p} = \begin{bmatrix} x_{11} & x_{12} & \cdots & x_{1p} \\ x_{21} & x_{22} & \cdots & x_{2p} \\ \vdots & \vdots & \vdots & \vdots \\ x_{n1} & x_{n2} & \cdots & x_{np} \end{bmatrix}$$

Factor analysis results. Factor analysis with Varimax rotation is conducted on the normalized data set (subtract mean and divide by standard deviation). There, factors are retained and these three factors account for 97.9 percent of total variance. The factor loadings represent the correlations between the factors and mechanical parameters. The higher the value, the more highly correlated is that factor with mechanical properties. It is illustrated in Table 5.11.

Factor interpretation.
Factor 1. The mechanical attributes that load high in factor 1 were compression work and rupture force at all experimental compression levels. Therefore, factor 1 relates to the hardness and toughness of the gels.

TABLE 5.11 Factor Loading after Varimax Rotation

Mechanical properties	Factor 1	Factor 2	Factor 3
Compression work			
Small	**0.886**	0.416	−0.195
Medium	**0.896**	0.369	−0.237
Large	**0.912**	0.298	−0.279
Rupture	**0.922**	0.200	−0.308
Resilience			
Small	−0.415	−0.684	0.460
Medium	−0.469	−0.625	0.594
Large	−0.479	−0.526	0.691
Rupture	−0.240	−0.566	**0.758**
Compressibility			
Small	0.291	**0.940**	−0.135
Medium	0.345	**0.897**	−0.268
Large	0.384	**0.846**	−0.346
Rupture	0.382	**0.777**	−0.479
Force			
Small	**0.925**	0.323	−0.199
Medium	**0.925**	0.323	−0.199
Large	**0.925**	0.323	−0.199
Rupture	**0.925**	0.323	−0.199

Factor contribution to variances: factor 1: 7.871; factor 2: 5.315; factor 3: 2.482.
High factor loadings are shown in bold.

Factor 2. The mechanical attributes that load high on factor 2 were compressibility at all experimental compression levels. Therefore factor 2 relates to fracturability.

Factor 3. Since resilience at rupture was loaded high at factor 3, factor 3 relates to elasticity.

For all 24 test samples, their factor scores are computed and listed in Table 5.12.

From Table 5.12 we can see that the gel formed at 100°C from a 20% protein concentration has the highest scores for both factor 1 and factor 2, which indicates this gel is the hardest, toughest and the most unfracturable gel. For factor 3, the gel formed at 80°C and 19% protein concentration has the highest score for factor 3, which means this gel is the most elastic.

5.8.2 Procrustes factor analysis for automobile body assembly process

Factor analysis is a multivariate tool that is very similar to PCA. Factor analysis is also used to condense a set of observed variables into a smaller number of transformed variables called components or factors.

TABLE 5.12 Factor Scores

Process conditions		Factor scores		
Protein concentration (%)	Temperature (°C)	Factor 1	Factor 2	Factor 3
Regular APP gel				
18	80	−0.756	0.353	0.556
18	85	−0.783	0.310	0.196
18	90	−0.854	0.269	−0.210
18	93	−0.810	−0.124	−0.960
18	96	−0.669	−0.189	−1.905
18	**100**	−0.263	0.163	−2.231
19	80	−0.513	0.802	**1.486**
19	85	−0.408	0.663	1.201
19	90	−0.552	0.535	0.327
19	93	−0.452	0.357	−0.518
19	96	0.214	0.644	−1.191
19	100	0.686	0.880	−0.919
20	80	−0.491	0.515	1.195
20	85	−0.419	0.577	1.088
20	90	−0.452	0.437	0.282
20	93	−0.385	0.530	−0.685
20	96	1.248	1.443	0.002
20	**100**	**2.709**	**1.637**	1.012
Low glycinin APP gel				
20	80	0.452	−2.056	0.943
20	85	0.435	−1.797	0.623
20	90	0.492	−1.575	0.801
20	93	0.457	−1.696	0.240
20	96	0.578	−1.307	−0.439
20	100	0.537	−1.371	−0.896

Both factor analysis and PCA are used to identify an underlying structure or pattern beneath a set of multivariate data. However, physical interpretation of PCA generated modes is sometimes difficult. This is especially true when PCA is applied to flexible parts which can have complex deformations. Although, in such circumstances, PCA can describe the variation pattern, the variation cannot be resolved to provide a better physical meaning to the pattern. Resolving the variation pattern is generating modes or factors of variation along a particular direction of rotation or translation (e.g., one factor describing variation along only x-direction translation and another factor describing variation along x-y rotation). Factors or modes generated by factor analysis can be rotated to resolve the variation pattern. Rotated factors can provide better physical interpretation and also generate clues as to causes of variation.

Procrustes factor analysis can be used very effectively for hypothesis testing (confirmatory analysis). Suspected causes of variation can be

156 Chapter Five

translated into a matrix, which is the matrix **O** in Eq. (5.59), that represents the factor loadings required to confirm a particular cause of variation. After the required or target factor matrix **O** is set up, factor analysis can be performed on the actual data to obtain the initial factor matrix, which is the factor loading matrix **L**. The actual factor matrix can then be "rotated" to obtain the fitted factor matrix. If we really can find a rotation **T** which can rotate **L** into a matrix that is very close to **O**, then the truth of the hypothesis is "validated." At least we can say that the hypothesis is very likely to be true. One case study is discussed to illustrate the use of procrustes factor rotation.

5.8.3 Hinge variation study using procrustes factor analysis

This section shows how factor analysis can be used to root-cause the hinge-variation problem discussed in Sec. 5.5. The problem was analyzed by using PCA. Figure 5.29 illustrates the problem.

The unrotated factors of variation are the same as the modes of variation generated by PCA. To perform procrustes (confirmatory) factor analysis (hypothesis testing), possible causes or hypotheses for the hinge variation have to be identified. These hypotheses are identified as follows:

1. Underbody dash-extension building approximately 1 mm outboard and the corresponding body-side region building inboard by about 2 mm.
2. Right body-side "boat-anchor" building approximately 3 mm inboard.

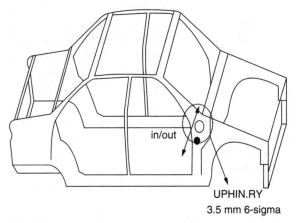

Figure 5.29 High cross-car variation at right UPHIN check point.

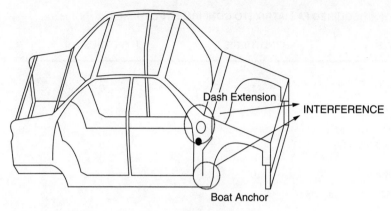

Figure 5.30 Interference at hinge region (hypothesis).

These two hypotheses are illustrated in Fig. 5.30. It was suspected that the conditions listed above were causing an interference in the hinge region.

Target factor matrix O. The hypotheses have to be first translated into a target factor matrix **O**. This matrix gives the target or required loadings for the selected points in each factor for the hypothesis to be true. For the case study being considered, the selected points were 5R-HGU, 5R-RKU, UPHIN.RY, and ROCK.RY. The first two points were on the underbody and the next two points on the full body. These points best represented the interference regions in the B-I-W.

The location of the points is shown in Fig. 5.31 and the target matrix is shown in Fig. 5.32.

All the points selected for the analysis were cross-car check points. This was because check points along the fore/aft and up/down direction did not exhibit any significant variation.

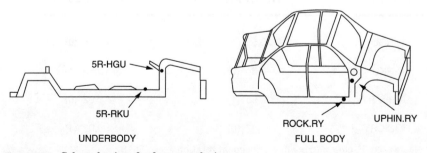

Figure 5.31 Selected points for factor analysis.

REQUIRED FA MATRIX (TO CONFIRM HYPOTHESES):

POINTS	HYPOTHESIS I	HYPOTHESIS II
5R-HGU	1	0
5R-RKU	0	1
UPHIN.RY	1	0
ROCK.RY	0	1

Figure 5.32 Target matrix O.

The ones represent maximum loading in a particular factor while the zeros represent minimum loading in a factor. The target matrix was developed to represent maximum variation at the interference regions.

Results of factor rotation. The factor matrix obtained by PCA was rotated orthogonally to best fit the target matrix. The factors were rotated orthogonally to maintain independence of the factors. Figure 5.33 shows the resultant matrix. The required or target loadings are shown in parentheses next to the actual loadings obtained.

The resultant matrix is quite close to the target matrix, thus confirming the hypothesis. Various factors contribute to the difference in the target matrix and the rotated matrix. For this case study, one reason could be the location of the points. Although the points were close to the interference regions, they were not exactly at the metal contact points.

FA MATRIX OBTAINED AFTER ROTATION OF MODES:

POINTS	HYPOTHESIS I	HYPOTHESIS II
5R-HGU	0.55 (1)	−0.18 (0)
5R-RKU	−0.24 (0)	0.50 (1)
UPHIN.RY	0.95 (1)	0.02 (0)
ROCK.RY	0.07 (0)	0.93 (1)

Figure 5.33 Factor matrix after orthogonal rotation.

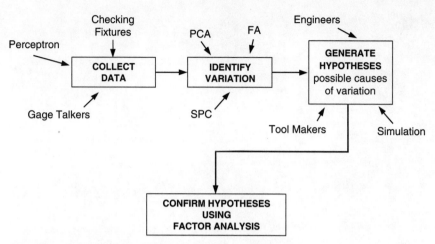

Figure 5.34 Using factor analysis for root-cause analysis.

Figure 5.34 illustrates the steps in using factor analysis for root-cause analysis. Generating hypotheses for possible causes of variation is critical to the successful implementation of confirmatory factor analysis. Engineering experience and simulation greatly assist in formulating the hypotheses. Usually more than one hypothesis has to be tested before the cause of variation can be found.

Chapter 6

Discriminant Analysis

6.1 Introduction

Discriminant analysis is a powerful multivariate statistical method that can classify individual objects or units into groups. Specifically, in discriminant analysis, there are many individual objects and each object has several variables associated with it, that is, $(x_1, x_2, ..., x_n)$. The whole population of objects can be partitioned into several groups with the values of variables strongly related to the group the object belongs to. Discriminant analysis is a method that classifies each object into groups depending on the values of the variables of each object. We give the following examples to illustrate this kind of a situation.

Example 6.1: People's Weight Every person can be classified into one of the three groups—underweight, normal weight, and overweight. In order to determine a person's weight category, we need to look into at least two variables, height (x_1) and weight (x_2). Table 6.1 gives a data set which has people's height and weight data in all the three weight categories.

TABLE 6.1 People's Weight Categories

Weight category	H (m) x_1	W (kg) x_2	Weight category	H (m) x_1	W (kg) x_2	Weight category	H (m) x_1	W (kg) x_2
N	1.52	48.5	O	1.53	61	U	1.54	45
N	1.61	59.6	O	1.60	64	U	1.63	47
N	1.65	65.3	O	1.66	82.6	U	1.67	47.4
N	1.71	64.3	O	1.72	82.8	U	1.70	46.2
N	1.73	71.8	O	1.74	87.8	U	1.74	60.5
N	1.80	74.5	O	1.81	98.2	U	1.81	60.1
N	1.92	77.4	O	1.87	90.9	U	1.91	65.7
N	1.83	80.4	O	1.90	101	U	1.65	51.7
N	1.63	58.5	O	1.64	86	U	1.69	45.7
N	1.77	65.8	O	1.58	67.4	U	1.78	53.9
N	1.75	67.4	O	1.76	77.4	U	1.74	51.5
N	1.82	76.2	O	1.81	88.5	U	1.80	58.3

Weight categories: N = normal, O = overweight, U = underweight, H = height, W = weight. In this example, discriminant analysis can be used to analyze people's height and weight data with known categories and develop criteria to classify any new person with height and weight data into appropriate weight categories.

Example 6.2: Remote Sensing Data for Crops A satellite remote sensing device picks up signals for four variables (x_1, x_2, x_3, x_4) on cropland. Table 6.2 gives some values for several known crop plots.

Discriminant analysis is able to use the available data to develop classification criteria. So in future, if new remote sensing data are received, discriminant analysis will be able to classify the kind of cropland it should be.

TABLE 6.2 Remote Sensing Data for Crops

Crops	x_1	x_2	x_3	x_4
Corn	16	27	31	33
Corn	15	23	30	30
Corn	16	27	27	26
Corn	18	20	25	23
Corn	15	15	31	32
Corn	15	20	32	28
Corn	12	18	25	34
Soybeans	20	23	23	25
Soybeans	24	24	25	32
Soybeans	21	25	23	24
Soybeans	27	45	24	12
Soybeans	12	13	15	42
Soybeans	22	32	31	43
cotton	31	32	33	34
cotton	29	24	26	28
cotton	34	32	28	45
cotton	26	25	23	24
cotton	53	48	75	26
cotton	34	35	25	78
Sugarbeets	22	23	25	42
Sugarbeets	25	25	24	26
Sugarbeets	34	25	16	52
Sugarbeets	54	23	21	54
Sugarbeets	25	43	32	15
Sugarbeets	26	54	2	54
Clover	87	54	61	21
Clover	51	31	31	16
Clover	96	48	54	62
Clover	31	31	11	11
Clover	56	13	13	71
Clover	32	13	27	32
Clover	36	26	54	32
Clover	53	8	6	54
Clover	32	32	62	16

In general, discriminant analysis is a step-by-step process described as follows:

6.1.1 Discriminant analysis steps

Step 1: Determine variables and groups. We need to determine the discriminant variables ($x_1, x_2, x_3, ..., x_p$). The key consideration is that these variables should contain the key variables that can discriminate groups based on their values. The variable selection can be based on previous experience or relevant research. The groups should be mutually exclusive; each object can only belong to one group.

Step 2: Obtain initial sample of objects. After the determination of variables and groups, we need to obtain an initial sample of objects. We should know what group each object belongs to, to obtain the values of variables for each object. The initial sample should have sufficient size. Usually, a portion of the initial sample is used to develop discriminant criteria, which is the rule for classification. This portion of the initial sample is often called the analysis sample. The remaining portion of the initial sample is often used to validate the discriminant criteria and estimate the classification errors.

Step 3: Develop discriminant criteria. There are several methods to develop discriminant criteria based on the situation. First of all, the discriminant criteria could consist of linear function(s) and cutoff score(s). In such a case, we often call it a linear discriminant function. The criteria could also be quadratic functions and cutoff scores as well as other forms. The type of discriminant criteria to be used also depends on whether there are two groups or multiple groups. Usually, for the two-group discriminant analysis situation, linear discriminant analysis is used. For multiple groups, the discriminant criteria based on Mahalanobis distance are the most popular method.

In subsequent sections, we will discuss the discriminant analysis for various situations. Discriminant analysis was first discussed by Fisher (1936). There is much excellent literature on discriminant analysis that provides more comprehensive discussions (Lachenbrush, 1975; Hand, 1981; Huberty, 1994).

6.2 Linear Discriminant Analysis for Two Normal Populations with Known Covariance Matrix

Classification Rule 6.1. We discuss the discriminant analysis with a somewhat theoretical case. Assume that we have two groups of objects. Group 1 objects follow a multivariate normal distribution $N(\mu_1, \Sigma)$, and group 2 objects follow another multivariate normal distribution $N(\mu_2, \Sigma)$. In this case, each group has different means, and both groups have the same covariance matrix.

For any new unknown object $\mathbf{x}$, we develop a linear discriminant function:

$$\mathbf{w}'\mathbf{x} = w_1 x_1 + w_2 x_2 + \cdots + w_p x_p$$

and a cutoff score K such that if $\mathbf{w'x} > K$, we classify $\mathbf{x}$ into group 1; otherwise, we classify $\mathbf{x}$ into group 2.

Theorem 6.1. *For the above discriminant classification,*

$$\mathbf{w'x} = w_1x_1 + w_2x_2 + \cdots + w_px_p = (\boldsymbol{\mu}_1 - \boldsymbol{\mu}_2)'\boldsymbol{\Sigma}^{-1}\mathbf{x} \quad (6.1)$$

$$K = \tfrac{1}{2}(\boldsymbol{\mu}_1 - \boldsymbol{\mu}_2)'\boldsymbol{\Sigma}^{-1}(\boldsymbol{\mu}_1 - \boldsymbol{\mu}_2) \quad (6.2)$$

Proof. From Chap. 2, the linear combination of a multivariate vector will follow a univariate normal distribution, that is,

$$\mathbf{w'x} \sim N(\mathbf{w'}\boldsymbol{\mu}, \mathbf{w'}\boldsymbol{\Sigma}\mathbf{w})$$

For a good discriminant function, the vector $\mathbf{w}$ should be chosen such that the distance from the centers of two populations, $\boldsymbol{\mu}_1$ and $\boldsymbol{\mu}_2$, is maximized:

$$\text{Maximize } \frac{(\mathbf{w'}\boldsymbol{\mu}_1 - \mathbf{w'}\boldsymbol{\mu}_2)^2}{\mathbf{w'}\boldsymbol{\Sigma}\mathbf{w}} \quad (6.3)$$

From the Cauchy-Schwarz inequality, for two vectors, $\mathbf{a}$ and $\mathbf{b}$,

$$(\mathbf{a'b})^2 \leq (\mathbf{a'a})(\mathbf{b'b})$$

If we let

$$\mathbf{a} = \frac{\boldsymbol{\Sigma}^{1/2}\mathbf{w}}{\sqrt{\mathbf{w'}\boldsymbol{\Sigma}\mathbf{w}}} \qquad \mathbf{b} = \boldsymbol{\Sigma}^{1/2}(\boldsymbol{\mu}_1 - \boldsymbol{\mu}_2)$$

then

$$\frac{(\mathbf{w'}\boldsymbol{\mu}_1 - \mathbf{w'}\boldsymbol{\mu}_2)^2}{\mathbf{w'}\boldsymbol{\Sigma}\mathbf{w}} = \frac{\left[\mathbf{w'}\boldsymbol{\Sigma}^{1/2}\boldsymbol{\Sigma}^{1/2}(\boldsymbol{\mu}_1 - \boldsymbol{\mu}_2)\right]^2}{\mathbf{w'}\boldsymbol{\Sigma}\mathbf{w}} = \left[\left(\frac{\boldsymbol{\Sigma}^{1/2}\mathbf{w}}{\sqrt{\mathbf{w'}\boldsymbol{\Sigma}\mathbf{w}}}\right)^T \boldsymbol{\Sigma}^{-1/2}(\boldsymbol{\mu}_1 - \boldsymbol{\mu}_2)\right]^2$$

$$\leq \left(\boldsymbol{\Sigma}^{-1/2}(\boldsymbol{\mu}_1 - \boldsymbol{\mu}_2)\right)^T \left(\boldsymbol{\Sigma}^{1/2}(\boldsymbol{\mu}_1 - \boldsymbol{\mu}_2)\right) = (\boldsymbol{\mu}_1 - \boldsymbol{\mu}_2)^T \boldsymbol{\Sigma}^{-1}(\boldsymbol{\mu}_1 - \boldsymbol{\mu}_2) \quad (6.4)$$

In Eq. (6.3), if we let $\mathbf{w} = \boldsymbol{\Sigma}^{-1}(\boldsymbol{\mu}_1 - \boldsymbol{\mu}_2)$, then

$$\frac{(\mathbf{w'}\boldsymbol{\mu}_1 - \mathbf{w'}\boldsymbol{\mu}_2)^2}{\mathbf{w'}\boldsymbol{\Sigma}\mathbf{w}} = \frac{\left[(\boldsymbol{\mu}_1 - \boldsymbol{\mu}_2)^T \boldsymbol{\Sigma}^{-1}(\boldsymbol{\mu}_1 - \boldsymbol{\mu}_2)\right]^2}{(\boldsymbol{\mu}_1 - \boldsymbol{\mu}_2)^T \boldsymbol{\Sigma}^{-1}\boldsymbol{\Sigma}\boldsymbol{\Sigma}^{-1}(\boldsymbol{\mu}_1 - \boldsymbol{\mu}_2)} = (\boldsymbol{\mu}_1 - \boldsymbol{\mu}_2)^T \boldsymbol{\Sigma}^{-1}(\boldsymbol{\mu}_1 - \boldsymbol{\mu}_2)$$

Clearly, $\mathbf{w} = \Sigma^{-1}(\mu_1 - \mu_2)$ is the vector that makes Eq. (6.4) an equation. It is also the vector that makes Eq. (6.3) achieve its maximum. Therefore

$$\mathbf{w}'\mathbf{x} = w_1 x_1 + w_2 x_2 + \cdots + w_p x_p = (\mu_1 - \mu_2)' \Sigma^{-1} \mathbf{x}$$

should be the linear discriminant function. Now, let us determine the cutoff score K. Recall the discriminant rule: if $\mathbf{w}'\mathbf{x} > K$, we classify $\mathbf{x}$ into group 1; otherwise, we classify $\mathbf{x}$ into group 2.

If an observation $\mathbf{x}$ is from group 1, then $\mathbf{w}'\mathbf{x} \sim N(\mathbf{w}'\mu_1, \mathbf{w}'\Sigma\mathbf{w})$, then the probability that we misclassify $\mathbf{x}$ into group 2 is

$$P(2|1) = P[\mathbf{w}'\mathbf{x} < K] = P\left[\frac{\mathbf{w}'\mathbf{x} - \mathbf{w}'\mu_1}{\sqrt{\mathbf{w}'\Sigma\mathbf{w}}} < \frac{K - \mathbf{w}'\mu_1}{\sqrt{\mathbf{w}'\Sigma\mathbf{w}}}\right]$$

$$= P\left[z < \frac{K - \mathbf{w}'\mu_1}{\sqrt{\mathbf{w}'\Sigma\mathbf{w}}}\right] = P\left[z \geq \frac{\mathbf{w}'\mu_1 - K}{\sqrt{\mathbf{w}'\Sigma\mathbf{w}}}\right] = \Phi\left(\frac{\mathbf{w}'\mu_1 - K}{\sqrt{\mathbf{w}'\Sigma\mathbf{w}}}\right) \quad (6.5)$$

On the other hand, if an observation $\mathbf{x}$ is from group 2, then $\mathbf{w}'\mathbf{x} \sim N(\mathbf{w}'\mu_2, \mathbf{w}'\Sigma\mathbf{w})$ and then the probability that we misclassify $\mathbf{x}$ into group 1 is

$$P(1|2) = P[\mathbf{w}'\mathbf{x} > K] = P\left[\frac{\mathbf{w}'\mathbf{x} - \mathbf{w}'\mu_2}{\sqrt{\mathbf{w}'\Sigma\mathbf{w}}} > \frac{K - \mathbf{w}'\mu_2}{\sqrt{\mathbf{w}'\Sigma\mathbf{w}}}\right] = P\left[z \geq \frac{K - \mathbf{w}'\mu_2}{\sqrt{\mathbf{w}'\Sigma\mathbf{w}}}\right]$$

$$= \Phi\left(\frac{K - \mathbf{w}'\mu_2}{\sqrt{\mathbf{w}'\Sigma\mathbf{w}}}\right) \quad (6.6)$$

where

$$\Phi(x) = \int_x^\infty \frac{1}{\sqrt{2\pi}} e^{-1/2 t^2} dt$$

For a good classification procedure, we have $P(2|1) = P(1|2)$, then

$$\frac{\mathbf{w}'\mu_1 - K}{\sqrt{\mathbf{w}'\Sigma\mathbf{w}}} = \frac{K - \mathbf{w}'\mu_2}{\sqrt{\mathbf{w}'\Sigma\mathbf{w}}}$$

By substituting $\mathbf{w} = \Sigma^{-1}(\mu_1 - \mu_2)$ and solving for K, we get

$$K = \tfrac{1}{2}(\mu_1 - \mu_2)\Sigma^{-1}(\mu_1 + \mu_2)$$

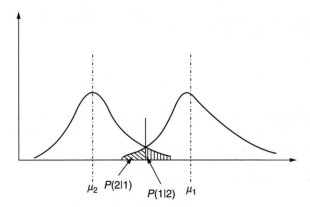

Figure 6.1 Misclassification probability of discriminant analysis for single variable.

Also

$$P(1|2) = P(2|1) = P\left[z > \tfrac{1}{2}(\boldsymbol{\mu}_1 - \boldsymbol{\mu}_2)\boldsymbol{\Sigma}^{-1}(\boldsymbol{\mu}_1 - \boldsymbol{\mu}_2)\right] \quad (6.7)$$

Remarks.

1. Equal misclassification probability can be intuitively explained by the following univariate discriminant analysis case, where $x_1 \sim N(\mu_1, \sigma^2)$, $x_2 \sim N(\mu_2, \sigma^2)$, $\mu_1 > \mu_2$. From Fig. 6.1, it is clear that making $P(1|2)$ and $P(2|1)$ equal will actually minimize the total misclassification error, $P(2|1) + P(1|2)$. In this case, the cutoff point, $K = (\mu_1 + \mu_2)/2$, is the midpoint between two means, μ_1 and μ_2. The classification rule is: if $x > K$, then classify x into group 1, otherwise x is classified into group 2.

2. Our linear 2 group discriminant classification rule states that, for any new multivariate observation $\mathbf{x}$, if $\mathbf{w}'\mathbf{x} - K > 0$, then $\mathbf{x}$ is classified into group 1; otherwise, $\mathbf{x}$ is classified into group 2.

$$\begin{aligned}\mathbf{w}'\mathbf{x} - K &= (\boldsymbol{\mu}_1 - \boldsymbol{\mu}_2)\boldsymbol{\Sigma}^{-1}\mathbf{x} - \tfrac{1}{2}(\boldsymbol{\mu}_1 - \boldsymbol{\mu}_2)^T\boldsymbol{\Sigma}^{-1}(\boldsymbol{\mu}_1 + \boldsymbol{\mu}_2) \\ &= \tfrac{1}{2}\left[(\mathbf{x} - \boldsymbol{\mu}_2)^T\boldsymbol{\Sigma}^{-1}(\mathbf{x} - \boldsymbol{\mu}_2) - (\mathbf{x} - \boldsymbol{\mu}_1)^T\boldsymbol{\Sigma}^{-1}(\mathbf{x} - \boldsymbol{\mu}_1)\right] \\ &= \tfrac{1}{2}\left[D_2^2 - D_1^2\right]\end{aligned} \quad (6.8)$$

where $D_1^2 = (\mathbf{x} - \boldsymbol{\mu}_1)^T\boldsymbol{\Sigma}^{-1}(\mathbf{x} - \boldsymbol{\mu}_1)$ is called the Mahalanobis squared distance between $\mathbf{x}$ and $\boldsymbol{\mu}_1$, and $D_2^2 = (\mathbf{x} - \boldsymbol{\mu}_2)^T\boldsymbol{\Sigma}^{-1}(\mathbf{x} - \boldsymbol{\mu}_2)$ is the Mahalanobis squared distance from $\mathbf{x}$ to $\boldsymbol{\mu}_2$.

From Eq. (6.8), the classification rule actually is the following:

> If the Mahalanobis distance from $\mathbf{x}$ to $\boldsymbol{\mu}_2$ is larger than that from $\mathbf{x}$ to $\boldsymbol{\mu}_1$, then $\mathbf{x}$ should be classified as in group 1; otherwise, $\mathbf{x}$ should be classified as in group 2.

3. If $\mathbf{x} \sim N(\boldsymbol{\mu}, \boldsymbol{\Sigma})$, then the Mahalanobis squared distance from $\mathbf{x}$ to $\boldsymbol{\mu}$ is $D^2 = (\mathbf{x} - \boldsymbol{\mu})^T \boldsymbol{\Sigma}^{-1} (\mathbf{x} - \boldsymbol{\mu})$.

Recall that the probability density function of the multivariate normal distribution is

$$f(\mathbf{x}) = (2\pi)^{-p/2} |\boldsymbol{\Sigma}|^{-1/2} \exp\left[-\tfrac{1}{2} (\mathbf{x} - \boldsymbol{\mu})^T \boldsymbol{\Sigma}^{-1} (\mathbf{x} - \boldsymbol{\mu})\right]$$

compared with the univariate normal density function

$$f(x) = \frac{1}{\sqrt{2\pi}\sigma} \exp\left[-\frac{1}{2}\left(\frac{x-\mu}{\sigma}\right)^2\right]$$

where $(x - \mu)/\sigma$ is the well-known z score of x. Clearly, the Mahalanobis squared distance D^2 is the equivalent parameter as a squared z score for multivariate normal random variables.

6.3 Linear Discriminant Analysis for Two Normal Populations with Equal Covariance Matrices

In practical application of discriminant analysis, the population multivariate parameters, such as $\boldsymbol{\mu}_1$, $\boldsymbol{\mu}_2$, and $\boldsymbol{\Sigma}$, are not known. In this case, $\boldsymbol{\mu}_1$, $\boldsymbol{\mu}_2$, and $\boldsymbol{\Sigma}$ must be estimated from the initial sample.

Specifically, there are N_1 multivariate observations from group 1 and N_2 observations from group 2. They are $\mathbf{x}_1, \ldots, \mathbf{x}_{N_1}, \mathbf{x}_{N_1+1}, \ldots, \mathbf{x}_{N_1+N_2}$ observations, where the first N_1 observations are from group 1 and the next N_2 observations are from group 2. Let

$$\bar{\mathbf{x}}_1 = \frac{1}{N_1} \sum_{i=1}^{N_1} \mathbf{x}_i \qquad \bar{\mathbf{x}}_2 = \frac{1}{N_2} \sum_{i=N_1+1}^{N_1+N_2} \mathbf{x}_i$$

and

$$\mathbf{S}_1 = \frac{1}{N_1 - 1} \sum_{i=1}^{N_1} (\mathbf{x}_i - \bar{\mathbf{x}}_1)(\mathbf{x}_i - \bar{\mathbf{x}}_1)^T \qquad \mathbf{S}_2 = \frac{1}{N_2 - 1} \sum_{i=N_1+1}^{N_1+N_2} (\mathbf{x}_i - \bar{\mathbf{x}}_2)(\mathbf{x}_i - \bar{\mathbf{x}}_2)^T$$

Clearly, $\bar{\mathbf{x}}_1$ and $\bar{\mathbf{x}}_2$ are statistical estimates of $\boldsymbol{\mu}_1$ and $\boldsymbol{\mu}_2$, and $\mathbf{S}_1$ and $\mathbf{S}_2$ are sample covariance matrices for population 1 and population 2, respectively. The statistical estimate of $\boldsymbol{\Sigma}$ is

$$\mathbf{S}_p = \frac{1}{N_1 + N_2 - 2} [(N_1 - 1)\mathbf{S}_1 + (N_2 - 1)\mathbf{S}_2]$$

By using statistical estimates of $\boldsymbol{\mu}_1$, $\boldsymbol{\mu}_2$, and $\boldsymbol{\Sigma}$ as defined above, we have the following linear discriminant classification rule.

Classification Rule 6.2. Assume that we have two groups of objects. Group 1 objects follow a multivariate normal distribution $N(\mu_1, \Sigma)$, and group 2 objects follow another multivariate normal distribution $N(\mu_2, \Sigma)$.

For any new unknown object $\mathbf{x}$, we develop a linear discriminant function:

$$\mathbf{w'x} = w_1 x_1 + w_2 x_2 + \cdots + w_p x_p = (\bar{\mathbf{x}}_1 - \bar{\mathbf{x}}_2)^T \mathbf{S}_p^{-1} \mathbf{x}$$

and a cutoff score $K = \frac{1}{2}(\bar{\mathbf{x}}_1 - \bar{\mathbf{x}}_2)^T \mathbf{S}_p^{-1}(\bar{\mathbf{x}}_1 + \bar{\mathbf{x}}_2)$ such that if $\mathbf{w'x} > K$, we classify $\mathbf{x}$ into group 1; otherwise, we classify $\mathbf{x}$ into group 2.

Misclassification probability (Srivastava, 2002). The misclassification probabilities for this rule are as follows:

$$P(2|1) \approx \Phi\left(\frac{1}{2}\Delta\right) + \frac{a_1}{N_1} + \frac{a_2}{N_2} + \frac{a_3}{N_1 + N_2 - 2}$$

$$P(1|2) \approx \Phi\left(\frac{1}{2}\Delta\right) + \frac{a_2}{N_1} + \frac{a_1}{N_2} + \frac{a_3}{N_1 + N_2 - 2}$$

where

$$\Delta = \sqrt{(\mu_1 - \mu_2)^T \Sigma^{-1}(\mu_1 - \mu_2)}$$

$$a_1 = \frac{\Delta^2 + 12(p-1)}{16\Delta}\phi(\Delta) \quad a_2 = \frac{\Delta^2 - 4(p-1)}{16\Delta}\phi(\Delta) \quad a_3 = \frac{\Delta}{4}(p-1)\phi(\Delta)$$

where

$$\phi(\Delta) = \frac{1}{\sqrt{2\pi}} e^{-1/2\Delta^2}$$

Example 6.3: Freezer Ownership The following data describes a survey data for freezer ownership of 10 families, where income (x_1) and family size (x_2) are variables, and the freezer ownership is the category variable. It has two exclusive groups—a family either has a freezer or does not have a freezer.

Income (1000 £) (x_1)	Family size (x_2)	Freezer ownership
2.5	1	No
3.0	2	No
4.0	2	No
4.5	5	Yes
5.0	4	Yes
5.5	2	No
6.0	4	Yes
6.5	4	Yes
8.5	2	Yes
10.0	4	Yes

Assume that group 1 corresponds to "no freezer," and group 2 corresponds to "have freezer."

$$S_1 = \frac{1}{N_1-1}\sum_{i=1}^{N_1}(\mathbf{x}_i-\bar{\mathbf{x}}_1)(\mathbf{x}_i-\bar{\mathbf{x}}_1)^T = \begin{bmatrix} 1.75 & 0.417 \\ 0.417 & 0.25 \end{bmatrix}$$

$$S_2 = \frac{1}{N_2-1}\sum_{i=N_1+1}^{N_1+N_2}(\mathbf{x}_i-\bar{\mathbf{x}}_2)(\mathbf{x}_i-\bar{\mathbf{x}}_2)^T = \begin{bmatrix} 4.475 & -1.15 \\ -1.15 & 0.9667 \end{bmatrix}$$

$$S_p = \frac{1}{N_1+N_2-2}[(N_1-1)S_1+(N_2-1)S_2]$$

$$= \frac{1}{8}\left\{3\begin{bmatrix} 1.75 & 0.417 \\ 0.417 & 0.25 \end{bmatrix}+5\begin{bmatrix} 4.475 & -1.15 \\ -1.15 & 0.9667 \end{bmatrix}\right\} = \begin{bmatrix} 3.453 & -0.563 \\ -0.563 & 0.698 \end{bmatrix}$$

The average income for group 1 $= \dfrac{2.5+3.0+4.0+5.5}{4} = 3.75$

The average family size for group 1 $= \dfrac{1+2+2+2}{4} = 1.75$

So, $\bar{\mathbf{x}}_1 = (3.75, 1.75)^T$. Similarly, $\bar{\mathbf{x}}_2 = (5.55, 3.75)^T$.

$$\mathbf{w}'\mathbf{x} = (\bar{\mathbf{x}}_1-\bar{\mathbf{x}}_2)^T S_p^{-1} \mathbf{x} = [(3.75,1.75)-(5.55,3.75)]\begin{bmatrix} 3.453 & -0.563 \\ -0.563 & 0.698 \end{bmatrix}^{-1}\begin{pmatrix} x_1 \\ x_2 \end{pmatrix}$$

$$= (-1.8,-2)\begin{bmatrix} 0.333 & 0.269 \\ 0.269 & 1.649 \end{bmatrix}\begin{pmatrix} x_1 \\ x_2 \end{pmatrix} = -1.137x_1 - 3.782x_2$$

$$K = \frac{1}{2}(\bar{\mathbf{x}}_1-\bar{\mathbf{x}}_2)^T S_p^{-1}(\bar{\mathbf{x}}_1+\bar{\mathbf{x}}_2) = \frac{1}{2}(-1.8,-2)\begin{bmatrix} 0.333 & 0.269 \\ 0.269 & 1.649 \end{bmatrix}\begin{pmatrix} 9.3 \\ 5.5 \end{pmatrix} = -15.687$$

Therefore, the classification rule for this problem is as follows:

If $-1.137x_1 - 3.782x_2 > -15.687$, or $1.137(\text{Income}) + 3.782(\text{family size}) < 15.687$, then classify it as "will not own freezer"; otherwise, classify it as "will own freezer."

If a family has an income of 6000 pounds, with family size 2, then $1.137 \times (6.0) + 3.782 \times (2) = 14.386$, we classify this family as "will not own a freezer."

6.4 Discriminant Analysis for Two Normal Population with Unequal Covariance Matrices

Assume that we have two groups of objects. Group 1 objects follow a multivariate normal distribution $N(\mu_1, \Sigma_1)$, and group 2 objects follow another multivariate normal distribution $N(\mu_2, \Sigma_2)$. Then the following quadratic discriminant function should be used:

$$g(\mathbf{x}) = (\mathbf{x} - \boldsymbol{\mu}_2)^T \Sigma_2^{-1}(\mathbf{x} - \boldsymbol{\mu}_2) - (\mathbf{x} - \boldsymbol{\mu}_1)^T \Sigma_1^{-1}(\mathbf{x} - \boldsymbol{\mu}_1) - \ln\left(\frac{|\Sigma_1|}{|\Sigma_2|}\right)$$

If the population distribution parameters $\boldsymbol{\mu}_1, \boldsymbol{\mu}_2, \Sigma_1, \Sigma_2$ are unknown, then we can replace them with the corresponding statistical estimates. We get the following quadratic discriminant function:

$$g(\mathbf{x}) = (\mathbf{x} - \bar{\mathbf{x}}_2)^T S_2^{-1}(\mathbf{x} - \bar{\mathbf{x}}_2) - (\mathbf{x} - \bar{\mathbf{x}}_1)^T S_1^{-1}(\mathbf{x} - \bar{\mathbf{x}}_1) - \ln\left(\frac{|S_1|}{|S_2|}\right)$$

Classification Rule 6.3. If $g(\mathbf{x}) > 0$, we classify $\mathbf{x}$ into group 1; otherwise, we classify $\mathbf{x}$ into group 2.

6.5 Discriminant Analysis for Several Normal Populations

Assume that there are k groups. For each group i, $i = 1, 2, \ldots, k$, its objects will follow the multivariate normal distribution $\mathbf{N}(\boldsymbol{\mu}_i, \Sigma)$. Also assume that $\bar{\mathbf{x}}_i$ is the sample mean for the initial analysis sample for the ith group, with sample size N_i, for $i = 1, 2, \ldots, k$. We further define

$$\mathbf{S}_p = \frac{1}{N_1 + \cdots + N_k - k}\left[\sum_i (\mathbf{x}_i - \bar{\mathbf{x}}_1)(\mathbf{x}_i - \bar{\mathbf{x}}_1)^T + \cdots + \sum_i (\mathbf{x}_i - \bar{\mathbf{x}}_k)(\mathbf{x}_i - \bar{\mathbf{x}}_k)^T\right]$$

as the pooled sample covariance matrix.

There are two discriminant classification rules. One is based on linear discriminant function and the other is based on the Mahalanobis squared distance.

6.5.1 Linear discriminant classification

Let

$$\mathbf{w}_i = \mathbf{x}' \mathbf{S}_p^{-1} \bar{\mathbf{x}}_i - \tfrac{1}{2} \bar{\mathbf{x}}_i^T \mathbf{S}_p^{-1} \bar{\mathbf{x}}_i$$

be the linear discriminant function for the ith group, for $i = 1, \ldots, k$. Then we have the following rule.

Classification Rule 6.4. Assign $\mathbf{x}$ to group i, if $w_i = \text{Max}(w_1, w_2, \ldots, w_k)$.

6.5.2 Discriminant classification based on the Mahalanobis squared distances

First, for each new object **x**, we compute the following Mahalanobis squared distances:

$$D_i^2 = (\mathbf{x} - \bar{\mathbf{x}}_i)^T \mathbf{S}_p^{-1}(\mathbf{x} - \bar{\mathbf{x}}_i)$$

then for $i = 1,\ldots, k$, we have the following rule.

Classification Rule 6.5. Assign **x** to group i if $D_i^2 = \min\{D_1^2,\ldots, D_k^2\}$

Remark. Rule 6.5 essentially means that we assign a new object to the group, which is closest to it, based on the Mahalanobis distance measure.

Example 6.4: (Continue with Example 6.1) Recall the weight category example (Table 6.1) and assume normal weight as group 1, overweight as group 2, and underweight as group 3. It can be computed as

$$\bar{\mathbf{x}}_1 = (1.73, 67.48), \quad \bar{\mathbf{x}}_2 = (1.72, 82.30), \quad \bar{\mathbf{x}}_3 = (1.72, 52.75)$$

$$\mathbf{S}_p = \begin{bmatrix} 0.012 & 0.9511 \\ 0.9511 & 98.63 \end{bmatrix}$$

So

$$\mathbf{S}_p^{-1} = \begin{bmatrix} 353.55 & -3.409 \\ -3.409 & 0.043 \end{bmatrix}$$

Because $\mathbf{w}_i = \mathbf{x}' \mathbf{S}_p^{-1} \bar{\mathbf{x}}_i - \tfrac{1}{2} \bar{\mathbf{x}}_i^T \mathbf{S}_p^{-1} \bar{\mathbf{x}}_i$, so

$$\mathbf{w}_1 = \mathbf{x}' \mathbf{S}_p^{-1} \bar{\mathbf{x}}_1 - \tfrac{1}{2} \bar{\mathbf{x}}_1^T \mathbf{S}_p^{-1} \bar{\mathbf{x}}_1$$

$$= (x_1, x_2) \begin{bmatrix} 353.55 & -3.409 \\ -3.409 & 0.043 \end{bmatrix} \begin{bmatrix} 1.73 \\ 67.48 \end{bmatrix} - \frac{1}{2}(1.73, 67.48)\begin{bmatrix} 353.55 & -3.409 \\ -3.409 & 0.043 \end{bmatrix}\begin{bmatrix} 1.73 \\ 67.48 \end{bmatrix}$$

$$= 390.68 x_1 - 3.08 x_2 - 233.59$$

$$\mathbf{w}_2 = \mathbf{x}' \mathbf{S}_p^{-1} \bar{\mathbf{x}}_2 - \tfrac{1}{2} \bar{\mathbf{x}}_2^T \mathbf{S}_p^{-1} \bar{\mathbf{x}}_2$$

$$= (x_1, x_2) \begin{bmatrix} 353.55 & -3.409 \\ -3.409 & 0.043 \end{bmatrix} \begin{bmatrix} 1.72 \\ 82.30 \end{bmatrix} - \frac{1}{2}(1.72, 82.30)\begin{bmatrix} 353.55 & -3.409 \\ -3.409 & 0.043 \end{bmatrix}\begin{bmatrix} 1.72 \\ 82.30 \end{bmatrix}$$

$$= 335.23 x_1 - 2.4 x_2 - 189.33$$

$$\mathbf{w}_3 = \mathbf{x}' \mathbf{S}_p^{-1} \bar{\mathbf{x}}_3 - \tfrac{1}{2} \bar{\mathbf{x}}_3^T \mathbf{S}_p^{-1} \bar{\mathbf{x}}_3$$

$$= (x_1, x_2) \begin{bmatrix} 353.55 & -3.409 \\ -3.409 & 0.043 \end{bmatrix} \begin{bmatrix} 1.72 \\ 52.75 \end{bmatrix} - \frac{1}{2}(1.72, 52.75)\begin{bmatrix} 353.55 & -3.409 \\ -3.409 & 0.043 \end{bmatrix}\begin{bmatrix} 1.72 \\ 52.75 \end{bmatrix}$$

$$= 439.75 x_1 - 3.71 x_2 - 280.81$$

Assume that we have two new people. The first one has height = 1.83, weight = 90.0 kg

The second person has height = 1.65, weight = 63 kg. Then for the first person,

$$W_1 = 390.68(1.83) - 3.08(90) - 233.59 = 204.15$$
$$W_2 = 335.23(1.83) - 2.40(90) - 189.33 = 208.14$$
$$W_3 = 439.75(1.83) - 3.71(90) - 280.81 = 190.02$$

Because $W_2 = 208.14$ is larger than both W_1 and W_3, the first person should belong to group 2, which is overweight.

For second person

$$W_1 = 390.68(1.65) - 3.08(63) - 233.59 = 216.99$$
$$W_2 = 335.23(1.65) - 2.40(63) - 189.33 = 212.60$$
$$W_3 = 439.75(1.65) - 3.71(63) - 280.81 = 211.05$$

Because W_1 is larger than both W_2 and W_3, the second person is in the "normal weight" group.

MINITAB will produce the following computer printout for linear discriminant functions:

Discriminant analysis: category versus height, weight

Linear Method for Response: Category

```
Predictors:  Height   Weight
Group           N        O        U
Count          12       12       12

Summary of Classification

Put into       ....True Group....
Group           N        O        U
N              12        2        0
O               0       10        0
U               0        0       12
Total N        12       12       12
N Correct      12       10       12
Proportion   1.000    0.833    1.000

N =   36      N Correct =   34     Proportion Correct = 0.944
```

The above computer printout indicates that there are two misclassification cases when discriminant function is used on the data.

```
Squared Distance Between Groups

              N         O         U
N        0.0000   10.7110    8.8383
O       10.7110    0.0000   38.9859
U        8.8383   38.9859    0.0000

Linear Discriminant Function for Group
              N         O         U
Constant -233.59   -189.33   -280.81
Height    390.68    335.23    439.75
Weight     -3.08     -2.40     -3.71
```

Variable	Pooled Mean	Means for Group		
		N	O	U
Height	1.7228	1.7283	1.7183	1.7217
Weight	67.508	67.475	82.300	52.750

Variable	Pooled StDev	StDev for Group		
		N	O	U
Height	0.1092	0.1113	0.1180	0.0973
Weight	9.931	9.195	12.770	6.947

Pooled Covariance Matrix

	Height	Weight
Height	0.0119	
Weight	0.9511	98.6270

Covariance Matrix for Group N

	Height	Weight
Height	0.012	
Weight	0.950	84.544

Covariance Matrix for Group O

	Height	Weight
Height	0.014	
Weight	1.325	163.071

Covariance Matrix for Group U

	Height	Weight
Height	0.009	
Weight	0.578	48.266

The above computer printout gives the Mahalanobis distances between the centers (means) of each pair of groups; the linear classification functions for all three groups and covariances matrices.

Summary of Misclassified Observations

Observation	True Group	Pred Group	Group	Squared Distance	Probability
14 **	O	N	N	3.382	0.650
			O	4.621	0.350
			U	20.487	0.000
23 **	O	N	N	2.486	0.577
			O	3.110	0.423
			U	20.572	0.000

The above computer printout indicates that there are two misclassification cases, objects 14 and 23, and they are in the overweight group but misclassified as in the normal group.

We can also use the Mahalanobis squared distance to do the classification. MINITAB will generate the following computer output:

Squared Distance Between Groups

	N	O	U
N	0.0000	10.7110	8.8383
O	10.7110	0.0000	38.9859
U	8.8383	38.9859	0.0000

Summary of Classified Observations

Observation	True Group	Pred Group	Group	Squared Distance	Probability
1	N	N	N	3.884	0.943
			O	17.488	0.001
			U	9.543	0.056

2	N	N	N	1.280	0.980
			O	9.657	0.015
			U	11.927	0.005
3	N	N	N	1.241	0.923
			O	6.244	0.076
			U	15.058	0.001
4	N	N	N	0.1569	0.964
			O	13.1849	0.001
			U	6.8417	0.034
5	N	N	N	0.7709	0.922
			O	5.7408	0.077
			U	14.8298	0.001
6	N	N	N	0.5060	0.984
			O	9.5400	0.011
			U	11.0571	0.005
7	N	N	N	4.337	0.767
			O	22.707	0.000
			U	6.723	0.233
8	N	N	N	1.885	0.894
			O	6.162	0.105
			U	16.837	0.001
9	N	N	N	0.8671	0.973
			O	12.9693	0.002
			U	8.1817	0.025
10	N	N	N	1.240	0.791
			O	18.868	0.000
			U	3.905	0.209
11	N	N	N	0.1818	0.963
			O	13.3987	0.001
			U	6.8007	0.035
12	N	N	N	0.7924	0.984
			O	9.7153	0.011
			U	11.4974	0.005
13	O	O	N	7.120	0.230
			O	4.706	0.770
			U	27.359	0.000
14 **	O	N	N	3.382	0.650
			O	4.621	0.350
			U	20.487	0.000
15	O	O	N	18.952	0.000
			O	1.360	1.000
			U	53.324	0.000
16	O	O	N	11.2174	0.004
			O	0.0061	0.996
			U	39.9512	0.000
17	O	O	N	16.5075	0.000
			O	0.6635	1.000
			U	49.5032	0.000
18	O	O	N	26.272	0.000
			O	3.942	1.000
			U	65.346	0.000
19	O	O	N	8.136	0.055
			O	2.463	0.945
			U	32.234	0.000
20	O	O	N	19.731	0.000
			O	3.546	1.000
			U	53.459	0.000
21	O	O	N	29.320	0.000
			O	4.851	1.000
			U	69.888	0.000
22	O	O	N	7.899	0.056
			O	2.261	0.944
			U	31.200	0.000

23 **	O	N	N	2.486	0.577
			O	3.110	0.423
			U	20.572	0.000
24	O	O	N	9.7976	0.011
			O	0.7581	0.989
			U	36.7958	0.000
25	U	U	N	5.414	0.418
			O	26.032	0.000
			U	4.754	0.582
26	U	U	N	7.8122	0.029
			O	35.6716	0.000
			U	0.8107	0.971
27	U	U	N	10.7188	0.005
			O	42.4665	0.000
			U	0.2902	0.995
28	U	U	N	15.926	0.001
			O	52.645	0.000
			U	1.059	0.999
29	U	U	N	2.752	0.379
			O	24.314	0.000
			U	1.762	0.621
30	U	U	N	9.0145	0.015
			O	38.8883	0.000
			U	0.6582	0.985
31	U	U	N	14.358	0.004
			O	47.649	0.000
			U	3.160	0.996
32	U	U	N	4.497	0.174
			O	28.135	0.000
			U	1.384	0.826
33	U	U	N	15.4895	0.001
			O	51.7848	0.000
			U	0.9822	0.999
34	U	U	N	13.9533	0.001
			O	48.9950	0.000
			U	0.8226	0.999
35	U	U	N	12.5440	0.002
			O	46.4377	0.000
			U	0.3506	0.998
36	U	U	N	10.1513	0.008
			O	41.3824	0.000
			U	0.5355	0.992

6.6 Case Study: Discriminant Analysis of Vegetable Oil by Near-Infrared Reflectance Spectroscopy

Near-infrared reflectance (NIR) (Bewig et al., 1994) can be used for rapid quantitative determination of moisture, liquid, protein, carbohydrates and fiber in cereals, grains, feeds, meats, and dairy products. The objective of this project is to use NIR technique to differentiate vegetable oil types and to classify unknown oil samples.

Four types of vegetable oils—peanut, soybean, cottonseed, and canola oil—are used. The near-infrared reflectance technique works in the following fashion.

The near-infrared light emitted from the light source is beamed at oil samples. For different types of oil, the reflectance to the near-infrared light

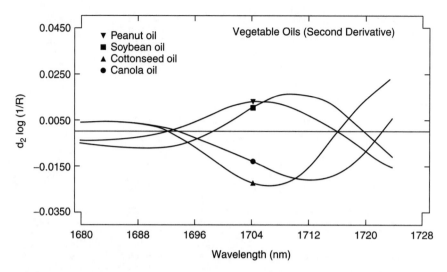

Figure 6.2 The reflectance of vegetable oils for wavelengths from 1680 to 1728 nm.

might be different, and here the reflectance is the fraction of radiant energy that is reflected from the oil sample surface. Log (1/reflectance) is often used as the measure of the magnitude of reflectance. It is found that at different wavelengths of the infrared light signal, the discrepancies in reflectance among these four kinds of oil are very different; see Figs. 6.2–6.4.

Figures 6.2 to 6.4 illustrate the differences in the log (1/reflectance) values, denoted by d_2 log (1/R), for these four types of oils at different near-infrared light wavelengths, because at the wavelengths 1704, 1802,

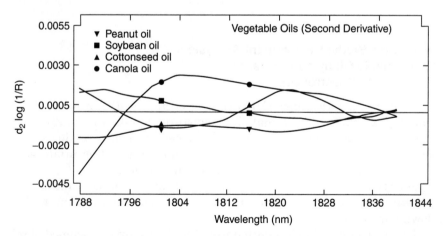

Figure 6.3 The reflectance of vegetable oils for wavelengths from 1788 to 1844 nm.

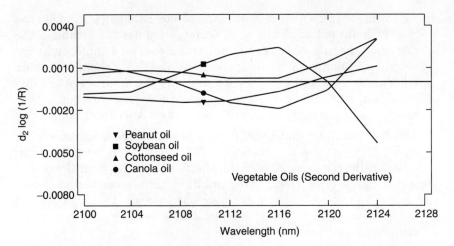

Figure 6.4 The reflectance of vegetable oils for wavelengths from 2100 to 2128 nm.

1816, and 2110 nm, the differences on $d_2 \log (1/R)$ are quite significant. So $d_2 \log (1/R)$ for these four wavelengths will be selected as our four discriminant variables (x_1, x_2, x_3, x_4). Figure 6.5 illustrates that if we use more than one variable as our discriminant criteria, we can classify unknown oil samples with quite a high accuracy.

Therefore, a discriminant analysis based oil sample identification procedure on near-infrared reflectance data is established as follows:

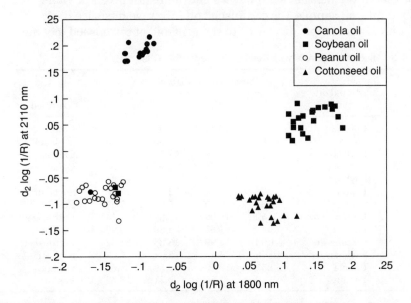

Figure 6.5 Reflectance plot for wavelength of 2110 vs. 1800 nm.

1. An initial 43 known oil samples are used to establish and verify the discriminant rules, of which 30 samples are used to establish the discriminant rule. Nine of them are cottonseed oil sample, eight are peanut oil, another eight are soybean oil, and five of them are canola oil. Thirteen oil samples are used to verify the accuracy of the discriminant rule. Three of them are cottonseed, four are peanut oil, another four are soybean oil and two of them are canola oil.
2. The discriminant variables (x_1, x_2, x_3, x_4) are the measured reflectance d_2 log $(1/R)$, at the wavelengths 1704, 1802, 1816, and 2110 nm. Specifically, for each oil sample, the near-infrared light beams of wavelengths 1704, 1802, 1816, and 2110 nm are beamed at the oil sample and the reflectance at these four wavelengths, (x_1, x_2, x_3, x_4) is recorded. Then we have the initial data set to establish the discriminant analysis rules.
3. In this project, we have four known groups; they are cottonseed, peanut, soybean, and canola oils. Mahalanobis distance discriminant rules are used. Specifically, we compute the pooled sample covariance matrix S_p and centroid vector of each group.
4. For each new sample, the measure of reflectances for the four wavelengths are obtained. Then we compute its Mahalanobis distances to each centroid. It will be classified by the oil type based on the smallest Mahalanobis distance.

For the 13 verification samples, we get the result given in Table 6.3. For example, oil sample 1 is cottonseed oil. After we measure its (x_1, x_2, x_3, x_4), its Mahalanobis distances to the centroid of cottonseed, peanut,

Table 6.3 Discriminant Analysis Results for Vegetable Oils

Sample number	Actual oil type	Mahalanobis distances				Classified as
		C	P	S	CL	
1	Cottonseed	1.49	16.54	17.07	31.08	Cottonseed
2	Soybean	14.58	15.16	13.99	19.76	Soybean
3	Soybean	32.06	18.39	17.74	1.51	Canola*
4	Canola	29.77	16.68	15.91	2.00	Canola
5	Soybean	24.45	12.99	12.14	25.24	Soybean
6	Peanut	18.75	5.86	2.60	17.96	Soybean*
7	Peanut	17.82	2.99	5.07	17.37	Peanut
8	Cottonseed	0.59	16.99	17.39	31.14	Cottonseed
9	Peanut	18.43	1.65	4.89	17.27	Peanut
10	Cottonseed	1.00	17.99	18.35	31.82	Cottonseed
11	Soybean	19.08	5.45	2.04	16.72	Soybean
12	Canola	30.58	17.40	16.53	1.61	Canola
13	Peanut	16.10	1.56	5.21	18.43	Peanut

C = cottonseed, P = peanut, S = soybean, CL = canola.
*Misclassification.

soybean, and canola oils are computed. They are 1.49, 16.54, 17.07, and 31.08. Since the Mahalanobis distance to the cottonseed centroid is the smallest, we classify this oil sample as cottonseed oil. For oil sample 3, it is actually a soybean oil sample. Again, after we measure its (x_1, x_2, x_3, x_4), its Mahalanobis distances to the centroid of cottonseed, peanut, soybean, and canola oils are computed. They are 32.06, 18.39, 17.74, and 1.51. Since its Mahalanobis distance to the centroid of canola oil is the smallest, we classify it as canola oil. Clearly, it is a mistake. In these 13 verification samples, we have two misclassifications. Overall, this discriminant analysis procedure has a reasonable accuracy.

Chapter 7

Cluster Analysis

7.1 Introduction

Cluster analysis is a multivariate statistical method that identifies hidden groups, in a large number of objects, based on their characteristics. Similar to discriminant analysis, each object has multiple characteristics, which can be expressed as a random vector $\mathbf{X} = (X_1, X_2, ..., X_p)$ with values that vary from object to object. The primary objective of cluster analysis is to identify similar objects on the basis of the characteristics they possess. Cluster analysis clusters similar objects into groups, so that the objects within a group are very similar and objects from different groups are significantly different in their characteristics. Unlike discriminant analysis, in which the number of groups and group names are known before the analysis, in cluster analysis, the number of groups and group features are unknown before the analysis, and they are determined after the analysis.

Let us look at the following example.

Example 7.1: Cereal Type Table 7.1 gives the nutritional contents of 25 types of cereals. Certainly, some brands are very similar in terms of nutritional content. Cluster analysis is able to identify hidden groups in the cereal brands, identify the features of each group and compare their differences.

In general, cluster analysis will work on the data given in Table 7.2. Cluster analysis usually consists of the following steps:

Step 1: Select cluster variables and distance measures. In the cereal type problem, Table 7.1 gives several variables about nutrition content. Actually, we can make the number of variables much larger, by adding additional items such as vitamin content. How many and which variables are to be selected will affect the analysis results. In cluster analysis, it is implicitly assumed that every variable is equally important.

TABLE 7.1 Nutrition Contents of Cereals

Brand	Calories (Cal/oz)	Protein (g)	Fat (g)	Na (mg)	Fiber (g)	Carbs (g)	Sugar (g)	K (mg)
Cheerios	110	6	2	290	2.0	17.0	1	105
Cocoa Puffs	110	1	1	180	0.0	12.0	13	55
Honey Nut Cheerios	110	3	1	250	1.5	11.5	10	90
Kix	110	2	1	260	0.0	21.0	3	40
Lucky Charms	110	2	1	180	0.0	12.0	12	55
Oatmeal Raisin Crisp	130	3	2	170	1.5	13.5	10	120
Raisin Nut Bran	100	3	2	140	2.5	10.5	8	140
Total Corn Flakes	110	2	1	200	0.0	21.0	3	35
Total Raisin Bran	140	3	1	190	4.0	15.0	14	230
Trix	110	1	1	140	0.0	13.0	12	25
Wheaties Honey Gold	110	2	1	200	1.0	16.0	8	60
All-Bran	70	4	1	260	9.0	7.0	5	320
Apple Jacks	110	2	0	125	1.0	11.0	14	30
Corn Flakes	100	2	0	290	1.0	21.0	2	35
Corn Pops	110	1	0	90	1.0	13.0	12	20
Mueslix Crispy Blend	160	3	2	150	3.0	17.0	13	160
Nut & Honey Crunch	120	2	1	190	0.0	15.0	9	40
Nutri Grain Almond Raisin	140	3	2	220	3.0	21.0	7	130
Nutri Grain Wheat	90	3	0	170	3.0	18.0	2	90
Product 19	100	3	0	320	1.0	20.0	3	45
Raisin Bran	120	3	1	210	5.0	14.0	12	240
Rice Krispies	110	2	0	290	0.0	22.0	3	35
Special K	110	6	0	230	1.0	16.0	3	55
Life	100	4	2	150	2.0	12.0	6	95
Puffed Rice	50	1	0	0	0.0	13.0	0	15

If this assumption is not true, then we may group variables, or only select a subset of variables that we think are about equally important. Distance measure is a numerical measure indicating how similar or different a pair of objects are. We will discuss distance measure in detail later.

TABLE 7.2 Typical Data Set in Cluster Analysis

Objects	Variables			
	1	2	...	p
1	x_{11}	x_{12}	...	x_{1p}
2	x_{21}	x_{22}	...	x_{2p}
⋮	⋮	⋮	...	⋮
N	x_{N1}	x_{N2}	...	x_{Np}

Step 2: Select cluster algorithm. Cluster algorithm is the procedure to determine "clusters," or "groups." There are two categories of cluster algorithms, hierarchical and nonhierarchical, that we will discuss in detail in Secs. 7.3 and 7.4.

Step 3: Perform cluster analysis. Cluster analysis will determine the cluster structure—specifically, which objects form a cluster, how many clusters, the features of clusters, etc.

Step 4: Interpretation. We need to explain what these clusters mean and how should we name and make sense of these clusters.

In this chapter, we provide an overview of commonly used cluster analysis methods. More comprehensive discussions on cluster analysis can be found in Everitt et al. (2001) and Aldenderfer and Blashfield (1985).

7.2 Distance and Similarity Measures

Cluster analysis clusters similar objects into the same group, and places dissimilar objects into different groups. Obviously, we need to define what is "similar" and how to quantify similarity before clustering can take place.

In cluster analysis, we usually use "distance" to represent how close each pair of objects are. There are different distance measures, and we will discuss some of the most frequently used distance measures in cluster analysis.

7.2.1 Euclidean distance

By referring to the data format in Table 7.2, the Euclidean distance between any two objects, that is, the distance between object i and object k, d_{ik}, is

$$d_{ik} = \sqrt{\sum_{j=1}^{N}(x_{ij} - x_{kj})^2} \qquad (7.1)$$

Some references call this Euclidean distance "ruler distance."

7.2.2 Standardized Euclidean distance

The numerical scale of different variables may vary greatly. If we look at the data in Table 7.1, we can see that the variable "calories" usually has the numerical scale around 100, but the variable "fat" is in the range of 0–2. Clearly, by using the Euclidean distance defined in Eq. (7.1), the distance will be dominated by the few variables with larger numerical scales. If we want each variable to have about an equal importance in deciding the distances between objects, than we need to standardize each variable so that each of them has about the same numerical scale. The standardization of each variable is done by subtracting the mean and dividing by the standard deviation, which is similar to standardizing normally distributed data. Specifically, we will transform each raw data x_{ij} into standardized data z_{ij} by

$$z_{ij} = \frac{x_{ij} - \bar{x}_{.j}}{s_j} \tag{7.2}$$

where

$$\bar{x}_{.j} = \frac{\sum_{k=1}^{N} x_{kj}}{N} \quad \text{and} \quad s_j = \sqrt{\frac{\sum_{k=1}^{N} (x_{kj} - \bar{x}_{.j})^2}{N-1}}$$

then the standardized Euclidean distance between any two objects i and k is

$$d_{ik} = \sqrt{\sum_{j=1}^{p} (z_{ij} - z_{kj})^2} \tag{7.3}$$

Standardized Euclidean distance is also called the Pearson distance.

7.2.3 Manhattan distance (city block distance)

Manhattan distance is defined by

$$d_{ik} = \sum_{j=1}^{p} |x_{ij} - x_{kj}| \tag{7.4}$$

If we use standardized variables, then the standardized Manhattan distance is

$$d_{ik} = \sum_{j=1}^{p} |z_{ij} - z_{kj}| \tag{7.5}$$

In cluster analysis, we need to compute the distances between each pair of objects. This result can be represented by the following distance matrix:

$$\mathbf{D} = \begin{bmatrix} 0 & d_{12} & d_{13} & \cdots & d_{1N} \\ d_{21} & 0 & d_{23} & \cdots & d_{2N} \\ d_{31} & d_{32} & 0 & \cdots & d_{3N} \\ \vdots & \vdots & \vdots & \vdots & \vdots \\ d_{N1} & d_{N2} & d_{N3} & \cdots & 0 \end{bmatrix} \tag{7.6}$$

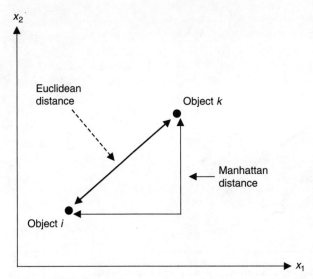

Figure 7.1 Euclidean distance and Manhattan distance between two objects.

Figure 7.1 illustrates the difference between Euclidean distance and Manhattan distance.

7.2.4 Distance between clusters and linkage method

In cluster analysis, it is desirable that the distances between objects within a cluster (group) are small and the distances between different clusters are large, as illustrated in Fig. 7.2. The distance between any

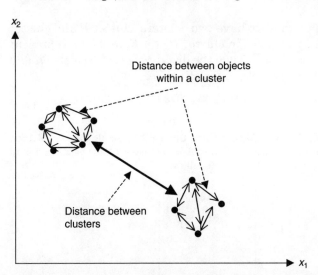

Figure 7.2 Within cluster distance and between cluster distance.

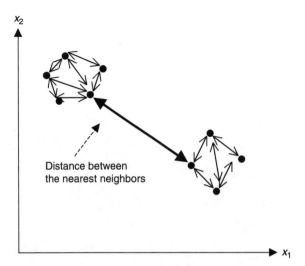

Figure 7.3 The distance between two clusters in single linkage method.

pair of objects has already been defined in the previous subsection. However, what is the distance between clusters? The definition of the distance between clusters depends on how we define the relationship between clusters. The relationship between clusters is called the linkage method. There are several different linkage methods which we will discuss as follows.

Single linkage method. In single linkage method, the distance between two clusters is defined to be the distance of the nearest neighbors, as illustrated in Fig. 7.3.

Specifically, assume that we have two clusters, cluster R and cluster S. Let r represent any element in cluster R, $r \in R$, and let s represent any element in cluster S, $s \in S$. Then the distance between cluster R and cluster S in single linkage method is defined as

$$d_{(R)(S)} = \min\{d_{rs} | r \in R, s \in S\} \tag{7.7}$$

Example 7.2: Distance Between Cluster 1 and Cluster 2 by Single Linkage Method
Assume that cluster 1 has objects 1, 2, and 3, and cluster 2 has objects 4, 5, and 6. A portion of distance matrix is given as follows:

$$\mathbf{D} = \begin{matrix} 1 \\ 2 \\ 3 \\ 4 \\ 5 \\ 6 \end{matrix} \begin{bmatrix} . & . & . & 10 & 8 & 6 \\ . & . & . & 6 & 9 & 5 \\ . & . & 0 & 13 & 11 & 8 \\ . & . & . & 0 & . & . \\ . & . & . & . & 0 & . \\ . & . & . & . & . & . \end{bmatrix}$$

According to Eq. (7.7), the distance between cluster 1 and cluster 2 is

$$d_{(1)(2)} = \min\{d_{14}, d_{15}, d_{16}, d_{24}, d_{25}, d_{26}, d_{34}, d_{35}, d_{36}\}$$
$$= \min\{10, 8, 6, 6, 9, 5, 13, 11, 8\} = 5 = d_{26}$$

Complete linkage method. The complete linkage method is actually the farthest neighbor method, as illustrated in Fig. 7.4. Specifically, assume that we have two clusters, cluster R and cluster S. Let r represent any element in cluster R, $r \in R$, and let s represent any element in cluster S, $s \in S$. Then the distance between cluster R and cluster S in complete linkage method is defined as

$$d_{(R)(S)} = \max\{d_{rs} | r \in R, s \in S\} \qquad (7.8)$$

Example 7.3: Distance Between Cluster 1 and Cluster 2 by Complete Linkage Method In this example, we use the same data as that of Example 7.2, According to Eq. (7.8), the distance between cluster 1 and cluster 2 in complete linkage method is

$$d_{(1)(2)} = \max\{d_{14}, d_{15}, d_{16}, d_{24}, d_{25}, d_{26}, d_{34}, d_{35}, d_{36}\}$$
$$= \max\{10, 8, 6, 6, 9, 5, 13, 11, 8\} = 13 = d_{34}$$

Average link method. In single linkage and complete linkage methods, the distance between two clusters depends on only one pair of objects; either it is the nearest neighbor or the farthest neighbor. In average

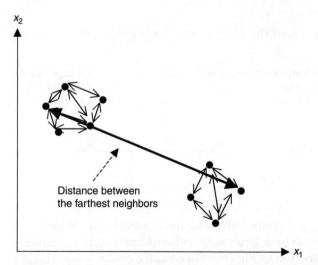

Figure 7.4 The distance between two clusters in complete linkage method.

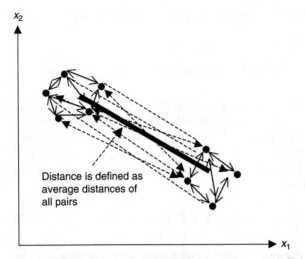

Figure 7.5 The distance between two clusters in average linkage method.

linkage method, all the distances between any pair of neighbors between these two clusters are used. The average of all these distances are used as the distance between the two clusters, as illustrated in Fig. 7.5. Specifically, the distance between cluster R and cluster S in average linkage method is defined as

$$d_{(R)(S)} = \frac{\sum_r \sum_s d_{rs}}{n_R n_S} \tag{7.9}$$

where n_R and n_S represent the number of objects in clusters R and S.

Example 7.4: Distance Between Cluster 1 and Cluster 2 by Average Linkage Method
In this example, we use the same data as that of Examples 7.2 and 7.3. According to Eq. (7.9), the distance between cluster 1 and cluster 2 in the average linkage method is

$$d_{(1)(2)} = \frac{d_{14} + d_{15} + d_{16} + d_{24} + d_{25} + d_{26} + d_{34} + d_{35} + d_{36}}{3 \times 3}$$

$$= \frac{10 + 8 + 6 + 6 + 9 + 5 + 13 + 11 + 8}{9} = 8.44$$

Centroid method. Again assume that cluster R contains n_R objects and cluster S contains n_S objects. The centroid is the gravitational center of each cluster, as illustrated in Fig. 7.6. Specifically, centroids for the two

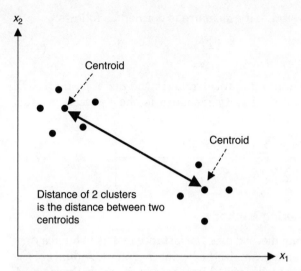

Figure 7.6 The distance between two clusters in the centroid method.

clusters then are

$$\overline{\mathbf{x}}_R = \frac{\sum_r \mathbf{x}_r}{n_R} = \begin{bmatrix} \overline{\mathbf{x}}_{r1} \\ \overline{\mathbf{x}}_{r2} \\ \vdots \\ \overline{\mathbf{x}}_{rp} \end{bmatrix} \quad \text{and} \quad \overline{\mathbf{x}}_S = \frac{\sum_s \mathbf{x}_s}{n_S} = \begin{bmatrix} \overline{\mathbf{x}}_{s1} \\ \overline{\mathbf{x}}_{s2} \\ \vdots \\ \overline{\mathbf{x}}_{sp} \end{bmatrix} \quad (7.10)$$

The distance between the two clusters will be

$$d_{(R)(S)} = \sqrt{(\overline{\mathbf{x}}_{r1} - \overline{\mathbf{x}}_{s1})^2 + \cdots + (\overline{\mathbf{x}}_{rp} - \overline{\mathbf{x}}_{sp})^2} \quad (7.11)$$

if Euclidean distance is used.

7.2.5 Similarity

Similarity is a measure of similarity between two objects or two clusters. Its maximum value is 1 (or 100%) and its minimum value is 0. The larger the similarity value, the more similar the two objects or clusters are. Higher similarity value indicates smaller distance. Specifically, given two objects $\mathbf{x}_r$ and $\mathbf{x}_s$, a similarity measure between $\mathbf{x}_r$ and $\mathbf{x}_s$ is denoted by s_{rs} which satisfies the following conditions:

1. $0 \leq s_{rs} \leq 1$
2. $s_{rs} = 1$ if and only if $\mathbf{x}_r = \mathbf{x}_s$
3. $s_{rs} = s_{sr}$

One commonly used similarity measure is defined as follows:

$$s_{rs} = 1 - \frac{d_{rs}}{d_{max}} \quad (7.12)$$

where d_{max} is the maximum distance value in the distance matrix **D**. Another commonly used similarity measure is the Pearson product moment correlation defined as

$$q_{rs} = \frac{\sum_{j=1}^{p}(x_{rj} - \bar{x}_{r.})(x_{sj} - \bar{x}_{s.})}{\left[\sum_{j=1}^{p}(x_{rj} - \bar{x}_{r.})^2 \sum_{j=1}^{p}(x_{sj} - \bar{x}_{s.})^2\right]} \quad (7.13)$$

7.3 Hierarchical Clustering Method

The hierarchical clustering method uses the distance matrix to build a tree-like diagram, called a dendogram. In the beginning, all individual objects are considered as clusters with one object, that is, itself. If there are N objects, we will have N clusters. Then, the two objects with the closest distance are selected and combined into a single cluster. Now the number of clusters are changed from n to $N - 1$. Then we compute the distances between other objects to the newly formed cluster. This distance calculation is based on one of the linkage methods, such as single linkage, complete linkage, etc. After this, we update the distance matrix. Now we have $n - 1$ clusters, so the new distance matrix will be a $(N-1) \times (N-1)$ matrix. We will continue this process. At each step of this process, the number of clusters is reduced by 1. Finally, all objects are combined into one cluster. Their mutual relationship is expressed by the dendogram. Then we examine the shape of dendogram and decide how many clusters are there in the whole population and which objects should be included in each cluster. Specifically, we have the following step-by-step procedure.

Step 0. Establish distance measure from object to object, then compute the distance matrix **D**, where

$$\mathbf{D} = \begin{bmatrix} 0 & d_{12} & d_{13} & \cdots & d_{1N} \\ d_{21} & 0 & d_{23} & \cdots & d_{2N} \\ d_{31} & d_{32} & 0 & \cdots & d_{3N} \\ \vdots & \vdots & \vdots & \cdots & \vdots \\ d_{N1} & d_{N2} & d_{N3} & \cdots & 0 \end{bmatrix}$$

Now, we have N clusters, each containing only one object.

Step 1. Find the minimum distance in the distance matrix **D**. Assume that the distance from object r to object s is, d_{rs}. Then object r and s are selected to form a single cluster (r, s).

Step 2. Now delete the rows and columns corresponding to object r and s in **D**. Then add a new row and column corresponding to cluster (r, s). So the numbers of rows and columns of **D** are reduced by 1. Compute the distance from other objects to cluster (r, s), by using one of the linkage methods, using these new distances to fill the row and column corresponding to cluster (r, s), we now have a new **D** matrix.

Step 3. Repeat steps 1 and 2 $N-1$ times until all objects form a single cluster. At each step, record merged clusters and the value of distances at dendogram.

We will use the following example to illustrate the hierarchical clustering method.

Example 7.5: February's Weather in a Small Town Table 7.3 contains a small town's weather for February from 1982 to 1993. The meaning of each variable is quite obvious. Can we find clusters in weather pattern, that is, are there any subgroups such that the weather pattern in those years in each subgroup are quite similar?

Step 0: Establish distance matrix. In this problem, because the variables are in different units and numerical scale, the standardized Euclidean distance is selected. That is, for each individual variable, we subtract the mean and divide by standard deviation. We get the following distance matrix:

$$D = \begin{bmatrix} 0.00 & 2.44 & 1.90 & 1.86 & 5.32 & 1.82 & 2.70 & 2.56 & 4.48 \\ 2.44 & 0.00 & 1.92 & 1.49 & 3.31 & 2.16 & 2.98 & 3.69 & 5.23 \\ 1.90 & 1.92 & 0.00 & 2.45 & 4.98 & 1.25 & 1.57 & 2.36 & 3.78 \\ 1.86 & 1.48 & 2.45 & 0.00 & 3.75 & 2.19 & 3.64 & 4.01 & 5.84 \\ 5.32 & 3.31 & 4.98 & 3.75 & 0.00 & 4.73 & 5.90 & 6.71 & 8.13 \\ 1.82 & 2.16 & 1.25 & 2.19 & 4.73 & 0.00 & 2.40 & 3.10 & 4.59 \\ 2.70 & 2.98 & 1.57 & 3.64 & 5.90 & 2.40 & 0.00 & 1.57 & 2.42 \\ 2.56 & 3.69 & 2.36 & 4.01 & 6.71 & 3.10 & 1.57 & 0.00 & 2.05 \\ 4.48 & 5.23 & 3.78 & 5.83 & 8.13 & 4.59 & 2.42 & 2.05 & 0.00 \end{bmatrix}$$

TABLE 7.3 February's Weather in a Small Town from 1982 to 1993

Year	x_1, Mean temp.	x_2, Max. temp.	x_3, Min. temp.	x_4, Soil temp. (@ 10 cm)	x_5, Monthly rainfall (mm)	x_6, Max. rain in a day	x_7, Days with snow
1982	4.2	13.3	−5.3	4.0	23	6	0
1983	1.0	7.8	−5.3	3.0	34	11	8
1984	2.9	11.4	−5.1	3.2	65	17	0
1985	1.6	10.2	−6.0	2.9	7	2	5
1986	−1.1	2.7	−9.0	1.5	22	5	24
1987	3.3	13.4	−7.3	2.7	46	15	2
1988	4.5	13.0	−2.9	3.7	89	22	4
1989	5.7	13.5	−2.7	5.2	92	16	0
1990	6.6	14.9	−0.6	5.5	131	29	0

Step 1: Find the minimum distance and create a new cluster. The smallest distance (nonzero distance) is the distance between object 3 and object 6 (year 1884 and 1987), $d_{3,6} = 1.25$; so object (3, 6) will be combined into a single cluster.

Step 2: Update the distances. Then we need to compute the distance between other objects to the cluster (3, 6). If we use the single linkage method, then

$$d_{1,(3,6)} = \min(d_{13}, d_{16}) = \min(1.90, 1.82) = 1.82$$
$$d_{2,(3,6)} = \min(d_{23}, d_{26}) = \min(1.92, 2.16) = 1.92$$
$$d_{4,(3,6)} = \min(d_{43}, d_{46}) = \min(2.45, 2.19) = 2.19$$
$$d_{5,(3,6)} = \min(d_{53}, d_{56}) = \min(4.98, 4.73) = 4.73$$
$$d_{7,(3,6)} = \min(d_{73}, d_{76}) = \min(1.57, 2.40) = 1.57$$
$$d_{8,(3,6)} = \min(d_{83}, d_{86}) = \min(2.36, 3.10) = 2.36$$
$$d_{9,(3,6)} = \min(d_{93}, d_{96}) = \min(3.78, 4.59) = 3.78$$

Step 3: Repeat steps 1 and 2 and establish dendogram. We illustrate step 3 by showing the following MINITAB steps for this example. Here is a portion of the MINITAB output.

```
Standardized Variables, Euclidean Distance, Single Linkage
Amalgamation Steps
```

Step	Number of clusters	Similarity level	Distance level	Clusters joined		New cluster	Number of obs. in new cluster
1	8	84.67	1.246	3	6	3	2
2	7	81.71	1.486	2	4	2	2
3	6	80.70	1.569	7	8	7	2
4	5	80.64	1.573	3	7	3	4
5	4	77.63	1.818	1	3	1	5
6	3	77.12	1.860	1	2	1	7
7	2	74.83	2.046	1	9	1	8
8	1	59.28	3.309	1	5	1	9

Clearly, at first, because the minimum distance in the original distance matrix is $d_{36} = 1.246$, so objects 3 and 6 are joined as a new cluster 3. After updating the distance matrix, the remaining minimum distance is $d_{24} = 1.486$. So objects 2 and 4 are joined to form a new cluster, called cluster 2. Again the distance matrix is updated. The next cluster is formed by joining objects 7 and 8, because $d_{78} = 1.569$ is the smallest. Then, after updating the distance matrix, a new cluster is formed by joining cluster 3 (objects 3 and 6), and cluster 7 (objects 7 and 8), where the distance between cluster 3 and cluster 7 is

$$d_{(3)(7)} = \min(d_{7,(3,6)}, d_{8,(3,6)}) = d_{7,(3,6)} = 1.573$$

And the process continues until all the objects are combined into a single cluster and the resulting dendogram is as shown in Fig. 7.7.

In a dendogram, the distances between clusters and the joining process are described very well. Of course, we usually do not want to form just one single cluster, we need to "cut" the dendogram to get several clusters. As we discussed earlier, a good clustering should be as follows:

1. The objects within a cluster should be similar, in other words, the distances between the objects within a cluster should be smaller.
2. The objects from different clusters should be dissimilar, or the distances between them should be large.

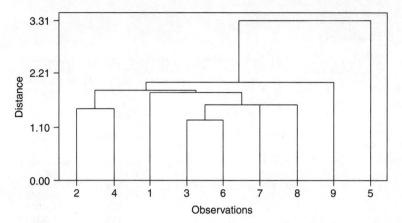

Figure 7.7 The distances between clusters and the joining process in a dendogram.

We may use visual observation, as in the above dendogram. We may think "four clusters may be good" [(2, 4) is a cluster, (1, 3, 6, 7, 8) is a cluster, (9) is a cluster, and (5) is a cluster]. Or we can try to "cut" the clusters by distance measures or the similarity measure. Here we use the similarity measure as defined by Eq. (7.12), that is, $s_{rs} = 1 - d_{rs}/d_{max}$. For example, if we cut the cluster by 80% similarity, that is, all objects within a cluster have similarity level of 80% or higher, then we have five clusters. By this cutting, it seems that there are too many clusters (Fig. 7.8).

If we cut off the cluster by 75% similarity, then we get three clusters. So how many clusters should we cut? It also depends on whether we can give good interpretation and characterization to clusters. For example, if we cut four clusters, by MINITAB, we have

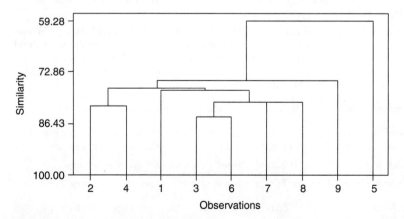

Figure 7.8 The dendogram based on distance.

194   Chapter Seven

```
Final Partition

Number of clusters:   4
```

	Number of observations	Within cluster sum of squares	Average distance from centroid	Maximum distance from centroid
Cluster1	5	9.634	1.358	1.707
Cluster2	2	1.104	0.743	0.743
Cluster3	1	0.000	0.000	0.000
Cluster4	1	0.000	0.000	0.000

Cluster Centroids

Variable	Cluster1	Cluster2	Cluster3	Cluster4	Grand centrd
x1	0.3863	-0.7836	-1.7792	1.4150	0.0000
x2	0.4686	-0.5596	-2.2120	0.9880	-0.0000
x3	0.0990	-0.2913	-1.6120	1.6997	0.0000
x4	0.1901	-0.4575	-1.6168	1.5812	0.0000
x5	0.1579	-0.8837	-0.8469	1.8245	-0.0000
x6	0.1771	-0.8275	-1.0007	1.7705	0.0000
x7	-0.4621	0.2224	2.4827	-0.6171	-0.0000

Distances Between Cluster Centroids

	Cluster1	Cluster2	Cluster3	Cluster4
Cluster1	0.0000	2.3580	5.3978	3.3416
Cluster2	2.3580	0.0000	3.4569	5.4898
Cluster3	5.3978	3.4569	0.0000	8.1279
Cluster4	3.3416	5.4898	8.1279	0.0000

It looks like the within–cluster distances are all smaller than 2 (standardized Euclidean distance), the between–cluster distances are all greater than 2 (from centroid to centroid), and the centroid vectors are significantly different from cluster to cluster.

Now we try to explain the features of each cluster:

1. Object 9 (year 1990) is cluster 4. We see that in 1990, rainfall is very high: it rains every day and it is warm. This is a very unique year, and we can call it a warm and rainy February.

2. Object 5 (year 1986) is cluster 3. We see that in 1986, it is very cold, and there are 24 snow days! This is also very unique. We can call it a cold snowy February.

3. Objects 1, 3, 6, 7, 8 (year 1982, 1984, 1987, 1988, 1989) form cluster 1. We can call it a typical February. However, what is the difference between cluster 1 and cluster 2 (year 1983 and 1985)? Actually, it is very difficult to describe their differences. So they might be merged into a single cluster. Then a three-cluster partition seems to make more sense.

Therefore, finally, we will have a three-cluster partition as follows:

Cluster 1: (Year 1982, 1983, 1984, 1985, 1987, 1988, 1989).   This cluster is named as typical February, that is, not too cold, not too warm, average rain/snow.

Cluster 2: (Year 1986).   It can be named as warm and rainy February.

Cluster 3: (Year 1990).   It can be named as cold and snowy February.

The following are MINITAB printout for a three-cluster partition for this example:

Final Partition

Number of clusters: 3

	Number of observations	Within cluster sum of squares	Average distance from centroid	Maximum distance from centroid
Cluster1	7	18.681	1.568	2.243
Cluster2	1	0.000	0.000	0.000
Cluster3	1	0.000	0.000	0.000

Cluster Centroids

Variable	Cluster1	Cluster2	Cluster3	Grand centrd
x1	0.0520	-1.7792	1.4150	0.0000
x2	0.1749	-2.2120	0.9880	-0.0000
x3	-0.0125	-1.6120	1.6997	0.0000
x4	0.0051	-1.6168	1.5812	0.0000
x5	-0.1397	-0.8469	1.8245	-0.0000
x6	-0.1100	-1.0007	1.7705	0.0000
x7	-0.2665	2.4827	-0.6171	-0.0000

Distances Between Cluster Centroids

	Cluster1	Cluster2	Cluster3
Cluster1	0.0000	4.8054	3.9309
Cluster2	4.8054	0.0000	8.1279
Cluster3	3.9309	8.1279	0.0000

7.4 Nonhierarchical Clustering Method (K-Mean Method)

In hierarchical clustering the number of clusters is unknown. The process is initiated with a distance matrix, and once an object is assigned to a cluster it is never changed. In nonhierarchical clustering method, the analyst has to specify the number of clusters first. If the analyst chooses k clusters, then either the analyst or the computer package provides k initial seeds, or the seeds are initial cluster centroids for these k clusters. After that, the other objects are assigned to the clusters whose centroid is closest in terms of the distance measure. Then the centroids are recalculated based on added objects and the objects that have long distances to their centroids are reassigned.

Specifically, the nonhierarchical clustering method will follow the following three-step process:

Step 1. Start with k initial seeds, each of them as initial clusters. Compute their centroids.

Step 2. Compute the distance of each object (which includes the objects that have already been assigned to their clusters), to the centroid of each cluster. Assign the object to the cluster which is closest to it. Reassign the object if necessary.

Step 3. Recompute the centroid based on reassignment and repeat step 2. Stop if no objects can be reassigned.

Some of the typical difficulties with the nonhierarchical clustering method include the following:

1. The composition of the clusters is very sensitive to the initial seeds selection. For different seeds, we could get very different kinds of clusters.
2. Sometimes it is very difficult to make a good choice on the number of clusters before analyzing the data.

However, hierarchical and nonhierarchical clustering methods may be combined to identify the seeds and number of clusters. Then the results can be input into nonhierarchical clustering method to refine the cluster solution.

Example 7.6: Weather Pattern Problem Revisited For the same data set, if we choose the number of clusters $k = 3$, and without specifying the seeds, then MINITAB will come up with the following results:

Cluster 1. 1982, 1984, and 1987

Cluster 2. 1983, 1985, and 1986

Cluster 3. 1988, 1989, and 1990

K-means cluster analysis: x_1, x_2, x_3, x_4, x_5, x_6, x_7

Standardized Variables

Final Partition

Number of clusters: 3

	Number of observations	Within cluster sum of squares	Average distance from centroid	Maximum distance from centroid
Cluster1	3	2.829	0.959	1.170
Cluster2	3	9.071	1.673	2.305
Cluster3	3	4.175	1.162	1.401

Cluster Centroids

Variable	Cluster1	Cluster2	Cluster3	Grand centrd
x1	0.1152	-1.1154	1.0002	0.0000
x2	0.4109	-1.1104	0.6995	-0.0000
x3	-0.3899	-0.7316	1.1214	0.0000
x4	-0.1777	-0.8439	1.0216	0.0000
x5	-0.2914	-0.8714	1.1628	-0.0000
x6	-0.1155	-0.8853	1.0007	0.0000
x7	-0.5310	0.9759	-0.4449	-0.0000

Distances Between Cluster Centroids

	Cluster1	Cluster2	Cluster3
Cluster1	0.0000	2.7549	2.8208
Cluster2	2.7549	0.0000	4.9375
Cluster3	2.8208	4.9375	0.0000

Clearly, this is very different from what we get from hierarchical clustering method. If we use the result from hierarchical clustering as the seed, that is,

Cluster 1. Year 1982, 1983, 1984, 1985, 1987, 1988, 1989
Cluster 2. Year 1986
Cluster 3. Year 1990

Then MINITAB will give the following final clustering result:

Cluster 1. Year 1982, 1983, 1984, 1985, and 1987 (average)
Cluster 2. Year 1986 (cold and snowy)
Cluster 3. Year 1988, 1989, and 1990 (warm and rainy)

K-means cluster analysis: $x_1, x_2, x_3, x_4, x_5, x_6, x_7$

```
Standardized Variables

Final Partition

Number of clusters:   3

                                Within cluster      Average         Maximum
                Number of       sum of              distance        distance
                observations    squares             from centroid   from centroid
Cluster1        5               7.849               1.251           1.314
Cluster2        1               0.000               0.000           0.000
Cluster3        3               4.175               1.162           1.401

Cluster Centroids

Variable        Cluster1        Cluster2        Cluster3        Grand centrd
x1              -0.2443         -1.7792          1.0002          0.0000
x2               0.0227         -2.2120          0.6995         -0.0000
x3              -0.3504         -1.6120          1.1214          0.0000
x4              -0.2896         -1.6168          1.0216          0.0000
x5              -0.5283         -0.8469          1.1628         -0.0000
x6              -0.4003         -1.0007          1.0007          0.0000
x7              -0.2296          2.4827         -0.4449         -0.0000

Distances Between Cluster Centroids

                Cluster1        Cluster2        Cluster3
Cluster1        0.0000          4.3037          3.2805
Cluster2        4.3037          0.0000          6.8739
Cluster3        3.2805          6.8739          0.0000
```

7.5 Cereal Brand Case Study

Now let us apply cluster analysis to the cereal-type data (Table 7.1) that we introduced at the beginning of this chapter. If we use standardized Euclidean distance and average linkage method, MINITAB will create the following computer output and dendogram (Fig 7.9):

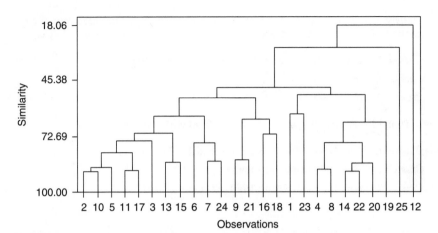

Figure 7.9 Cluster analysis of cereal data.

Cluster analysis of observations: calories, protein, fat, Na, fiber, carbs, suga

```
Standardized Variables, Euclidean Distance, Average Linkage

Amalgamation Steps

Step  Number of  Similarity  Distance  Clusters  New      Number of obs.
      clusters   level       level     joined    cluster  in new cluster
  1     24        90.22       0.752    14   22    14           2
  2     23        90.16       0.757     2   10     2           2
  3     22        89.50       0.807    11   17    11           2
  4     21        89.03       0.844     4    8     4           2
  5     20        88.02       0.921     2    5     2           3
  6     19        86.00       1.077    14   20    14           3
  7     18        85.34       1.128    13   15    13           2
  8     17        85.05       1.150     7   24     7           2
  9     16        84.28       1.209     9   21     9           2
 10     15        80.65       1.489     2   11     2           5
 11     14        76.11       1.837     4   14     4           5
 12     13        75.86       1.856     6    7     6           3
 13     12        74.93       1.928     2    3     2           6
 14     11        71.89       2.162    16   18    16           2
 15     10        71.44       2.196     2   13     2           8
 16      9        66.00       2.615     4   19     4           6
 17      8        64.20       2.753     9   16     9           4
 18      7        62.72       2.867     2    6     2          11
 19      6        61.99       2.923     1   23     1           2
 20      5        53.68       3.563     2    9     2          15
 21      4        52.21       3.675     1    4     1           8
 22      3        48.39       3.970     1    2     1          23
 23      2        28.93       5.466     1   25     1          24
 24      1        18.06       6.302     1   12     1          25
```

First, if we cut off at similarity level of higher than 65%, we will get more than nine clusters, which will be difficult to explain and characterize. Then we set the cutoff similarity score at 60% and we get six clusters, which are as follows:

Cluster 1. Cheerois and Special *K*

Cluster 2. Cocoa Puffs, Honey Nut Cheerios, Lucky Charms, Oatmeal Raisin Crisp Raisin Nut Bran, Trix, Wheaties Honey Gold, Apple Jacks, Corn Pops, Nut & Honey Crunch, Life

Cluster 3. Kix, Total Corn Flakes, Corn Flakes, Nutri-Grain Wheat, Product 19, Rice Krispies

Cluster 4. Total Raisin Bran, Mueslix Crispy Blend, Nutri-Grain Almond Raisin, Raisin Bran

Cluster 5. All-Bran

Cluster 6. Puffed Rice

The relevant MINITAB is as follows:

```
Final Partition
Number of clusters:    6

                                 Within cluster       Average           Maximum
                   Number of        sum of            distance          distance
                  observations     squares          from centroid     from centroid
Cluster1              2             4.273              1.462             1.462
Cluster2             11            28.222              1.526             2.230
Cluster3              6             9.803              1.214             2.023
Cluster4              4             9.280              1.498             1.839
Cluster5              1             0.000              0.000             0.000
Cluster6              1             0.000              0.000             0.000

Cluster Centroids

Variable      Cluster1    Cluster2    Cluster3    Cluster4    Cluster5
calories       0.0188      0.0614     -0.2941      1.4267     -1.8585
Protein        2.5266     -0.3791     -0.2638      0.2435      1.0045
Fat            0.1053      0.2251     -0.7726      0.7638      0.1053
Na             0.9001     -0.4318      0.8300     -0.0463      0.9001
Fiber         -0.0973     -0.3625     -0.4214      0.9968      3.5497
Carbs          0.2979     -0.6499      1.2908      0.3599     -2.0603
Sugar         -1.1714      0.6429     -1.0268      0.8894     -0.5206
K             -0.1368     -0.3127     -0.5669      1.2825      2.9598

Variable      Cluster6   Grand centrd
calories      -2.7972      0.0000
Protein       -1.2785     -0.0000
Fat           -1.2115      0.0000
Na            -2.7453     -0.0000
Fiber         -0.8266      0.0000
Carbs         -0.5709     -0.0000
Sugar         -1.6053     -0.0000
K             -0.9754      0.0000

Distances Between Cluster Centroids

           Cluster1    Cluster2    Cluster3    Cluster4    Cluster5
Cluster1    0.0000      3.8112      3.1554      3.9983      5.8919
Cluster2    3.8112      0.0000      3.0577      2.8572      6.0620
Cluster3    3.1554      3.0577      0.0000      4.0394      6.6727
Cluster4    3.9983      2.8572      4.0394      0.0000      5.4661
Cluster5    5.8919      6.0620      6.6727      5.4661      0.0000
Cluster6    6.2936      4.7023      4.9399      6.8453      7.6909
```

```
              Cluster6
Cluster1      6.2936
Cluster2      4.7023
Cluster3      4.9399
Cluster4      6.8453
Cluster5      7.6909
Cluster6      0.0000
```

From the centroid of each cluster, it seems that we can characterize the clusters as follows:

Cluster 1. High protein and low sugar cereals

Cluster 2. Common cereals

Cluster 3. Low sugar hydrocarbons

Cluster 4. High energy, high potash, medium fiber, and sugar

Cluster 5. Light and low hydrocarbon, fiber loaded, and high potash

Cluster 6. Just rice, low on everything

We can refine these six clusters by using the above clustering as the seeds for a K-means analysis. However, another MINITAB run by using K-means yields the same clustering classification, and of course, the same centroids and interpretations of the clusters, as we obtained by using the hierarchical cluster analysis that we just conducted.

Chapter

8

Mahalanobis Distance and Taguchi Method

8.1 Introduction

Dr. Genichi Taguchi is a well-known Japanese quality expert who pioneered the idea of robust engineering (Taguchi and Wu, 1979; Taguchi, 1993). The robust engineering method is a comprehensive quality strategy that builds robustness into a product/process during its design stage. Robust engineering is a combination of engineering design principles and Taguchi's version of design of experiments, which are called orthogonal array experiments. Taguchi and coworkers [Taguchi (1981), Taguchi, Elsayed, and Hsiang (1989)] also developed a system of online quality control methods, which is featured by quality loss function and Taguchi tolerance design. The system of robust engineering, Taguchi online quality control methods, and other methods developed by Taguchi are usually called the Taguchi method.

The recent addition to the Taguchi method is the Mahalanobis-Taguchi system (MTS). (Taguchi et al., 2001; Taguchi and Jugulum, 2002). The Mahalanobis-Taguchi system is a multivariate data-based pattern recognition and diagnosis system. A significant number of case studies have been recorded, based on the MTS, which range from health diagnosis, electronic manufacturing, chemical industry, fire prevention, etc. There are more than 70 published case applications in a wide variety of companies such as Nissan, Fuji, Xerox, Ford, Sony, Delphi, Yamaha, Minolta, Konica, and several medical firms.

In this approach, a large number of multivariate data sets are collected, and then the data sets for healthy or normal groups are partitioned from other data sets for abnormal or unhealthy groups. For example, in liver disease diagnosis, for each potential patient, many kinds of medical tests are conducted. Therefore, for each patient, a complete set of test records is a multivariate data set. In the MTS approach, the test records of a large number of people are collected. The people who are known to be healthy are classified in "healthy groups"; their test data sets are used to create a baseline metric for a healthy population. This baseline metric for the healthy population is based on the Mahalanobis distance. In MTS, this Mahalanobis distance is scaled so that the average length for the Mahalanobis distance is approximately equal to 1 for the known healthy group. Taguchi calls this distance a unit space. For the objects in the abnormal group, the scaled Mahalanobis distances are supposed to be significantly larger than 1. In the MTS approach, it is recommended that people start with a large number of variables in each multivariate data set so that the chance for the multivariate data set to contain important variables is good. The important variables in this case are those variables which make the scaled Mahalanobis distance large for abnormal objects. Then Taguchi's orthogonal array experiment is used to screen this initial variable set for important variables. After the orthogonal experiment, a smaller set of important variables is selected as the monitoring variables to detect future abnormal conditions. In the liver disease example, we may start with 40 to 60 kinds of tests. After a large number of patients' test data are collected and the baseline Mahalanobis distance is established for the healthy group, orthogonal array experiments are conducted to screen these 40 to 60 variables. After experimental data analysis, only important variables, that is, the variables which make Mahalanobis distance large for the abnormal group, are selected in future liver disease tests. We may end up with 5 to 10 variables (tests) after this selection. These 5 to 10 tests will be selected as the future key diagnostic tests for liver disease.

Taguchi also applied a similar approach to screen important variables for quality monitoring, employee screening tests selection, and many others.

However, Taguchi's MTS approach has been debated by many people in the statistical community (Woodall et al., 2003), and alternative approaches have been proposed.

In this chapter we will give a full description of MTS. Examples and case studies are provided. Comments and proposed alternative approaches by other researchers are also presented.

8.2. Overview of the Mahalanobis-Taguchi System (MTS)

The Mahalanobis-Taguchi system is a four-stage process. The first three stages are baseline Mahalanobis space creation, abnormal group testing, and variable screening. Stage 4 is the establishment of a threshold

TABLE 8.1 Raw Data Format in the Mahalanobis-Taguchi System

Objects	Variables (characteristics)					
	X_1	X_1	...	X_i	...	X_{p-1}	X_p
1	x_{11}	x_{12}	...	x_{1i}	...	$x_{1,p-1}$	x_{1p}
2	x_{21}	x_{22}	...	x_{2i}	...	$x_{2,p-1}$	x_{2p}
⋮	⋮	⋮	...	⋮	...	⋮	⋮
k	x_{k1}	x_{k2}	...	x_{ki}	...	$x_{k,p-1}$	x_{kp}
⋮	⋮	⋮	...	⋮	...	⋮	⋮
N	x_{Nk}	x_{N2}	...	x_{Ni}	...	$x_{N,k-1}$	x_{Np}
Average	$\bar{X}_1$	$\bar{X}_2$	...	$\bar{X}_i$	...	$\bar{X}_{p-1}$	$\bar{X}_p$
Standard deviation	s_1	s_2	...	s_i	...	s_{p-1}	s_p

by the quality loss function and the establishment of a future data monitoring system. We are going to describe these four stages in detail.

8.2.1 Stage 1: Creation of a baseline Mahalanobis space

In this stage, raw multivariate data are collected for known healthy or normal objects. The purpose of this stage is to create a baseline measurement scale for the healthy population. For example, in medical diagnosis, we collect test data of many known healthy people. In general, the raw data that we collect has the format illustrated in Table 8.1.
Here

$$\bar{X}_i = \frac{1}{N}\sum_{k=1}^{N} x_{ki} \quad \text{and} \quad s_i = \sqrt{\frac{\sum_{k=1}^{N}(x_{ki}-\bar{X}_i)^2}{N-1}} \quad \text{for } i=1,\ldots,p.$$

By subtracting the mean and dividing by the standard deviation, we get the standardized data as illustrated in Table 8.2.

TABLE 8.2 Standardized Data

Objects	Standardized variables (characteristics)						
	Z_1	Z_2	...	Z_i	...	Z_{p-1}	Z_p
1	z_{11}	z_{12}	...	z_{1i}	...	$z_{1,p-1}$	z_{1p}
2	z_{21}	z_{22}	...	z_{2i}	...	$z_{2,p-1}$	z_{2p}
⋮	⋮	⋮	...	⋮	...	⋮	⋮
k	z_{k1}	z_{k2}	...	z_{ki}	...	$z_{k,p-1}$	z_{kp}
⋮	⋮	⋮	...	⋮	...	⋮	⋮
N	z_{Nk}	z_{N2}	...	z_{Ni}	...	$z_{N,k-1}$	z_{Np}

Here

$$z_{ki} = \frac{x_{ki} - \bar{X}_i}{s_i} \quad \text{for all } k = 1, \ldots, N, i = 1, \ldots, p$$

Clearly, this is a typical normalization process for multivariate data analysis.

Then, the sample correlation matrix is computed for the standardized variables for the healthy group, that is,

$$\mathbf{R} = \begin{bmatrix} 1 & r_{12} & \cdots & r_{1p} \\ r_{21} & 1 & \cdots & r_{2p} \\ \vdots & \vdots & \vdots & \vdots \\ r_{p1} & r_{p2} & \cdots & 1 \end{bmatrix}$$

where

$$r_{ij} = \frac{1}{N-1} \sum_{k=1}^{N} z_{ki} z_{kj} \qquad (8.1)$$

Then for any multivariate observation,

$$\mathbf{x}_0 = (x_{01}, x_{02}, \ldots, x_{0i}, \ldots, x_{0p})$$

We can compute its scaled Mahalanobis distance as follows:

Step 1. Normalize $\mathbf{x}_0 = (x_{01}, x_{02}, \ldots, x_{0i}, \ldots, x_{0p})^T$ by subtracting $\bar{\mathbf{X}} = (\bar{X}_1, \bar{X}_2, \ldots, \bar{X}_i, \ldots, \bar{X}_p)^T$, and dividing by $s_1, s_2, \ldots, s_i, \ldots, s_p$ from the healthy group data set. So we get the normalized new observation as follows:

$$\mathbf{z}_0 = (z_{01}, z_{02}, \ldots, z_{0i}, \ldots, z_{0p})^T = \left(\frac{x_{01} - \bar{X}_1}{s_1}, \ldots, \frac{x_{0i} - \bar{X}_i}{s_i}, \ldots, \frac{x_{0p} - \bar{X}_p}{s_p} \right)^T$$
(8.2)

Step 2. Compute the scaled Mahalanobis distance (MD) for this observation:

$$MD_0 = \frac{1}{p} \mathbf{z}_0^T \mathbf{R}^{-1} \mathbf{z}_0 \qquad (8.3)$$

Remark. The scaled Mahalanobis distance is the regular Mahalanobis distance for normalized variables divided by p. That is for a new normalized observation z_0, the regular Mahalanobis distance is

$$D_0 = z_0^T R^{-1} z_0 \qquad (8.4)$$

The reason for this is that for any healthy observation z, the average scaled Mahalanobis distance is approximately equal to 1, that is,

$$E(\text{MD}) = E\left(\frac{1}{p} z^T R^{-1} z\right) \approx 1 \qquad (8.5)$$

Actually, Tracy, Young, and Mason (1992) and Woodall et al. (2003) stated that

$$E(\text{MD}) = \frac{N}{N-1} \qquad (8.6)$$

where N is the number of observations in the initial healthy group. Taguchi and Jugulum (2002) call the scaled Mahalanobis domain for initial healthy group the unit space, or the baseline Mahalanobis space.

8.2.2 Stage 2: Test and analysis of the Mahalanobis measure for abnormal samples

Test the sensitivity of the Mahalanobis measure. The purpose of the Mahalanobis-Taguchi system is to effectively identify new abnormal observations by the magnitude of the scaled Mahalanobis distance. It is very desirable that the scaled Mahalanobis distance is significantly larger for any abnormal observation. After establishing the initial baseline for the Mahalanobis space, we need to test if this scaled Mahalanobis measure is sensitive to abnormal observations.

In this stage, we collect multivariate data for known abnormal objects. For example, in the liver disease case, we purposely conduct tests on known liver disease patients and obtain test results. After new data sets for abnormal samples are collected, we use Eq. (8.3) to calculate the scaled Mahalanobis distances for each abnormal observation. It is certainly desirable that these newly calculated scaled Mahalanobis distances are significantly larger than 1—actually, the larger the better.

What if the newly calculated scaled Mahalanobis distances are not much larger than 1 for abnormal samples? If that happens, it often indicates that the variables that we originally selected are not able to discriminate between the abnormal ones and healthy ones. In such a case, we will have to add new variables hoping that these new variables will have better discriminating power. In the liver disease example, it means that the tests that we did are not sensitive enough to differentiate healthy ones from abnormal ones. Then we may have to add new tests, hoping that these new tests will be able to discriminate healthy people from abnormal people.

After we add new variables, we have to go back to Stage 1, measure the healthy objects with new variables, and augment the raw data set described by Table 8.1 and recalculate the standardized variables and reestablish the baseline Mahalanobis space. We will have to repeat this process until the discrepancy between the scaled Mahalanobis distances for the healthy group and that for the abnormal group become sufficiently large. Taguchi did not explicitly specify how large is sufficiently large. However, from the working procedure of the Mahalanobis-Taguchi system, if there is an overlap region, that is, if the high end of MD for healthy group objects is larger than the low end of MD for unhealthy group objects, it will create a lot of difficulties for MTS.

Analysis of abnormality. For an abnormal observation, if the scaled Mahalanobis distance is too large, it does not always mean that this observation is a "bad" object. For example, in a graduate students' admission process for an ordinary campus, the typical applicant might have reasonable GPA (grade point average) and GRE scores. A particular application may have an extremely high GPA and GRE and so its scaled Mahalanobis distance is high because it is "out of norm." However, this kind of abnormality is not harmful and it should be considered as "good abnormal." Therefore, when a new observation has a high scaled Mahalanobis distance, we should not sound the alarm right away. We should look into the variable or variables that makes the scaled Mahalanobis distance high and also into the kind of abnormality it is, before taking any action.

8.2.3 Stage 3 variable screening by using Taguchi orthogonal array experiments

After stages 1 and 2, we should have already established that the baseline scaled Mahalanobis space and the sensitivity of scaled Mahalanobis distance towards abnormal conditions is at a satisfactory level. However, we may have too many variables. In the future monitoring process, more variables means higher cost of monitoring and prevention. For example, in the liver disease diagnosis process, if we identify a large number of tests which can effectively detect liver disease by using scaled Mahalanobis measure, it is an effective diagnosis system, but is too expensive. Usually, not every variable plays an equal role in discriminating the abnormality. Some variables will be more important than others. It is very necessary to screen the variables that contribute less to the discriminating process so that the future monitoring and prevention cost is lower.

We will use the following example to illustrate how orthogonal arrays can be used to screen variables.

Example 8.1: A Five-Variable Problem Assume that we have five variables X_1, X_2, X_3, X_4, and X_5. By using the observations in the normal group, we have established the baseline Mahalanobis space. Specifically, we have already established **R** and

$\mathbf{R}^{-1}$ matrices, $\overline{\mathbf{X}} = (\overline{X}_1, \overline{X}_2, \overline{X}_3, \overline{X}_4, \overline{X}_5)^T$, and s_1, s_2, s_3, s_4, s_5 from the normal group data set. In addition, we have also obtained additional n multivariate data sets from the abnormal group. Specifically, we have the following data sets from known abnormal observations:

$$\mathbf{x}_1 = (x_{11}, x_{12}, x_{13}, x_{14}, x_{15})$$

$$\mathbf{x}_2 = (x_{21}, x_{22}, x_{33}, x_{24}, x_{35})$$

$$\vdots$$

$$\mathbf{x}_k = (x_{k1}, x_{k2}, x_{k3}, x_{k4}, x_{k5})$$

$$\vdots$$

$$\mathbf{x}_n = (x_{n1}, x_{n2}, x_{n3}, x_{n4}, x_{n5})$$

We will select a two-level orthogonal array with more than five columns, so that all five variables can be accommodated. In this example, we selected an $L_8(2^7)$ array and the screening experiment is illustrated in Table 8.3.

In orthogonal array layout, level 1 means "inclusion" of the corresponding variable, and level 2 means "exclusion" of the corresponding variable. Empty columns indicate that no variables are assigned to them.

For the orthogonal array experiment described in Table 8.3, for the first experimental run, the orthogonal array setting is 1-1-1-1-1, for X_1, X_2, X_3, X_4, and X_5, which means that all five variables will be used to compute scaled Mahalanobis distances for n abnormal objects.

For the second experimental run, the orthogonal array setting is 1-1-1-2-2 for X_1, X_2, X_3, X_4, and X_5, which means that only X_1, X_2, X_3 will be used to compute scaled Mahalanobis distances for all n abnormal objects. X_4 and X_5 will not be used.

For the third experimental run, the orthogonal array setting is 1-2-2-1-1 for X_1, X_2, X_3, X_4, and X_5, which means that only X_1, X_4, X_5 will be used to compute scaled Mahalanobis distances for all n abnormal objects. X_2 and X_3 will not be used. In general, in each run of the orthogonal array experiment, according to the experimental layout in each run, a subset of variables will be selected to compute the scaled Mahalanobis distances for all abnormal observations.

Like other Taguchi orthogonal array experiments, signal-to-noise ratio (S/N) is used as the metric to select important variables. In the setup described in Table 8.3, for each experimental run i, $i = 1, \ldots, 8$, n scaled Mahalanobis distances are

TABLE 8.3 A Typical Orthogonal Array Layout for Variable Screening

	L_8 Array columns							Scaled Mahalanobis distances				S/N
	1	2	3	4	5	6	7					
Expt. run no.	X_1	X_2	X_3	X_4	X_5			1	2	...	n	η
1	1	1	1	1	1	1	1	MD_{11}	MD_{12}	...	MD_{1n}	η_1
2	1	1	1	2	2	2	2	MD_{21}	MD_{22}	...	MD_{2n}	η_2
3	1	2	2	1	1	2	2					
4	1	2	2	2	2	1	1					
5	2	1	2	1	2	1	2					
6	2	1	2	2	1	2	1					
7	2	2	1	1	2	2	1					
8	2	2	1	2	1	1	2	MD_{81}	MD_{82}	...	MD_{8n}	η_8

computed for all n abnormal objects with selected subset of variable, that is, MD_{i1}, $MD_{i2}, \ldots, MD_{in}$. By Taguchi's terminology, the scaled Mahalanobis distance has the "larger-the-better" characteristic, because for abnormal observations, the larger the scaled Mahalanobis distance, the more easily the sensitively scaled Mahalanobis distance measure can pick up the abnormality. So the following signal-to-noise ratio will be used:

$$\eta_i = -\log_{10}\left[\frac{1}{n}\sum_{j=1}^{n}\left(\frac{1}{MD_{ij}}\right)^2\right] \quad \text{for all } i = 1,\ldots, 8. \quad (8.7)$$

By using Taguchi's experimental data analysis, the important variables will be selected. We will use Example 8.2 to show the data analysis procedure.

Example 8.2: Liver Disease Diagnosis (adapted from Taguchi and Jugulum, 2002) In liver disease diagnosis, the following five tests are often used:

Variables	Tests
X_1	Lactate dehydrogenase
X_2	Alkanline phosphatase
X_3	r-Glutamyl transpeptidase
X_4	Leucine aminopeptidase
X_5	Total cholesterol

We will use the test results as variables to establish our Mahalanobis-Taguchi system metric. First, we obtain the test results of 50 healthy people. It is listed in Table 8.4.

By using the test data of the healthy group, we can calculate the correlation matrix as follows:

$$R = \begin{bmatrix} 1 & 0.330 & 0.035 & 0.130 & 0.223 \\ 0.330 & 1 & -0.010 & 0.062 & 0.360 \\ 0.035 & -0.010 & 1 & 0.780 & 0.173 \\ 0.130 & 0.062 & 0.780 & 1 & 0.071 \\ 0.223 & 0.360 & 0.173 & 0.071 & 1 \end{bmatrix}$$

The inverse matrix for correlation matrix is

$$R^{-1} = \begin{bmatrix} 1.167 & -0.306 & 0.201 & -0.277 & -0.165 \\ -0.306 & 1.268 & 0.265 & -0.216 & -0.419 \\ 0.201 & 0.265 & 2.772 & -2.172 & -0.466 \\ -0.277 & -0.216 & -2.172 & 2.722 & 0.322 \\ -0.165 & -0.419 & -0.466 & 0.322 & 1.245 \end{bmatrix}$$

The mean test results for health samples are as follows:

$$\overline{X}_1 = 185.34, \quad \overline{X}_2 = 5.504, \quad \overline{X}_3 = 20.5, \quad \overline{X}_4 = 316.04, \quad \overline{X}_5 = 188.26$$

TABLE 8.4 Test Data of 50 Healthy People

Sample no.	X_1	X_2	X_3	X_4	X_5
1	190	7.7	12	275	220
2	217	6.3	23	340	225
3	220	9.2	16	278	220
4	204	4.6	15	289	171
5	198	5.2	13	312	192
6	188	6.1	16	304	216
7	187	6.1	13	272	215
8	190	7.7	12	278	169
9	195	3.6	13	319	160
10	230	7	35	441	197
11	206	6.6	12	320	224
12	211	6.1	14	241	213
13	181	4.8	17	282	193
14	205	5.1	11	218	181
15	206	4.3	19	280	210
16	184	5.7	13	309	181
17	205	6.4	23	328	175
18	180	5.2	13	258	181
19	154	4.3	15	255	220
20	217	7.7	47	430	213
21	184	5.1	13	281	206
22	167	6.9	14	346	197
23	152	4.8	40	355	177
24	176	3.7	28	337	180
25	150	4.6	16	317	165
26	155	6.4	20	382	177
27	184	5.3	38	388	217
28	178	5.9	53	416	213
29	196	7.6	14	327	184
30	157	8.4	11	279	182
31	178	4.9	9	233	186
32	171	4.3	10	297	190
33	173	4.5	19	302	159
34	183	4.2	12	288	221
35	144	4	11	245	151
36	182	3.8	14	317	155
37	163	6	12	268	164
38	213	5.2	12	318	187
39	205	4.2	13	295	136
40	169	6	27	334	247
41	163	3.7	21	356	171
42	145	5.1	33	327	186
43	181	5.7	16	286	178
44	154	3.9	22	312	170
45	187	6.2	14	262	139
46	209	5.5	45	365	176
47	173	4.3	44	417	172
48	223	5.4	14	374	164
49	180	4.8	38	413	172
50	204	5.1	40	336	215

TABLE 8.5 Test Data for Abnormal Group

Sample no.	X_1	X_2	X_3	X_4	X_5
1	336	13.7	192	856	144
2	365	25.8	274	1027	113
3	255	10.6	123	704	141
4	723	24.3	152	841	125
5	281	14.4	96	613	89
6	280	13.6	156	1059	109
7	234	13.2	123	823	148
8	206	14.7	98	612	116
9	384	13.4	150	626	112
10	265	21.8	173	1155	150
11	1070	16.8	94	760	91
12	478	30.2	248	1137	145
13	338	9.3	241	734	98
14	200	12.9	92	627	158
15	354	17.2	178	679	100

The standard deviations for the test results of healthy samples are as follows:

$$s_1 = 21.92, \quad s_2 = 1.30, \quad s_3 = 11.53, \quad s_4 = 52.37, \quad s_5 = 24.68$$

We also obtained 15 sets of test results of 15 people from the abnormal group. Their test data are listed in Table 8.5. Then we can use Eqs. (8.2) and (8.3) to calculate the scaled Mahalanobis distance. For example, for the first sample in the abnormal group, we first calculate the normalized test result:

$$\mathbf{z}_0 = (z_{01}, z_{02}, \ldots, z_{0i}, \ldots, z_{0p})^T = \left(\frac{x_{01} - \overline{X}_1}{s_1}, \ldots, \frac{x_{0i} - \overline{X}_i}{s_i}, \ldots, \frac{x_{0p} - \overline{X}_p}{s_p} \right)^T$$

$$= \left(\frac{336 - 185.34}{21.92}, \frac{13.7 - 5.50}{1.30}, \frac{192 - 20.5}{11.53}, \frac{856 - 316.04}{52.37}, \frac{144 - 188.26}{24.68} \right)^T$$

$$= (6.87, 6.31, 14.87, 10.31, -1.79)^T$$

The scaled Mahalanobis distance is

$$MD_0 = \frac{1}{p} \mathbf{z}_0^T \mathbf{R}^{-1} \mathbf{z}_0$$

$$= \frac{1}{5}(6.87, 6.31, 14.87, 10.31, -1.79) \begin{bmatrix} 1.167 & -0.306 & 0.201 & -0.277 & -0.165 \\ -0.306 & 1.268 & 0.265 & -0.216 & -0.419 \\ 0.201 & 0.265 & 2.772 & -2.172 & -0.466 \\ -0.277 & -0.216 & -2.172 & 2.722 & 0.322 \\ -0.165 & -0.419 & -0.466 & 0.322 & 1.245 \end{bmatrix}$$

$$\times \begin{bmatrix} 6.87 \\ 6.31 \\ 14.87 \\ 10.31 \\ -1.79 \end{bmatrix} = 73.70$$

TABLE 8.6 Scaled Mahalanobis Distances for Abnormal Group

Sample no.	Abnormal	Sample no.	Abnormal	Sample no.	Abnormal
1	73.699	6	57.457	11	373.578
2	209.460	7	31.411	12	204.913
3	25.899	8	30.215	13	140.176
4	191.393	9	70.738	14	19.081
5	34.277	10	92.097	15	99.433

Similarly, we can calculate the scaled Mahalanobis distances for people from both abnormal and normal groups. The results are summarized in Tables 8.6 and 8.7.

The average scaled Mahalanobis distance is 0.98 for the healthy group. Figure 8.1 illustrates the comparison of scaled Mahalanobis distances between the normal group and the abnormal group. Clearly, the gap between the scaled Mahalanobis distances of the healthy group and those of the abnormal group is quite large.

Now, we want to know if we can reduce the number of tests so that we have a more economical test procedure. We will do this by using Taguchi's orthogonal array. Since we currently have five variables, the L_8 array is used for this purpose. The experimental layout is in Table 8.8. The scaled Mahalanobis distances of the abnormal group, based on the subset of variables specified by each run of L_8, is given in Table 8.9. For example, the level settings for the first run of L_8 are all 1s. Here "1" means the corresponding variable is selected. So the first run means all five variables are selected. The 15 scaled Mahalanobis distances are computed using all five variables. In the second run, the level selection is 1-1-1-2-2, which means that X_1, X_2, and X_3 are selected, so the corresponding Mahalanobis distances are computed by using only these three variables.

TABLE 8.7 Scaled Mahalanobis Distances for Normal Group

Sample no.	Normal	Sample no.	Normal	Sample no.	Normal
1	0.885	18	0.286	35	1.235
2	0.755	19	1.391	36	0.686
3	1.975	20	1.803	37	0.674
4	0.505	21	0.319	38	0.682
5	0.314	22	1.223	39	1.414
6	0.342	23	1.228	40	1.510
7	0.417	24	0.464	41	0.861
8	1.136	25	0.707	42	1.078
9	0.947	26	1.701	43	0.131
10	1.909	27	0.744	44	0.554
11	0.949	28	1.751	45	1.576
12	1.078	29	0.795	46	1.829
13	0.181	30	2.133	47	1.189
14	1.343	31	0.556	48	1.911
15	0.957	32	0.591	49	0.860
16	0.167	33	0.337	50	1.364
17	0.410	34	1.149		

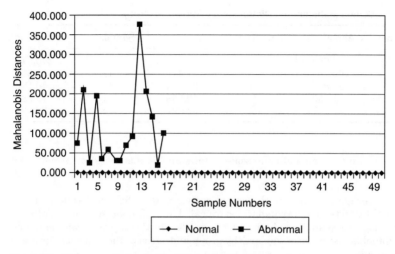

Figure 8.1 The comparison of scaled Mahalanobis distances between normal and abnormal groups.

The signal-to-noise ratio is computed based on the larger-the-better formula, because if a subset of variables can make the scaled Mahalanobis distances for the abnormal group larger, then it will be easier for scaled Mahalanobis distance to detect abnormal items. Therefore we use the following signal-to-noise ratio calculation formula:

$$\eta_i = -\log_{10}\left[\frac{1}{n}\sum_{j=1}^{n}\left(\frac{1}{MD_{ij}}\right)^2\right]$$

for each run; for example, for the first run of L_8 array,

$$\eta_1 = -\log_{10}\left[\frac{1}{15}\left(\frac{1}{73.699^2} + \frac{1}{209.46^2} + \cdots + \frac{1}{93.433^2}\right)\right] = 17.354$$

The signal-to-noise ratio calculations for each run is at the end of Table 8.9.

TABLE 8.8 L_8 Experimental Layout

Expt. no.	X_1	X_2	X_3	X_4	X_5
1	1	1	1	1	1
2	1	1	1	2	2
3	1	2	2	1	1
4	1	2	2	2	2
5	2	1	2	1	2
6	2	1	2	2	1
7	2	2	1	1	2
8	2	2	1	2	1

TABLE 8.9 Scaled Mahalanobis Distances for Abnormal Samples in L_8 Experiment

Expt. no.	1	2	3	4	5	6	7	8	9	10	11	12	13	14	15	S/N
1	73.699	209.460	25.899	191.393	34.277	57.457	31.411	30.215	70.738	92.097	373.578	204.913	140.176	19.081	99.433	17.354
2	94.211	245.932	32.553	253.687	31.392	60.594	38.295	32.957	71.192	111.816	558.887	266.815	135.475	24.548	96.444	17.938
3	50.722	85.787	22.968	240.766	24.736	77.145	33.807	14.667	44.929	89.606	606.541	135.338	43.837	12.688	42.478	15.764
4	47.242	67.179	10.099	601.651	19.045	18.649	4.928	0.888	82.139	13.207	1628.852	178.261	48.504	0.447	59.204	5.807
5	69.137	200.852	33.403	145.649	37.073	114.873	60.948	38.493	33.741	194.675	69.136	284.757	34.776	31.734	60.590	17.548
6	29.193	164.074	13.971	138.356	47.364	36.305	25.511	42.021	34.277	99.069	66.129	221.447	16.952	22.211	66.807	15.647
7	112.725	257.983	39.785	66.634	21.834	101.235	49.134	22.794	73.400	129.004	36.014	194.689	244.405	20.767	111.032	17.162
8	120.425	265.851	45.653	75.633	35.136	83.222	44.684	31.214	76.122	95.043	33.431	208.371	207.840	21.948	111.467	17.615

TABLE 8.10 Average S/N for Different Levels of Variables

	Level 1	Level 2	Gain
X_1	14.2158	16.99281	−2.77701
X_2	17.12164	14.08697	3.034678
X_3	17.51717	13.69143	3.825739
X_4	16.95697	14.25164	2.705329
X_5	16.5949	14.61371	1.981188

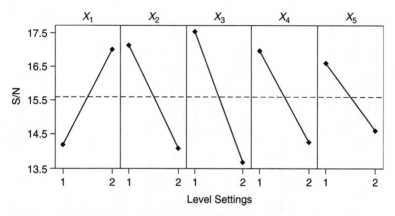

Figure 8.2 Main effect plot for Example 8.2.

Figure 8.2 shows the main effect plot of signal-to-noise ratios for each variables. Table 8.10 lists the comparison of signal-to-noise ratios for different levels of variables. Clearly, for variable X_1, level 2 (not including X_1) actually has larger signal-to-noise ratio than that of level 1 (including X_1). So having X_1 actually reduces the gap between the scaled Mahalanobis distances of the abnormal group versus that of the normal group. We also notice that for all other variables, X_2, X_3, X_4 and X_5, the signal-to-noise ratios at level 1 are quite significantly larger than those at level 2, which means that inclusion of X_2, X_3, X_4, and X_5 will significantly increase scaled Mahalanobis distances for the abnormal group, which is better. Therefore, the final variable selection will be X_2, X_3, X_4, and X_5.

8.2.4 Stage 4: Establish a threshold value (a cutoff MD) based on Taguchi's quality loss function and maintain a multivariate monitoring system

In the Mahalanobis-Taguchi system, the threshold value for the scaled Mahalanobis distance, T, is established by Dr. Taguchi's quality loss function and the Taguchi tolerance design approach. His tolerance design approach is a cost-based tolerance design. The most important consideration in cost is the quality loss due to the deviation from the ideal level.

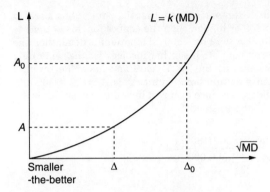

Figure 8.3 Quality loss function of MTS.

The quality loss function for the Mahalanobis-Taguchi system is

$$L = k(\text{MD}) \tag{8.8}$$

The quality loss will be zero, if MD = 0. We illustrate the quality loss function for the Mahalanobis-Taguchi system in Fig. 8.3.

In Fig. 8.3, Δ_0 is the functional limit and A_0 is the cost incurred when we reach the functional limit. From the equation of loss function, Eq. (8.8),

$$A_0 = k\Delta_0^2$$

So

$$k = \frac{A_0}{\Delta_0^2}$$

Therefore

$$L = \frac{A_0}{\Delta_0^2} \text{MD} \tag{8.9}$$

In Fig. 8.3, A is the internal cost, that is, the cost of dealing with the situation when the threshold value for MD $\geq T$. By using Eqs. (8.8) and (8.9) we have

$$A = \frac{A_0}{\Delta_0^2} T \tag{8.10}$$

So

$$T = \frac{A\Delta_0^2}{A_0} \tag{8.11}$$

Here T is the threshold value for the scaled Mahalanobis distance.

Example 8.3 In the liver disease example, assume that if a patient has a scaled Mahalanobis distance of MD = 800 or higher, then the chance for this patient to die is greater. The functional limit Δ_0 will be $\sqrt{800}$. If a patient is dead, then the loss to his or her family is at least equal to his or her incomes for the remaining life. Assume that his or her remaining life is 27 years and his or her annual income is $50,000. Then the loss at functional limit Δ_0 is $A_0 = 27 \times 50000 = \$1,350,000$. If the average medical procedure to cure a liver disease patient is $A = \$8000$, then by using Eq. (8.10), the threshold

$$T = \frac{A\Delta_0^2}{A_0} = \frac{8000 \times 800}{1,350,000} = 4.74$$

8.3 Features of the Mahalanobis-Taguchi System

1. In the Mahalanobis-Taguchi system, no particular probability distribution is assumed. The cutoff score is not based on misclassification probability or any probabilistic measures.

2. Unlike discriminant analysis, where the number of groups is known, MTS only assumes that there is a healthy or normal population. There is no separate unhealthy population, because each unhealthy situation is unique, according to Taguchi and Jugulum (2002). Hawkins (2003) evaluated data from Taguchi's liver disease case study and found that the sample covariance matrix of the healthy sample is vastly different from that of the abnormal group. Therefore the pooling of covariance matrices commonly used in linear discriminant analysis will not be appropriate. In a practical situation, this assumption (abnormal observations may not form a well-defined population) may often be true. The closest traditional multivariate statistical method to MTS might be the multivariate T^2 control chart. However, the usual practice of the T^2 chart does not include the practice of variables' screening, and maximization of Mahalanobis distances for abnormal group objects by adding variables.

3. The screening of variables by using Taguchi's orthogonal array experiment is a feature of MTS. The goal of variables' screening is to reduce the cost of monitoring. Many researchers agree with the necessity of the variables' screening, but they do not agree with the usage of an orthogonal array in this application (Woodall, 2003).

8.4. The Mahalanobis-Taguchi System Case Study

In this section a case study, related to inspection of automobile clutch discs, is discussed. This study was conducted by Japan Technical Software Co., Ltd.

Figure 8.4 An example of a clutch disc.

8.4.1 Clutch disc inspection

Clutch discs are designed in different shapes and dimensions for various torque transformation requirements. A typical shape of clutch is shown in Fig. 8.4. The outside diameter of such a disc is 180 mm, and it has a width of 15 mm. The clutch is glued with a special frictional material on both sides, and a slit-type pattern is engraved on the surface. It is important to maintain high standards of quality for these discs since they determine the driving performance.

Because of higher production quantities and inconsistencies in human judgment, it has been decided to automate the inspection process by summarizing the multidimensional information. The characteristics of the system include differential and integral characteristics produced by various patterns. These patterns are obtained by rotating a clutch disc at a constant speed. The pictures of these patterns are taken with the help of a charge-coupled device camera. These patterns are inputted to a personal computer with specially designed image input boards. The image data represent the concentration of the light. Based on this concentration, the condition of the product is identified. Figure 8.5 shows different types of product conditions. In this figure the x axis represents the rotating position of the disc and the y axis represents the concentration of the light. From this figure, it is clear that the patterns do not show any difference for the cases of no defects and small defects. In order to distinguish such cases, differential and integral characteristics are used.

Differential and integral characteristics. Let $Y(t)$ represent one of the wave patterns, obtained from one of the rotations of the disc. As shown in Fig. 8.6, p parallel lines on t axis are drawn at the same intervals. The number of cross sections, the wave intersects on each line is known as the differential characteristic. The range between cross sections of each line is known as the integral characteristic. The differential characteristics indicate the wave changes, similar to the concept of frequency, and are observed for each amplitude. The integral characteristics indicate the magnitude of the range of each amplitude. Since these characteristics

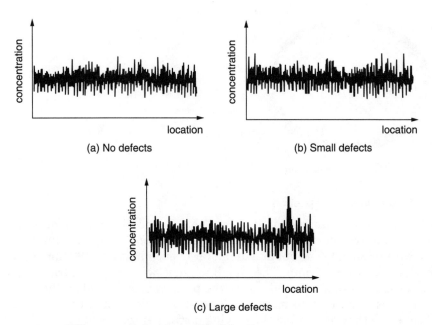

Figure 8.5 Different conditions (patterns) of the disc.

include information about frequency, amplitude, and distribution, they can be used for quick and efficient data processing. Besides these characteristics, there are other characteristics used for inspection.

Construction of the Mahalanobis space. From normal clutch discs, 1000 sets of data are taken to show images. These clutches have been found to be defect free. Each data set contains information on 40 variables

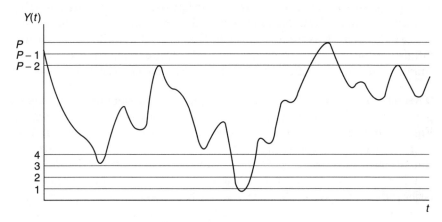

Figure 8.6 Differential and integral characteristics.

(including differential and integral characteristics). For all these observations, MDs are computed. These MDs constitute the Mahalanobis space. This is the reference point for the measurement scale.

Validation of measurement scale. To ensure the accuracy of this measurement scale, data sets were collected on the discs having defects (abnormal data). For these data sets MDs were computed. The values of these MDs were large enough to ensure the accuracy of the scale.

Identifying the variables of importance (dimensionality reduction). In order to identify the variables of importance, an L_{64} (2^{63}) array is selected. The variables are allocated to the first 40 columns of the array. These variables are numbered as $X_1, X_2, ..., X_{40}$. If the element of a column is 1, then the corresponding variable is considered as part of the system and if the element is 2, then that variable is not considered as part of the system. For each combination, S/N ratios are computed based on MDs of abnormal data. In this case larger-the-better type S/N ratios are used, since for abnormal data the measurement scale should give larger MDs. The S/N ratios (in decibel units) are shown in Table 8.11.

As explained in Sec. 8.2, for each run, gain in S/N ratios is computed. The gains are shown in Fig. 8.7. From this figure, it can be seen that 20 variables $X_1, X_3, X_4, X_{10}, X_{11}, X_{12}, X_{13}, X_{14}, X_{15}, X_{16}, X_{17}, X_{18}, X_{19}, X_{20}, X_{21}, X_{22}, X_{24}, X_{25}, X_{33}$, and X_{34} have positive gains. Hence these variables are considered important.

TABLE 8.11 S/N Ratios (Decibel Units) for 64 Runs of L_{64} (2^{63}) Array

Run no.	S/N ratio	Run no.	S/N ratio	Run no.	S/N ratio	Run no.	S/N ratio
1	1.54	17	2.12	33	2.18	49	2.13
2	1.63	18	2.1	34	2.17	50	2.02
3	1.98	19	2.21	35	2.21	51	2.14
4	2.08	20	2.22	36	2.19	52	2.07
5	2.2	21	2.1	37	2.17	53	2.16
6	2.3	22	2.14	38	2.18	54	2.11
7	2.35	23	2.07	39	2.16	55	2.12
8	2.38	24	2.11	40	2.16	56	2.08
9	2.26	25	2.2	41	2.26	57	2.21
10	2.38	26	2.19	42	2.25	58	2.15
11	2.33	27	2.22	43	2.21	59	2.22
12	2.48	28	2.2	44	2.25	60	2.16
13	1.9	29	2.17	45	2.19	61	2.15
14	2.06	30	2.17	46	2.23	62	2.11
15	1.87	31	2.23	47	2.17	63	2.17
16	2.03	32	1.93	48	2.23	64	2.14

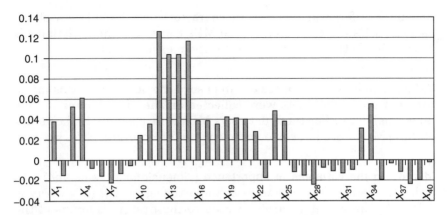

Figure 8.7 Gain in S/N ratios corresponding to the 40 variables.

Confirmation run. With a useful set of variables, a confirmation run is conducted by constructing a measurement scale. The Mahalanobis space is developed with the help of the inverse of correlation matrix of size 20 × 20. The same abnormal data sets are used to ensure the accuracy of the scale. It is found that the accuracy of this scale is very high. Moreover, this scale has better discrimination power than the scale with 40 variables, indicating that unimportant variables affect the discrimination power. This can be easily seen from Fig. 8.8. For normal data (defect-free discs), this difference is not important because the average MD of Mahalanobis space is unity irrespective of the number of variables, and all distances would be close to the average.

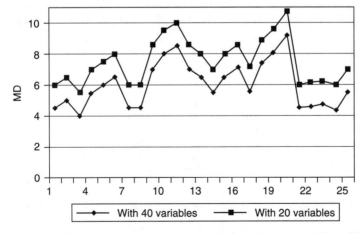

Figure 8.8 Discrimination power before and after selecting variables of importance.

8.5 Comments on the Mahalanobis-Taguchi System by Other Researchers and Proposed Alternative Approaches

Like other approaches proposed by Dr. Taguchi, the Mahalanobis-Taguchi system generated much discussion in the statistical community. Woodall et al. (2003) made the following comments on the Mahalanobis-Taguchi system:

1. The methods of the MTS are considered ad hoc in the sense that they are not developed using any underlying statistical theory. There is no distributional assumption in the MTS. In MTS, the normal item is the only population to be considered; however, the baseline normal group is only a sample. Because there are no distribution properties defined for either normal or abnormal items, the decision rules and sampling rules in MTS do not have a solid basis.

2. The cutoff value for MD is not clearly specified. From MTS descriptions it seems that the cutoff value of MD should be higher than the largest MD value in normal group. However, with no distribution assumption, when there is a mixed region (the lowest MD for abnormal items is lower than that of highest MD in normal group), the misclassification probability cannot be quantified.

3. Using fractional factorial array (Taguchi's orthogonal arrays) to screen important variables is not the best optimization algorithm for this purpose.

On the other hand, the comments made by Kenett (2003) are rather positive. Kenett credited the Taguchi-Mahanalobis system with developing methods yielding reduction in dimensions and presenting several new ideas and opens up new fields of research.

8.5.1 Alternative approaches

MYT decomposition. As we discussed earlier, the closest traditional multivariate statistical method to MTS is the T^2 control chart. One important element of MTS is the variable screening by using Taguchi's orthogonal array experiment. The purpose of this screening is to select a small subset of variables that are most sensitive to abnormal conditions in terms of Mahalanobis distance. Mason, Young, and Tracy (MYT) developed a procedure that is called MYT decomposition (Mason and Young, 2002). The purpose of this procedure is to locate the most significant variables for out-of-control conditions. MYT decomposition is an orthogonal decomposition procedure that can quantitatively evaluate the contribution of each variable, or a group of variables, to the T^2 statistics. Theoretically, it should not be too difficult to integrate the procedures such as MYT decomposition, with other optimization algorithms to develop excellent variable screening routines. More research in this area is definitely needed.

Stepwise discriminant analysis. Stepwise discriminant analysis is a special kind of discriminant analysis procedure. In the beginning of this procedure, many variables should be included in the discriminant function in order to include the variables that discriminate between groups. For example, an educational researcher interested in predicting high school graduates' choices for further education would probably include as many measures of personality, achievement motivation, academic performance, etc., as possible in order to learn which one(s) offer the best prediction. Then stepwise discriminant analysis will adopt a similar variable selection procedure as that of stepwise regression. Specifically, at each step all variables are reviewed and evaluated to determine which one will contribute the most to the discrimination between groups. One would only keep the important variables in the model, that is, those variables that contribute the most to the discrimination between groups.

Although there are differences between MTS and discriminant analysis in terms of their underlying assumptions, one common feature between MTS and stepwise discriminant analysis is the need for variable selection. This stepwise variable selection procedure is certainly worth further research.

Chapter 9

Path Analysis and the Structural Model

9.1 Introduction

In many quality improvement projects, the linear regression method and the design of experiments method (DOE) are very important tools to model relationships between dependent variables and independent variables. In industrial applications, the dependent variables are often key products or process performance characteristics, and the independent variables are often design parameters or process variables. We often use a P diagram to illustrate this relationship; see Fig. 9.1. For example, in a chemical process, the dependent variables are often the key performance characteristics $(Y_1, Y_2,..., Y_m)$ such as yield, purity, etc. The independent variables $(X_1, X_2,..., X_n)$ are often process factors such as temperature, pressure, air flow rate, etc. Clearly, if we understand the relationship between these dependent variables and independent variables, we can certainly adjust and control the independent variables, to achieve superior performance. The independent variables are often design parameters or process variables, which can be modeled as dependent variables.

In the design of experiments (DOE) method, we select a set of independent variables and systematically change their settings and measure the values of the dependent variables at these settings. We then construct empirical models relating independent variables and dependent variables. The design of experiments method is a very powerful method. It often effectively finds the causes and relationships among independent variables and dependent variables. It is a powerhouse for quality improvements, especially in the Six Sigma movement. However, when we

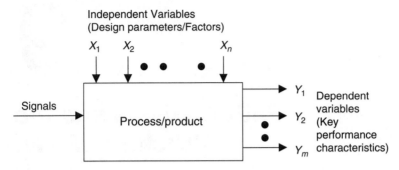

Figure 9.1 A P diagram.

use DOE on existing industrial facilities, we have the following limitations (Gerth and Hancock, 1995):

1. In the design of experiment method, the independent variables have to be set at different prespecified levels. This can be disrupting to the production process, and on many occasions it may not be possible to set variables at the prescribed levels.
2. The number of variables that can be studied at one time is limited. The practical high limit is 18, with 7 to 11 being typical for Taguchi methods and 3 or 4 for experimental design when interactions are of interest.

On the other hand, in a regular production process, a large number of process variables (usually independent variables) are monitored and recorded, though they are not artificially set and changed. Process performances (usually dependent variables) are also monitored and recorded. It seems that if we can develop an effective modeling scheme which can model the relationships among existing online data, we also can obtain a cause and effect model of independent and dependent variables.

Linear regression is a fairly well-established method for analyzing relationships among dependent variables and independent variables. However, the data collected online in manufacturing environments are usually featured by highly complex relationships among multiple output variables and correlated independent variables. In such a case, regular linear regression will suffer many vulnerabilities. Linear regression cannot deal well with the issues of multicolinearity, which is the mutual correlative relationship among independent variables. It cannot deal well with multiple dependent variables which may influence each other.

Path analysis with the structural model is a multivariate statistical method that can deal well with both correlated independent variables and correlated multiple dependent variables. Researchers in biology, social science, and medicine have been developing and applying path analysis since the 1920s. The method of path analysis was developed by the geneticist Sewell Wright to explain the relations in population genetics. He also applied path analysis to model the relationship between corn and hog prices (Johnson and Wishern, 1982). Duncan (1966) introduced path analysis in the field of sociology. Since then, many papers have appeared in the research journals of most behavioral sciences. Karl Joreskog (1977b) made a significant contribution toward making general path analysis techniques accessible to research communities. Today, with the development and increasing availability of computer programs, path analysis has become a well-established and respected data analysis method. Path analysis software packages, such as AMOS (Arbuckle, 1995), LISREL (Joreskog and Sorbom, 1993), and EQS (Bentler and Wu, 1993), handle a variety of ordinary least square regression designs as well as complex structural models.

However, the application case studies in business and industry have been very sparse. We believe that path analysis and the structural model is an untapped gold mine in quality engineering.

In this chapter, we describe what path analysis is and how it works, in Sec. 9.2. Section 9.3 will discuss the advantages and disadvantages of path analysis. Two industrial case studies will be presented in Sec. 9.4.

9.2 Path Analysis and the Structural Model

The goal of path analysis is to provide plausible explanations for the relationships among variables by constructing structural relationship models. In path analysis, there are two kinds of relationships among variables: one is called causation relationship, or cause and effect relationship, and the other is called casual correlation relationship.

Causation is defined as the cause and effect relationship. The existence of a cause and effect relationship of two variables requires the following:

1. Existence of a sufficient degree of correlation between two variables
2. One variable occurs before the other
3. One variable is clearly the outcome of the other
4. There are no other reasonable causes for the outcome

If two variables have a cause and effect relationship, say X causes Y to change, then we can also say that X structurally influences Y. Casual correlation relationship is defined as a civariant relationship among variables, but it is unclear if there is a cause and effect relationship.

In path analysis, three kinds of variables will be identified and studied:

1. Independent (exogenous) variables, X_i's, which are not influenced by other variables
2. Dependent (endogenous) variables, Y_i's, which are influenced by other variables in the model
3. Other variables or residuals (errors) represent those factors that are not actually measured but affect the dependent (endogenous) variables

In path analysis, path diagrams are used to graphically display the hypothesized relationships among variables in a structural model. For two variables, X and Y, the following types of relationships are possible:

1. $X \longrightarrow Y$, where X might structurally influence Y, but not vice versa.
2. $X \longleftarrow Y$, where Y might structurally influence X, but not vice versa.
3. $X \rightleftarrows Y$, where X might structurally influence Y, and Y might structurally influence X.
4. $X \smile Y$, where no structural relation is hypothesized between X and Y, but the variables may covary.

Structures that include relationships 1, 2, and 4 are called recursive models and are the focus of this chapter. Models that involve variables that structurally influence each other, that is, relationship 3, are called nonrecursive and they are not discussed in this book.

We will show how path diagram works in the following two examples.

Example 9.1: Path Diagram of Simple Linear Regression For the simple linear regression model $Y = \beta_1 X_1 + \beta_2 X_2 + \beta_3 X_3 + \varepsilon$. We can use the path diagram shown in Fig. 9.2.

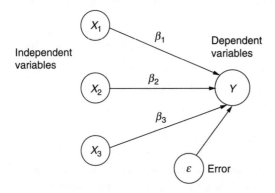

Figure 9.2 Path diagram for linear regression.

In the path diagram, the value by the arc is the coefficient to be multiplied with, for example, if the circle before the arc is X_1, and the value by the arc is β_1, then the value at the end of arc is $\beta_1 X_1$.

$$\left(X_1\right) \xrightarrow{\beta_1} \beta_1 X_1$$

If there is no value by the arc, the default value is 1.

In the above path diagram, it is also assumed that X_1, X_2, X_3 will all structurally influence Y, which means that they all have a cause and effect relationship with Y. Also, the above path diagram implies that X_1, X_2, X_3 are all independent of each other. In many practical cases, this assumption may not be true. X_1, X_2, X_3 may be correlated with each other, and this is called multicollinearity in linear regression modeling.

The next example shows how a path diagram represents a linear regression model with multicollinearity.

Example 9.2: Path Diagram of Linear Regression with Multicollinearity For the simple linear regression model

$$Y = \beta_1 X_1 + \beta_2 X_2 + \beta_3 X_3 + \varepsilon \tag{9.1}$$

in which X_1, X_2, X_3 are correlated with each other, and

$$\text{Cor}(X_1, X_2) = \rho_{12} \quad \text{Cor}(X_1, X_3) = \rho_{13} \quad \text{Cor}(X_2, X_3) = \rho_{23}$$

We can use the path diagram linear regression model with multicollinearity, as shown in Fig. 9.3. In path analysis, we usually use normalized variables for both dependent variables Y_i's and independent variables X_i's. By normalized variables we mean that

$$y = \frac{Y - \mu_Y}{\sigma_Y} \tag{9.2}$$

$$x_i = \frac{X_i - \mu_i}{\sigma_i} \tag{9.3}$$

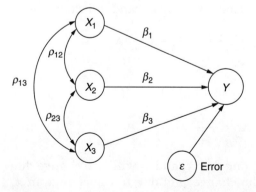

Figure 9.3 Path diagram for linear regression with multicollinearity.

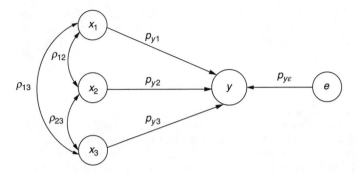

Figure 9.4 Path diagram for linear regression with multicolinearity with standardized variables.

By using normalized variables, Eq. (9.1) can be normalized to

$$\frac{Y-\mu_Y}{\sigma_Y} = \beta_1 \frac{\sigma_1}{\sigma_Y}\left(\frac{X_1-\mu_1}{\sigma_1}\right) + \beta_2 \frac{\sigma_2}{\sigma_Y}\left(\frac{X_2-\mu_2}{\sigma_2}\right) + \beta_3 \frac{\sigma_3}{\sigma_Y}\left(\frac{X_3-\mu_3}{\sigma_3}\right) + \frac{\sigma_\varepsilon}{\sigma_Y}\left(\frac{\varepsilon}{\sigma_\varepsilon}\right) \quad (9.4)$$

By using Eqs. (9.2) and (9.3), the original variables X_1, X_2, X_3, and Y are converted into standardized variables x_1, x_2, x_3, and y. We will get the standardized structural mode for Eq. (9.1) as

$$y = p_{y1}x_1 + p_{y2}x_2 + p_{y3}x_3 + p_{y\varepsilon}e_\varepsilon \quad (9.5)$$

where the path coefficients $p_{yi} = \beta_i(\sigma_i/\sigma_Y)$ are standardized coefficients for standardized independent variables, and $p_{y\varepsilon} = \sigma_\varepsilon/\sigma_Y$ is the coefficient relating error to Y. The path diagram based on standardized variables is shown in Fig. 9.4.

9.2.1 How to use the path diagram and structural model

How is the relationship specified by the path diagram and structural model? We can use a very simple example. Assume that the original variable X_1 is changed to $X_1 + \Delta X_1$. Then by using a standardized variable,

$$x_1 + \Delta x_1 = \frac{X_1 + \Delta X_1 - \mu_1}{\sigma_1} = \frac{X_1 - \mu_1}{\sigma_1} + \frac{\Delta X_1}{\sigma_1} \quad (9.6)$$

So

$$\Delta x_1 = \frac{\Delta X_1}{\sigma_1} \quad (9.7)$$

Because of the correlation among x_i's, x_2 and x_3 will also change due to the change in x_1. Then, the change in y due to the change in x_1

is as follows:

$$\Delta x_2 = \rho_{12} \Delta x_1 \tag{9.8}$$

$$\Delta x_3 = \rho_{13} \Delta x_1 \tag{9.9}$$

Therefore,

$$\Delta y = p_{y1} \Delta x_1 + p_{y2} \Delta x_2 + p_{y3} \Delta x_3 = p_{y1} \Delta x_1 + p_{y2} \rho_{12} \Delta x_1 + p_{y3} \rho_{13} \Delta x_1 \tag{9.10}$$

where $p_{y1} \Delta x_1$ is the change in y due to the direct effect of Δx_1, $p_{y2} \rho_{12} \Delta x_1 = p_{y2} \Delta x_2$ is the change in y due to the indirect effect of Δx_1 through the change in x_2, and $p_{y3} \rho_{13} \Delta x_1 = p_{y3} \Delta x_3$ is the change in y due to the indirect effect of Δx_1 through the change in x_3.

If $\Delta x_1 = 1$, or $\Delta X_1 = \sigma_1$, that is, the original variable is increased by one standard deviation, then by Eq. (9.10), the change in y is

$$\Delta y = p_{y1} + p_{y2} \rho_{12} + p_{y3} \rho_{13} \tag{9.11}$$

However, for any pair of standardized variables, say (y, x_1), when x_1 is changed by 1, the change in the other, that is y, is equal to $\rho_{y,x_1} = \text{Cor}(y, x_1)$ (see Hines and Montgomery, 1990). Therefore,

$$\rho_{y,x_1} = p_{y1} + p_{y2} \rho_{12} + p_{y3} \rho_{13} \tag{9.12}$$

Similarly,

$$\rho_{y,x_2} = p_{y1} \rho_{12} + p_{y2} + p_{y3} \rho_{23} \tag{9.13}$$

$$\rho_{y,x_3} = p_{y1} \rho_{13} + p_{y2} \rho_{23} + p_{y3} \tag{9.14}$$

Equations (9.12) to (9.14) are called the normal equations for the structural model illustrated in Fig. 9.4. Normal equation is the basis for the parameter estimation in path analysis and the structural model. For the situation illustrated by Example 9.2, the corresponding raw data set will appear as in Table 9.1.

TABLE 9.1 Data Set for the Structural Model in Example 9.2

	Y	X_1	X_2	X_3
1	y_1	x_{11}	x_{12}	x_{13}
2	y_2	x_{21}	x_{22}	x_{23}
⋮	⋮	⋮	⋮	⋮
i	y_i	x_{i1}	x_{i2}	x_{i3}
⋮	⋮	⋮	⋮	⋮
N	y_N	x_{N1}	x_{N2}	x_{N3}

By using the raw data, we can calculate the sample correlation matrix for the data set illustrated in Table 9.1 as follows:

$$\mathbf{R} = \begin{bmatrix} 1 & r_{y,x_1} & r_{y,x_2} & r_{y,x_3} \\ r_{y,x_1} & 1 & r_{12} & r_{13} \\ r_{y,x_2} & r_{21} & 1 & r_{23} \\ r_{y,x_3} & r_{31} & r_{32} & 1 \end{bmatrix} \quad (9.15)$$

By substituting the population correlation coefficients with sample correlation coefficients, we have the following normal equations to estimate the path coefficients:

$$r_{y,x_1} = p_{y1} + p_{y2}r_{12} + p_{y3}r_{13} \quad (9.16)$$

$$r_{y,x_2} = p_{y1}r_{12} + p_{y2} + p_{y3}r_{23} \quad (9.17)$$

$$r_{y,x_3} = p_{y1}r_{13} + p_{y2}r_{23} + p_{y3} \quad (9.18)$$

Clearly, after we calculate the sample correlation matrix (9.15), we can use Eqs. (9.16) to (9.18) to find path coefficients p_{y1}, p_{y2}, and p_{y3}.

Example 9.3: Estimating Path Coefficients Assume, in a linear regression with multicolinearity case as illustrated by Example 9.2, we have the following sample correlation matrix:

$$\mathbf{R} = \begin{bmatrix} 1 & r_{y,x_1} & r_{y,x_2} & r_{y,x_3} \\ r_{y,x_1} & 1 & r_{12} & r_{13} \\ r_{y,x_2} & r_{21} & 1 & r_{23} \\ r_{y,x_3} & r_{31} & r_{32} & 1 \end{bmatrix} = \begin{bmatrix} 1 & 0.7 & 0.6 & 0.9 \\ 0.7 & 1 & 0.5 & 0.2 \\ 0.6 & 0.5 & 1 & 0.4 \\ 0.9 & 0.2 & 0.4 & 1 \end{bmatrix}$$

Then the corresponding normal equations (9.16) to (9.18) will be

$$p_{y1} + 0.5p_{y2} + 0.2p_{y3} = 0.7$$
$$0.5p_{y1} + p_{y2} + 0.4p_{y3} = 0.6$$
$$0.2p_{y1} + 0.4p_{y2} + p_{y3} = 0.9$$

By solving these equations we get

$$p_{y1} = 0.534 \qquad p_{y2} = 0.011 \qquad p_{y3} = 0.805$$

The corresponding path diagram will be

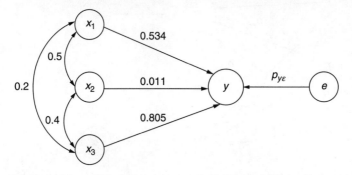

If X_1 increases by one standard deviation, then $\Delta x_1 = \Delta X/\sigma_1 = 1$ and the change in y will be

$$\Delta y = p_{y1} + p_{y2}r_{12} + p_{y3}r_{13} = 0.534 + 0.011 \times 0.5 + 0.805 \times 0.2$$
$$= 0.534 + 0.0055 + 0.161 = 0.7005$$

standard deviation, in which 0.534 standard deviation will be the direct effect from x_1, 0.0055 (standard deviation) will be the indirect effect of x_1 through x_2; and 0.161 (standard deviation) will be the indirect effect of x_1 through x_3.

Similarly, if X_2 increases by one standard deviation, then $\Delta x_2 = \Delta X_2/\sigma_2 = 1$ and the change in y will be

$$\Delta y = p_{y1}r_{12} + p_{y2} + p_{y3}r_{23} = 0.534 \times 0.5 + 0.011 + 0.805 \times 0.4$$
$$= 0.267 + 0.011 + 0.322 = 0.6$$

standard deviation, in which 0.011 standard deviation will be the direct effect from x_2, 0.267 (standard deviation) will be the indirect effect of x_2 through x_1; and 0.322 (standard deviation) will be the indirect effect of x_2 through x_3.

Finally, if X_3 is increasing by one standard deviation, then $\Delta x_3 = \Delta X_3/\sigma_3 = 1$ and the change in y will be

$$\Delta y = p_{y1}r_{13} + p_{y2}r_{23} + p_{y3} = 0.534 \times 0.2 + 0.011 \times 0.4 + 0.805$$
$$= 0.107 + 0.0044 + 0.805 = 0.9164$$

standard deviation, in which 0.805 standard deviation will be the direct effect from x_3, 0.107 (standard deviation) will be the indirect effect of x_3 through x_1; and 0.0044 (standard deviation) will be the indirect effect of x_3 through x_2.

Example 9.4: Other Example of Path Diagram Figure 9.5 illustrates a multiple regression case of two dependent variables, y_1 and y_2; and three independent variables, x_1, x_2, and x_3, with multicollinearity.

The regression equations based on standardized variables are

$$y_1 = p_{y1,x1}x_1 + p_{y1,x2}x_2 + p_{y1,x3}x_3 + p_{y1,e1}e_1$$

$$y_2 = p_{y2,x1}x_1 + p_{y2,x2}x_2 + p_{y2,x3}x_3 + p_{y2,e2}e_2$$

232 Chapter Nine

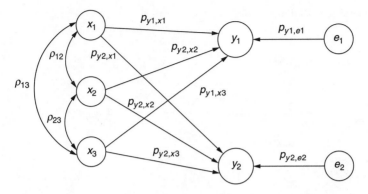

Figure 9.5 Path diagram for three independent variables and two dependent variables.

The raw data for this case will look like

	Y_1	Y_2	X_1	X_2	X_3
1	y_{11}	y_{21}	x_{11}	x_{12}	x_{13}
2	y_{21}	y_{22}	x_{21}	x_{22}	x_{23}
⋮	⋮	⋮	⋮	⋮	⋮
i	y_{i1}	y_{i2}	x_{i1}	x_{i2}	x_{i3}
⋮	⋮	⋮	⋮	⋮	⋮
N	y_{N1}	y_{N2}	x_{N1}	x_{N2}	x_{N3}

By using the raw data, we can obtain the following correlation matrix:

$$\mathbf{R} = \begin{bmatrix} 1 & 0 & r_{y1,x1} & r_{y1,x2} & y_{y1,x3} \\ 0 & 1 & r_{y2,x1} & r_{y2,x2} & r_{y2,x3} \\ . & . & 1 & r_{12} & r_{13} \\ . & . & . & 1 & r_{23} \\ . & . & . & . & 1 \end{bmatrix}$$

The normal equation to estimate the parameters of the structural model will be

$$r_{y1,x1} = p_{y1,x1} + p_{y1,x2}r_{12} + p_{y1,x3}r_{13}$$
$$r_{y1,x2} = p_{y1,x1}r_{12} + p_{y1,x2} + p_{y1,x3}r_{23}$$
$$r_{y1,x3} = p_{y1,x1}r_{13} + p_{y1,x2}r_{23} + p_{y1,x3}$$
$$r_{y2,x1} = p_{y2,x1} + p_{y2,x2}r_{12} + p_{y2,x3}r_{13}$$
$$r_{y2,x2} = p_{y2,x1}r_{12} + p_{y2,x2} + p_{y2,x3}r_{23}$$
$$r_{y2,x3} = p_{y2,x1}r_{13} + p_{y1,x2}r_{23} + p_{y2,x3}$$

In this example, it is assumed that y_1 and y_2 are mutually independent. In a more general case, path analysis can also deal with the cases where y_i's are mutually dependent.

Example 9.5: (Young and Sarle, 1983) Table 9.2 shows weight (lb), waist (in) and pulse, for 20 middle-aged men in a fitness club. It also illustrates their corresponding sports performance, namely, situps and jumps. In this example, we treat weight (x_1), waist (x_2), and pulse (x_3) as independent variables and situps (Y_1) and jumps (Y_2) as dependent variables. We use path analysis to find the direct and indirect effects of weight (x_1), waist (x_2), and pulse (x_3) to the dependent variables situps (Y_1) and jumps (Y_2).

The sample correlation matrix can be computed by MINITAB and is as follows:

Correlations: weight, waist, pulse, situps, jumps

	weight	waist	Pulse	Situps
waist	0.870			
Pulse	-0.266	-0.322		
Situps	-0.493	-0.646	0.288	
Jumps	-0.226	-0.191	0.080	0.669

We can use the path diagram in Fig. 9.5 and the normal equation to estimate the parameters of the structural model:

$$-0.493 = p_{y1,x1} + 0.87 p_{y1,x2} - 0.266 p_{y1,x3}$$

$$-0.646 = 0.87 p_{y1,x1} + p_{y1,x2} - 0.322 p_{y1,x3}$$

$$0.288 = -0.266 p_{y1,x1} - 0.322 p_{y1,x2} + p_{y1,x3}$$

$$-0.226 = p_{y2,x1} + 0.87 p_{y2,x2} - 0.266 p_{y2,x3}$$

$$-0.191 = 0.87 p_{y2,x1} + p_{y2,x2} - 0.322 p_{y2,x3}$$

$$0.08 = -0.266 p_{y2,x1} - 0.322 p_{y1,x2} + p_{y2,x3}$$

TABLE 9.2 Data for 20 Middle-aged Men

X_1 weight (lb)	X_2 waist (in)	X_3 pulse	Y_1 situps	Y_2 jumps
191	36	50	162	60
189	37	52	110	60
193	38	58	101	101
162	35	62	105	37
189	35	46	155	58
182	36	56	101	42
211	38	56	101	38
167	34	60	125	40
176	31	74	200	40
154	33	56	251	250
169	34	50	120	38
166	33	52	210	115
154	34	64	215	105
247	46	50	50	50
193	36	46	70	31
202	37	62	210	120
176	37	54	60	25
157	32	52	230	80
156	33	54	225	73
138	33	58	110	43

By solving the above equations, we get

$$p_{y1,x1} = 0.279 \quad p_{y1,x2} = -0.861 \quad p_{y1,x3} = 0.848$$
$$p_{y2,x1} = -0.248 \quad p_{y2,x2} = 0.323 \quad p_{y2,x3} = 0.245$$

Then we have the following path diagram:

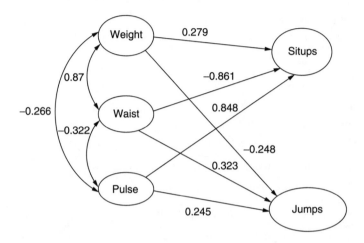

We then can summarize the direct and indirect effect of independent variables on dependent variables in Tables 9.3 and 9.4. Where the direct effects are the linkage coefficients from independent variables to dependent variables. For example, the direct effect of weight on situps is $p_{y1,x1} = 0.279$. The indirect effect of weight on situps is

$$r_{12}p_{y1,x2} + r_{13}p_{y1,x3} = 0.87 \times (-0.861) - 0.266 \times 0.848 = -0.975$$

Total effect = direct effect + indirect effect = 0.279 − 0.975 = − 0.696. That is, when weight increases by one standard deviation, the situps will decrease by 0.696 standard deviation by both direct and indirect effects.

TABLE 9.3 Analysis Correlation Decomposition for Situps

Variables	Direct effect	Indirect effect	Total effect
Weight	0.279	−0.975	−0.696
Waist	−0.861	0.030	−0.831
Pulse	0.848	0.203	1.051

TABLE 9.4 Analysis Correlation Decomposition for Jumps

Variables	Direct effect	Indirect effect	Total effect
Weight	−0.248	0.216	−0.032
Waist	0.323	−0.295	−0.028
Pulse	0.245	−0.038	0.207

Now we are very clear that only a sample correlation matrix is needed when we estimate the coefficients in structural models.

In general, it takes the following steps to establish the structural models:

Step 1. Develop a theoretically based cause-effect model that will specify the types of relationships between each pairs of variables

Step 2. Construct a path diagram to illustrate the relationship among variables

Step 3. Establish normal equations for estimating model parameters

Step 4. Collect data, compute the sample correlation matrix, and estimate structural model parameters

Step 5. Evaluate goodness of fit

Step 6. Interpret the model

All structural model software, such as AMOS, LISREL, and EQS, can perform all the calculations needed from Step 1 to Step 5 satisfactorily.

9.3 Advantages and Disadvantages of Path Analysis and the Structural Model

9.3.1 Advantages

One of the main advantages of path analysis is that it enables us to measure the direct and indirect effects that one variable has upon another. We can then compare the magnitude of the direct and indirect effects which would identify the relationships that characterize the process. In addition, path analysis enables us to decompose the correlation between any two variables into a sum of simple and compound paths. The real utility of path analysis comes in what we do with the path estimates once they are determined.

The decomposition of the correlation is extremely important since it yields information about the causal processes. Path analysis is superior to ordinary regression analysis since it allows us to move beyond the estimation of direct effects, the basic output of regression analysis. Rather, path analysis allows us to examine the processes underlying the observed relationships and to estimate the relative importance of alternative paths of influence. The model testing permitted by path analysis further encourages a more explicit approach in the search for explanations of the phenomena under investigation. The case study in this chapter will help in making the utility of this technique clearer. The underlying statistical assumptions, equations, and interpretations of coefficients in classical path analysis are similar to those in multiple regressions, but structural relationships are specified a priori. One major advantage of path analysis is that, in addition to direct structural effects, indirect effects through intervening variables can also be estimated. The determination of structural effect components in larger, more general structural equation models can become a complex and difficult task since all possible combinations of paths between a particular pair of variables must be identified. Fortunately, SEM programs, such as LISREL and AMOS, have incorporated algorithms to determine the various effects. AMOS can apply path analysis to basic techniques that are in current common use in structural modeling such as conventional linear regression, recursive and nonrecursive models, and factor analysis. Path analysis can also be used to analyze advanced techniques such as multiple regression models, simultaneous analysis of data from several different populations, estimation of means and additive constants in regression equations, and maximum likelihood estimation in the presence of missing data.

9.3.2 Disadvantages

A serious problem during the estimation of coefficients in structural equation models is that of model underidentification. The identification of a path model refers to the question of whether or not the researcher has sufficient variance and covariance information from the observed variables to estimate the unknown coefficients. A model is said to be identified if all unknown parameters are identified. An individual parameter is said to be identified if it can be expressed as a function of the variances and covariances of the observed variables in the model. More specifically, a structural equation model can have one of the three identification conditions:

1. *Just identified.* A system of equations of the model-implied parameters and the variances and covariances of observed variables can be uniquely solved for the unknown model parameters. This leads to a unique set of parameter estimates once sample variances and covariances of observed variables are available.

2. *Overidentified.* The system of equations can be solved for the model-implied parameters (in terms of the variances and covariances of observed variables), but for at least one such parameter there is no unique solution. Rather, based on the variances and covariances among observed variables, there exist at least two solutions for the same parameter.

3. *Underidentified.* The system of equations cannot be solved for all model parameters due to an insufficient number of variances and covariances of observed variables. Now, some parameters cannot be estimated solely on the basis of sample data from the observed variables.

9.4 Path Analysis Case Studies

9.4.1 Path analysis model relating plastic fuel tank characteristics with its hydrocarbon permeation (Hamade, 1996)

Plastic fuel tanks in automobiles are gaining popularity because of their many desirable properties, such as case of configuration, chemical and corrosion resistance, and light weight. Figure 9.6 shows some different designs of plastic fuel tanks.

Hydrocarbon permeation of a fuel tank is the emission of hydrocarbon compounds, mostly the light ingredients in the gasoline, to the air. The emission of hydrocarbon is considered as a volatile organic compound (VOC) and it reacts with air to form ground level ozone. This is associated with urban smog and causes respiratory problems. It is estimated that evaporative losses of hydrocarbons from vehicles contribute to about 10% of the human-made source of VOC in the atmosphere. Both EPA (Environmental Protection Agency) and CARB

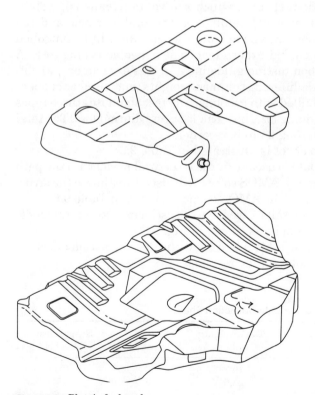

Figure 9.6 Plastic fuel tanks.

(California Air Resource Board) initiated requirements to reduce evaporative emissions of hydrocarbons from all gasoline fueled motor vehicles.

Hydrocarbon emission from the fuel tanks is a major source of emission from automobiles. Following are the three major types of hydrocarbon emissions from fuel tanks and related fuel systems.

1. *Running losses.* These are the losses that occur during sustained vehicle operation.
2. *Hot soak losses.* These are the losses that occur immediately following the engine shutoff. This is due to engine heat soaking back into the fuel system when a fully warmed-up vehicle is stationary.
3. *Diurnal losses.* These are the losses that occur when the vehicle is stationary for an extended period of time, with the engine switched off, equivalent to successive days of parking in hot weather.

This case study is to study the relationship among the three characteristics of the plastic fuel tank, which are volume (gallons), weight (pounds), and the number of openings on the tank surface, and the amount of hydrocarbon emissions from the fuel tank in a controlled experiment. This controlled experiment is designed according to EPA and CARB hydrocarbon testing guidelines. In this experiment, 21 different fuel tanks, with different volumes, weights, and number of openings, are drained and filled with gasoline. The test simulates three types of emission losses, running loss, hot soak loss, and diurnal loss. The total hydrocarbon emission (mg/24 h) is recorded.

The data collection sheet is illustrated in Table 9.5

The experimental data is used to fit the structural model with the path diagram shown in Fig. 9.7. AMOS software is used to estimate the structural model parameters. The AMOS output is listed in Table 9.6.

By using the output of AMOS, we have the estimated structural model for fuel tank emission study, as shown in Fig. 9.8.

It is convenient to use Table 9.7 to show the decomposition effects of the independent variables on the response.

TABLE 9.5 Fuel Tank Emission Data Collection Sheet

Tank no.	Tank volume (gallons)	Tank weight (lb)	No. of openings	HC permeations (mg/24 h)
1	19.0	19.7	6	169
2	36.0	22.5	5	—
⋮	⋮	⋮	⋮	⋮
21	19.0	19.7	6	149

TABLE 9.6 AMOS Output for Fuel Tank Emission Study
The Model is *Recursive*: There are no loops in the path diagram

		No. of openings	Weight	Volume	Permeation
Sample covariances					
No. of openings		0.522			
Weight		−0.268	3.317		
Volume		−0.691	10.570	50.493	
Permeation		20.297	48.190	320.755	5373.134
Sample correlations					
No. of openings		1.000			
Weight		−0.203	1.000		
Volume		−0.135	0.817	1.000	
Permeation		0.383	0.361	0.616	1.000

Maximum Likelihood Estimates

		Estimate (coefficient)	S.E. (standard error)	C.R. (critical ratio)
Regression weights				
Permeation	Volume	9.511	2.435	3.906*
Permeation	Weight	−12.127	9.615	−1.261
Permeation	No. of openings	45.305	14.119	3.209*
	Intercept	−70.243	187.062	−0.376
Standardized regression weights				
Permeation	Volume	0.922		
Permeation	Weight	−0.301		
Permeation	No. of openings	0.446		
Covariances				
Volume	Weight	10.570	3.736	2.829*
Weight	No. of openings	−0.268	0.300	−0.892
Volume	No. of openings	−0.691	1.158	−0.597
Squared multiple correlations (R^2)				
Permeation		0.630		

* Statistical Significance at .05 level

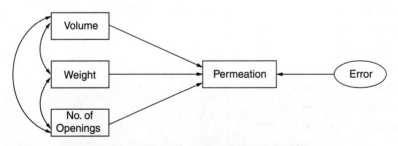

Figure 9.7 Path diagram format for fuel tank emission study.

TABLE 9.7 Path Analysis Correlation Decomposition for the Permeation Model

Independent variables	Indirect effect	Direct effect	Total effect
X_1—Volume	−0.304	0.92	0.616
X_2—Weight	0.661	−0.30	0.361
X_3—No. of openings	−0.067	0.45	0.383

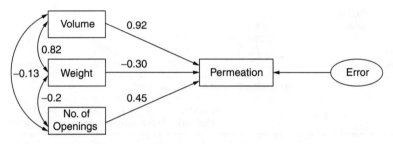

Figure 9.8 Path analysis results for the estimated model of tank permeation.

The direct effect is the path coefficient from independent variable to dependent variable. For example, the path coefficient from volume to permeation is 0.92. So the direct effect is 0.92, and the indirect effect for volume is obtained by

$$\text{Indirect effect} = (0.82)(-0.3) + (-0.13)(0.45) = -0.304$$

$$\text{Total effect} = \text{direct effect} + \text{indirect effect}$$

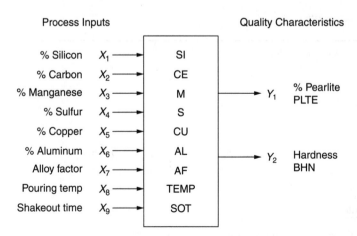

Figure 9.9 Casting process.

TABLE 9.8 A Sample of Data Collected in Casting Process

No.	BHN	PLTE	SI	CE	MN	S	CU	AL	AF	SOT	TEMP
1	223	82	1.86	4.31	0.78	0.011	0.12	0.007	1.26	1.00	2543
2	228	83	1.84	4.43	0.86	0.011	0.13	0.009	1.26	2.00	2541
2	228	97	1.77	4.38	0.77	0.010	0.13	0.013	1.16	1.10	2544
4	223	91	1.81	4.29	0.83	0.010	0.13	0.013	1.25	1.65	2620
5	223	72	1.79	4.35	0.80	0.012	0.14	0.012	1.17	0.95	2560

We can summarize the analysis results as follows:

1. For this model, the strongest effect on permeation is volume, followed by number of openings, then weight, based on total effect.
2. Volume and weight are correlated strongly. Increasing weight should make tank thicker to prevent permeation, but in current design, the volume usually also increases as weight increases. Larger volume makes larger areas to permeate. So the direct effect of weight on permeation is negative, but its indirect effect is positive. Overall, in current design practice, increase in volume and weight all make permeation increase.

9.4.2 Path analysis of a foundry process (Price and Barth, 1995)

This case study is from a foundry process in a casting plant. The plant produces three models of crankshafts. In this case study, there is a need to understand the relationships among input variables such as chemistries (for example, % silicon, % carbon equivalent, etc.); process settings (for example, temperature and shake out time); and output or response variables, which represent the quality of castings, for example, % pearlite and hardness. Figure 9.9 describes the problem.

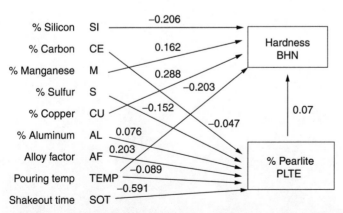

Figure 9.10 Path diagram and significant structural model coefficients.

Table 9.8 shows a sample of data collected for this study. LISREL software is used to analyze the data and estimate the parameters in a structural model. Figure 9.10 illustrates the path diagram and path coefficients computed by LISREL.

Clearly, manganese and copper contents positively affect the hardness; silicon content and temperature negatively affect the hardness. Important factors that affect percent pearlite include shakeout time, alloy factor, and sulfur content. Percent pearlite will slightly influence the hardness.

Chapter

10

Multivariate Statistical Process Control

10.1 Introduction

Statistical process control is the application of statistical techniques to control a process. In 1924, Walter. A. Shewhart of Bell Telephone Laboratories developed a statistical control chart to control important production variables in the production process. This chart is considered to be the beginning of statistical process control (SPC). It is one of the first quality assurance methods introduced in modern industry.

Statistical process control is based on a comparison of "what is a normal process," which is based on the data from a period of normal operation, with "what is happening now," which is based on a sample of data from current operation. The data collected from the normal operation condition is used to build control charts and control limits. The control charts and limits are built based on relevant statistical theory. The control limits are so designed that if the current operation is not much different from the normal operation, the statistic calculated from current data is within control limits. If the current operation is significantly different from the normal operation, then the statistic calculated from the data is outside the control limits. We treat this as an out-of-control situation. In statistical process control theory, the out-of-control situation is usually caused by assignable causes, or special causes, such as a sudden change in incoming material, degradation or malfunction of the machine, change of operators, etc. Usually, the production is stopped and investigation is conducted to find and eliminate assignable causes.

For univariate cases, that is, when there is only one variable to be monitored and controlled, there are many control charts available. For attribute variables, the popular control charts include the fraction defective chart (p chart), and the count chart (c chart). For continuous variables, the most popular ones include the X-bar and R charts and the X-bar and S charts. In a continuous variable case, at each sampling period, a sample of measurements on the variable, say X, are taken, and the average $\overline{X}$ is computed. The purpose of monitoring $\overline{X}$ is to detect if there is any significant change in population mean μ for the process. The range of the measurements in the sample R or the sample standard deviation S is also computed in order to detect if there is a significant change in the process dispersion, or process standard deviation σ. There are numerous books that discuss univariate statistical process control (Montgomery, 2000; Grant and Leavenworth, 1980).

However, in many real industrial situations, the variables to be controlled in the process are multivariate in nature. For example, in an automobile body assembly operation, the dimensions of subassemblies and body-in-white are multivariate and highly correlated. In the chemical industry (Mason and Young, 2002), many process variables, such as temperature, pressure, and concentration, are also multivariate and highly correlated.

Unfortunately, the current practice in industry toward these multivariate and highly correlated variables is usually to have one set of univariate control charts for each variable. This approach creates many control charts that could easily overwhelm the operator. Also, this approach produces misleading results. We illustrate this in Fig. 10.1.

Figure 10.1 shows that if (X_1, X_2) are correlated random variables, then an overwhelming portion of the observations fall in the eclipse region. The observations that fall outside the eclipse region are out of control. However, if we use two separate control limits based on univariate statistical theory, we can only detect some out-of-control observations; we miss some out-of-control observations, as Fig. 10.1 indicated.

Hotelling (1947) introduced a statistic which uniquely lends itself to plotting multivariate observations. This statistic is called Hotelling's T^2 and is discussed in Chap. 2. The T^2 statistic can be used to develop multivariate process control charts. However, because computing the T^2 statistic requires a lot of computations and requires some knowledge of matrix algebra, acceptance of multivariate control charts by industry was slow and hesitant.

Nowadays, with rapid progress in sensor technology and computing power, we are getting more and more data in production, manufacturing, and business operation. Most of these data are correlated multivariate data. The need to implement multivariate process control is growing. Also, with the increasing capability of modern computers, most of the laborious computational work can be accomplished in a split second, and it is getting easier and easier to implement multivariate process control.

This chapter tries to give a fair amount of coverage to multivariate process control techniques. Important literature on multivariate process

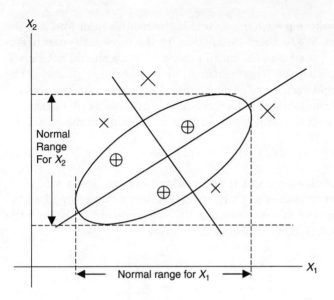

Figure 10.1 Univariate approach to multivariate variables.

control include Jackson (1959, 1985), Alt (1982), Doganaksoy et al. (1991), and Tracy, Mason, and Young (1992). Lowry and Montgomery (1995) wrote an excellent literature review on multivariate control charts. Extensive discussions on multivariate statistical process control can be found in Mason and Young (2002), as well as in Fuchs and Kenett (1998).

In this chapter, Sec. 10.2 quickly reviews the T^2 statistic and discusses the decomposition of the Hotelling T^2, and control charts based on known targets. Section 10.3 discusses a two-phase T^2 control chart for subgroups. Section 10.4 discusses a two-phase T^2 control chart for individual observations. Section 10.5 discusses the principal component control chart.

10.2. Multivariate Control Charts for Given Targets

The process control situation discussed in this chapter involves a multivariate random vector $\mathbf{X} = (X_1, X_2, \ldots, X_p)^T$. This multivariate random vector $\mathbf{X}$ characterizes the status of the process operation. In this chapter, we assume that $\mathbf{X}$ follows the multivariate normal distribution, that is, $\mathbf{X} \sim \mathbf{N}_p(\boldsymbol{\mu}, \boldsymbol{\Sigma})$ where $\boldsymbol{\mu} = (\mu_1, \mu_2, \ldots, \mu_p)^T$ is the mean vector and $\boldsymbol{\Sigma}$ is the covariance matrix. We further assume that for the process to operate satisfactorily, it is required that the mean process vector $\boldsymbol{\mu}$ should not significantly deviate from a given target vector $\boldsymbol{\mu}_0$, where $\boldsymbol{\mu}_0 = (\mu_{01}, \mu_{02}, \ldots, \mu_{0p})$; the values of the target vector are derived from the process requirements.

Clearly, the judgment about whether the process is at normal or abnormal operation becomes the following statistical hypothesis test:

$$H_0: \boldsymbol{\mu} = \boldsymbol{\mu}_0 \qquad H_1: \boldsymbol{\mu} \neq \boldsymbol{\mu}_0 \qquad (10.1)$$

in which $\boldsymbol{\mu} = \boldsymbol{\mu}_0$ indicates that the process is at normal operation.

In order to test hypothesis (10.1), we can collect a subgroup of multivariate data from the process. Assume the subgroup size is n, then the subgroup data set that we collected is as follows:

$$\mathbf{X}_{n \times p} = \begin{bmatrix} x_{11} & x_{12} & x_{13} & \cdots & x_{1p} \\ x_{21} & x_{22} & x_{23} & \cdots & x_{2p} \\ x_{31} & x_{32} & x_{33} & \cdots & x_{3p} \\ \vdots & \vdots & \vdots & \vdots & \vdots \\ x_{n1} & x_{n2} & x_{n3} & \cdots & x_{np} \end{bmatrix}$$

We can compute the sample mean vector for the subgroup as

$$\overline{\mathbf{X}} = \begin{bmatrix} \overline{X}_1 \\ \overline{X}_2 \\ \vdots \\ \overline{X}_j \\ \vdots \\ \overline{X}_p \end{bmatrix}$$

where

$$\overline{X}_j = \frac{\sum_{i=1}^n x_{ij}}{n} \qquad i = 1, \ldots, n, \ j = 1, \ldots, p$$

where $\overline{\mathbf{X}}$ is the statistical estimator of $\boldsymbol{\mu}$. We can compute

$$T_M^2 = n(\overline{\mathbf{X}} - \boldsymbol{\mu}_0)^T \mathbf{S}^{-1} (\overline{\mathbf{X}} - \boldsymbol{\mu}_0) \qquad (10.2)$$

where T_M^2 is the T^2 statistic for $\overline{\mathbf{X}}$, $\mathbf{S}$ is a sample covariance matrix computed from a known data set that follows $\mathbf{N}_p(\boldsymbol{\mu}, \boldsymbol{\Sigma})$, and it is a statistical estimate of $\boldsymbol{\Sigma}$.

From the results of Chap. 2, we know that if $\mu = \mu_0$, then the statistic T_M^2 has a distribution (Jackson 1985) given by

$$F_0 = \frac{N-p}{p(N-1)} T_M^2 \sim F_{p,N-p} \qquad (10.3)$$

where N is the number of observations to calculate the sample covariance matrix $\mathbf{S}$. When $\mu \neq \mu_0$, F_0 will have a noncentral F distribution and its value will be significantly larger, and then it is very likely that $F_0 > F_{\alpha,p,N-p}$. Therefore, the upper control limit for T_M^2 is

$$\text{UCL} = \frac{p(N-1)}{N-p} F_{\alpha,p,N-p} \qquad (10.4)$$

If the value of T_M^2 exceeds UCL in Eq. (10.4), then we reject H_0 in hypothesis (10.1) and assignable causes affecting the process mean are investigated.

Example 10.1: (Mason and Young, 2002) An Electrolysis Process In an electrolysis process, electrical current passes through a concentration of brine solution, where anode and cathode are separated by a porous diaphragm. The chlorine is displaced as a gas and the remaining water/brine solution contains the caustic. The unit performing this work is referred to as a cell, and several of these are housed together as a unit to form an electrolyzer. Many variables are measured and used as indicators of cell performance. These variables include caustic (NaOH) X_1, salt (NaCl) X_2, two kinds of impurities (I_1, I_2), X_3, X_4, and cell gases (Cl$_2$) X_5, (O$_2$) X_6.

Assume that the target values for the mean of $\mathbf{X} = (X_1, X_2, X_3, X_4, X_5, X_6)$ are

$$\mathbf{\mu_0} = \begin{bmatrix} \mu_{01} \\ \mu_{02} \\ \mu_{03} \\ \mu_{04} \\ \mu_{05} \\ \mu_{06} \end{bmatrix} = \begin{bmatrix} 142.66 \\ 197.20 \\ 0.14 \\ 2.40 \\ 98.25 \\ 1.30 \end{bmatrix}$$

Assume that we obtained 10 sets of measurement data as follows:

NaOH	NaCl	I_1	I_2	Cl$_2$	O$_2$
134.89	203	0.05	4	98.37	1.17
129.3	203.1	0.06	1.9	98.37	1.17
145.5	208.6	0.17	6.1	98.23	1.42
143.8	188.1	0.11	0.4	98.44	1.12
146.3	189.1	0.22	0.5	98.44	1.11
141.5	196.19	0.16	3.5	98.26	1.35
157.3	185.3	0.09	2.9	98.23	1.4
141.1	209.1	0.16	0.5	98.69	0.86
131.3	200.8	0.17	3.8	97.95	1.64
156.6	189	0.19	0.5	97.97	1.62

We can compute that the mean vector $\overline{X}$ for this data set is

$$\overline{X} = \begin{bmatrix} \overline{X}_1 \\ \overline{X}_2 \\ \overline{X}_3 \\ \overline{X}_4 \\ \overline{X}_5 \\ \overline{X}_6 \end{bmatrix} = \begin{bmatrix} 142.76 \\ 197.23 \\ 0.138 \\ 2.41 \\ 98.3 \\ 1.29 \end{bmatrix}$$

From past normal production data of $N = 20$ multivariate observations, the sample covariance matrix is

$$S = \begin{bmatrix} 72.27 & -44.94 & 0.10 & -5.50 & 0.06 & 0.09 \\ -44.94 & 90.78 & -0.23 & 9.40 & 0.52 & -0.36 \\ 0.10 & -0.23 & 0.005 & -0.047 & -0.008 & 0.008 \\ -5.50 & 9.40 & -0.047 & 3.47 & -0.033 & 0.037 \\ 0.06 & 0.52 & -0.008 & -0.033 & 0.053 & -0.052 \\ 0.09 & -0.36 & 0.008 & 0.037 & -0.052 & 0.055 \end{bmatrix}$$

$$T_M^2 = n(\overline{X} - \mu_0)^T S^{-1} (\overline{X} - \mu_0)$$

$$= 10 \times (142.76 - 142.66, 197.23 - 197.20, 0.138 - 0.14, 2.41 - 2.40,$$

$$98.30 - 98.25, 1.29 - 1.30)$$

$$\times \begin{bmatrix} 72.27 & -44.94 & 0.10 & -5.50 & 0.06 & 0.09 \\ -44.94 & 90.78 & -0.23 & 9.40 & 0.52 & -0.36 \\ 0.10 & -0.23 & 0.005 & -0.047 & -0.008 & 0.008 \\ -5.50 & 9.40 & -0.047 & 3.47 & -0.033 & 0.037 \\ 0.06 & 0.52 & -0.008 & -0.033 & 0.053 & -0.052 \\ 0.09 & -0.36 & 0.008 & 0.037 & -0.052 & 0.055 \end{bmatrix}^{-1} \begin{bmatrix} 142.76 - 142.66 \\ 197.23 - 197.20 \\ 0.138 - 0.14 \\ 2.41 - 2.40 \\ 98.3 - 98.25 \\ 1.29 - 1.30 \end{bmatrix}$$

$$= 10.2168$$

Comparing the critical value

$$\text{UCL} = \frac{p(N-1)}{N-p} F_{\alpha,p,N-p} = \frac{6(20-1)}{20-6} F_{0.01,6,14} = \frac{114}{16} 4.46 = 31.78$$

Clearly, T_M^2 is less than UCL, so it is in control.

10.2.1 Decomposition of the Hotelling T^2

In a subgroup of n observations specified by

$$\mathbf{X}_{n \times p} = \begin{bmatrix} x_{11} & x_{12} & x_{13} & \cdots & x_{1p} \\ x_{21} & x_{22} & x_{23} & \cdots & x_{2p} \\ x_{31} & x_{32} & x_{33} & \cdots & x_{3p} \\ \vdots & \vdots & \vdots & \vdots & \vdots \\ x_{n1} & x_{n2} & x_{n3} & \cdots & x_{np} \end{bmatrix}$$

For each multivariate observation, $\mathbf{X}_{i.} = (x_{i1}, x_{i2}, ..., x_{ip})^T$, for $i = 1, ..., n$, the following T^2 statistic

$$T_O^2 = \sum_{i=1}^{n}(X_{i.} - \mu_0)S^{-1}(X_{i.} - \mu_0) \qquad (10.5)$$

is called overall T^2 for all observations, T_O^2 has an asymptotic χ^2 distribution with $n \times p$ degrees of freedom (Jackson, 1985).

Therefore the control limit for T_O^2 is

$$\text{UCL} = \chi^2_{a,np} \qquad (10.6)$$

In this subgroup, the sample mean vector for the subgroup is

$$\mathbf{X} = \begin{bmatrix} \overline{X}_1 \\ \overline{X}_2 \\ \vdots \\ \overline{X}_j \\ \vdots \\ \overline{X}_p \end{bmatrix}$$

where

$$\overline{X}_j = \frac{\sum_{i=1}^{n} x_{ij}}{n} \qquad i = 1, ..., n, \ j = 1, ..., p$$

The following T^2 statistic

$$T_D^2 = \sum_{i=1}^{n}(X_{i.} - \overline{X})S^{-1}(X_{i.} - \overline{X}) \qquad (10.7)$$

is called T^2 for internal variability within the subgroup. T_D^2 has an asymptotic χ^2 distribution with $(n-1)p$ degrees of freedom (Jackson, 1985). Therefore, the control limit for T_D^2 is

$$\text{UCL} = \chi^2_{a,(n-1)p} \qquad (10.8)$$

From basic algebra we have

$$T_O^2 = T_M^2 + T_D^2 \qquad (10.9)$$

Equation (10.7) indicates that the overall variation around the process target, in a subgroup, is denoted by T_O^2, which can be decomposed into two parts, T_M^2 and T_D^2. T_M^2 is a measure of deviation of subgroup mean

$\overline{\mathbf{X}}$ with the process target $\boldsymbol{\mu}_0$, T_D^2 is a measure of variation of individual observations $\mathbf{X}_{i\cdot}$, for $i = 1, \ldots, n$, around the subgroup mean $\overline{\mathbf{X}}$. The role of T_M^2 in multivariate process control is similar to that of $\overline{X}$ chart, because when T_M^2 is out of control, it means that there is a significant deviation of subgroup mean $\overline{\mathbf{X}}$ with the process target $\boldsymbol{\mu}_0$. The role of T_D^2 in multivariate process control is similar to that of R chart or S chart, because if T_D^2 is out of control, it means that the variation of individual observations of the subgroup mean $\overline{\mathbf{X}}$ becomes excessive.

Example 10.2: T_D^2 and T_O^2 for the Data in Example 10.1 Based on the data set in Example 10.1, we have

$$T_D^2 = \sum_{i=1}^{n}(X_{i\cdot} - \overline{X})S^{-1}(X_{i\cdot} - \overline{X})$$

$= (134.89 - 142.76, 203 - 197.23, 0.05 - 0.138, 4 - 2.41, 98.37 - 98.3, 1.17 - 1.29)$

$$\times \begin{bmatrix} 72.27 & -44.94 & 0.10 & -5.50 & 0.06 & 0.09 \\ -44.94 & 90.78 & -0.23 & 9.40 & 0.52 & -0.36 \\ 0.10 & -0.23 & 0.005 & -0.047 & -0.008 & 0.008 \\ -5.50 & 9.40 & -0.047 & 3.47 & -0.033 & 0.037 \\ 0.06 & 0.52 & -0.008 & -0.033 & 0.053 & -0.052 \\ 0.09 & -0.36 & 0.008 & 0.037 & -0.052 & 0.055 \end{bmatrix}^{-1} \begin{bmatrix} 134.89 - 142.76 \\ 203 - 197.23 \\ 0.05 - 0.138 \\ 4 - 2.41 \\ 98.37 - 98.3 \\ 1.17 - 1.29 \end{bmatrix}$$

$+\cdots$

$+ (156.6 - 142.76, 189 - 197.23, 0.19 - 0.138, 0.5 - 2.41, 97.97 - 98.3, 1.62 - 1.29)$

$$\times \begin{bmatrix} 72.27 & -44.94 & 0.10 & -5.50 & 0.06 & 0.09 \\ -44.94 & 90.78 & -0.23 & 9.40 & 0.52 & -0.36 \\ 0.10 & -0.23 & 0.005 & -0.047 & -0.008 & 0.008 \\ -5.50 & 9.40 & -0.047 & 3.47 & -0.033 & 0.037 \\ 0.06 & 0.52 & -0.008 & -0.033 & 0.053 & -0.052 \\ 0.09 & -0.36 & 0.008 & 0.037 & -0.052 & 0.055 \end{bmatrix}^{-1} \begin{bmatrix} 156.6 - 142.76 \\ 189 - 197.23 \\ 0.19 - 0.138 \\ 0.5 - 2.41 \\ 97.97 - 98.3 \\ 1.62 - 1.29 \end{bmatrix}$$

$= 3.1973 + 5.5164 + 8.0139 + 2.7559 + 4.0121 + 1.9169 + 5.8459$

$+ 10.3165 + 4.4746 + 8.3223$

$= 54.3718$

The upper control limit for T_D^2 is

$$\text{UCL} = \chi^2_{\alpha,(n-1)p} = \chi^2_{0.01,54} = 81.04$$

So T_D^2 is in control.

$$T_O^2 = T_M^2 + T_D^2 = 10.2168 + 54.3718 = 64.5886$$

Its upper control limit is

$$\text{UCL} = \chi^2_{\alpha,np} = \chi^2_{0.01,60} = 88.38$$

So T_O^2 is in control.

10.3 Two-Phase T^2 Multivariate Control Charts with Subgroups

In real production processes, the target values for normal operation are unknown. In this situation, we often use a two-phase process to establish the process control procedure. The purpose of phase I is to obtain a baseline control limit based on a reference sample. The reference sample is the data collected from a period of known normal operation. Then we compute appropriate control limits based on the data from the reference sample. Phase II is the operational phase of process control in which subgroups of data from current production are collected and their appropriate T^2 statistic is computed and compared with control limits established in phase I.

This procedure is very similar to the practice of univariate process control. During stage I, the data collected in start-up sample are examined and out-of-control points are discarded, and only the observations that are not out of control are used to compute the control limits. We will discuss how to discover and eliminate out-of-control observations later in this section.

10.3.1 Reference sample and new observations

Assuming that we have obtained a reference sample,

$$\mathbf{Y}_{n_1, p} = \begin{bmatrix} \mathbf{Y}_{1.} \\ \mathbf{Y}_{2.} \\ \vdots \\ \mathbf{Y}_{i.} \\ \vdots \\ \mathbf{Y}_{n_1.} \end{bmatrix} = \begin{bmatrix} y_{11} & y_{12} & y_{13} & \cdots & y_{1p} \\ y_{21} & y_{22} & y_{23} & \cdots & y_{2p} \\ y_{31} & y_{32} & y_{33} & \cdots & y_{3p} \\ \vdots & \vdots & \vdots & \vdots & \vdots \\ y_{n_1 1} & y_{n_1 2} & y_{n_1 3} & \cdots & y_{n_1 p} \end{bmatrix}$$

The reference sample can be assumed as the observations from a multivariate random vector $\mathbf{Y} \sim \mathbf{N}_p(\boldsymbol{\mu}_1, \boldsymbol{\Sigma}_1)$. Let us further assume that any of the observations from current production are from the multivariate normal population, $\mathbf{X} \sim \mathbf{N}_p(\boldsymbol{\mu}_2, \boldsymbol{\Sigma}_2)$. Specifically, we take a new sample from the current production with the following format:

$$X_{n_2, p} = \begin{bmatrix} X_{1.} \\ X_{2.} \\ \vdots \\ X_{i.} \\ \vdots \\ X_{n_2.} \end{bmatrix} = \begin{bmatrix} x_{11} & x_{12} & x_{13} & \cdots & x_{1p} \\ x_{21} & x_{22} & x_{23} & \cdots & x_{2p} \\ x_{31} & x_{32} & x_{33} & \cdots & x_{3p} \\ \vdots & \vdots & \vdots & \vdots & \vdots \\ x_{n_2 1} & x_{n_2 2} & y_{n_2 3} & \cdots & y_{n_2 p} \end{bmatrix}$$

We assume that $\boldsymbol{\Sigma}_1 = \boldsymbol{\Sigma}_2 = \boldsymbol{\Sigma}$.

Let $\overline{X}$ and $\overline{Y}$ be the vector averages for these two multivariate normal samples. From the results in Chap. 2, we have

$$\overline{Y} \sim N_p\left(\mu_1, \frac{1}{n_1}\Sigma\right) \quad \text{and} \quad \overline{X} \sim N_p\left(\mu_2, \frac{1}{n_2}\Sigma\right)$$

and

$$\overline{X} - \overline{Y} \sim N_p\left[\mu_2 - \mu_1, \left(\frac{1}{n_1} + \frac{1}{n_2}\right)\Sigma\right] \quad (10.10)$$

In the process control of means, we really try to compare the mean of current production process, μ_2, with the process mean of the reference sample, μ_1. In other words, we conduct the following hypothesis testing:

$$H_0: \mu_2 = \mu_1 \quad H_1: \mu_2 \neq \mu_1 \quad (10.11)$$

If Σ is known, then

$$T_M^2 = \frac{n_1 n_2}{n_1 + n_2}(\overline{X} - \overline{Y})^T \Sigma^{-1}(\overline{X} - \overline{Y}) \quad (10.12)$$

will follow χ_p^2 if $H_0: \mu_2 = \mu_1$ is true. However, usually Σ is unknown and it has to be estimated from the sample. Because $\mathbf{Y}$ is the reference sample from normal operation, we often use the sample covariance matrix of the reference sample, $\mathbf{S}_Y$. In this case, the test statistic for $H_0: \mu_2 = \mu_1$ is

$$T_M^2 = n_2(\overline{X} - \overline{Y})^T \mathbf{S}_Y^{-1}(\overline{X} - \overline{Y}) \quad (10.13)$$

The upper control limit (UCL) in this case is

$$\text{UCL} = \frac{p(n_1 - 1)(n_1 + n_2)}{n_1(n_1 - p)} F_{\alpha, p, n_1 - p} \quad (10.14)$$

Similarly, we can also compute the other T^2 statistic

$$T_D^2 = \sum_{i=1}^{n_2}(\mathbf{X}_{i.} - \overline{\mathbf{X}})^T \mathbf{S}_Y^{-1}(\mathbf{X}_{i.} - \overline{\mathbf{X}}) \quad (10.15)$$

Its upper control limit is

$$\text{UCL} = \chi_{\alpha,(n_2-1)p}^2 \quad (10.16)$$

$$T_O^2 = \sum_{i=1}^{n_2}(\mathbf{X}_{i.} - \overline{\mathbf{Y}})^T \mathbf{S}_Y^{-1}(\mathbf{X}_{i.} - \overline{\mathbf{Y}}) \quad (10.17)$$

Its upper control limit is

$$\text{UCL} = \chi^2_{\alpha, n_2 p} \tag{10.18}$$

Example 10.3: Electrolysis Process with a Reference Sample Recall Example 10.1, assume that we have the following reference sample from the normal production:

NaOH	NaCl	I_1	I_2	Cl_2	O_2
134.89	203	0.05	4	98.37	1.17
129.3	203.1	0.06	1.9	98.37	1.17
145.5	208.6	0.17	6.1	98.23	1.42
143.8	188.1	0.11	0.4	98.44	1.12
146.3	189.1	0.22	0.5	98.44	1.11
141.5	196.19	0.16	3.5	98.26	1.35
157.3	185.3	0.09	2.9	98.23	1.4
141.1	209.1	0.16	0.5	98.69	0.86
131.3	200.8	0.17	3.8	97.95	1.64
156.6	189	0.19	0.5	97.97	1.62
135.6	192.8	0.26	0.5	97.65	1.94
128.39	213.3	0.07	3.6	98.43	1.23
138.1	198.3	0.15	2.7	98.12	1.36
140.4	186.1	0.3	0.3	98.15	1.37
139.3	204	0.25	3.8	98.02	1.54
152.39	176.3	0.19	0.9	98.22	1.3
139.69	186.1	0.15	1.6	98.3	1.25
130.3	190.5	0.23	2.6	98.08	1.37
132.19	198.6	0.09	5.7	98.3	1.16
134.8	196.1	0.17	4.9	97.98	1.5

We can compute that

$$\overline{Y} = (139.94, 195.72, 0.162, 2.535, 98.21, 1.344)^T$$

$$S_Y = \begin{bmatrix} 72.27 & -44.94 & 0.10 & -5.50 & 0.06 & 0.09 \\ -44.94 & 90.78 & -0.23 & 9.40 & 0.52 & -0.36 \\ 0.10 & -0.23 & 0.005 & -0.047 & -0.008 & 0.008 \\ -5.50 & 9.40 & -0.047 & 3.47 & -0.033 & 0.037 \\ 0.06 & 0.52 & -0.008 & -0.033 & 0.053 & -0.052 \\ 0.09 & -0.36 & 0.008 & 0.037 & -0.052 & 0.055 \end{bmatrix}$$

Clearly, $n_1 = 20$.

In current production, we obtained a new sample as follows:

NaOH	NaCl	I_1	I_2	Cl_2	O_2
126.33	190.12	0.20	6.32	98.10	1.47
137.83	188.68	0.22	14.28	98.10	1.47
124.30	199.23	0.34	15.56	98.10	1.47
131.72	191.0	0.27	7.55	98.10	1.47

We can compute that

$$\overline{X} = (130.04, 192.26, 0.26, 10.93, 98.1, 1.47)^T \qquad n_2 = 4$$

$$T_M^2 = n_2(\overline{\mathbf{X}} - \overline{\mathbf{Y}})^T \mathbf{S}_Y^{-1}(\overline{\mathbf{X}} - \overline{\mathbf{Y}})$$

$$= 4 \times (130.04 - 139.94, 192.26 - 195.72, 0.26 - 0.162, 10.93 - 2.535,$$
$$98.1 - 98.21, 1.47 - 1.344)$$

$$\times \begin{bmatrix} 72.27 & -44.94 & 0.10 & -5.50 & 0.06 & 0.09 \\ -44.94 & 90.78 & -0.23 & 9.40 & 0.52 & -0.36 \\ 0.10 & -0.23 & 0.005 & -0.047 & -0.008 & 0.008 \\ -5.50 & 9.40 & -0.047 & 3.47 & -0.033 & 0.037 \\ 0.06 & 0.52 & -0.008 & -0.033 & 0.053 & -0.052 \\ 0.09 & -0.36 & 0.008 & 0.037 & -0.052 & 0.055 \end{bmatrix}^{-1} \begin{bmatrix} 130.04 - 139.94 \\ 192.26 - 195.72 \\ 0.26 - 0.162 \\ 10.93 - 2.535 \\ 98.1 - 98.21 \\ 1.47 - 1.344 \end{bmatrix}$$

$$= 255.9704$$

$$\text{UCL} = \frac{p(n_1 - 1)(n_1 + n_2)}{n_1(n_1 - p)} F_{a,p,n_1-p} = \frac{6 \times 19 \times 24}{20 \times 14} F_{0.01,6,14} = \frac{2736}{280} 4.46 = 43.58$$

Clearly, $T_M^2 = 255.9704 > \text{UCL} = 43.58$, so T_M^2 is out of control.

$$T_D^2 = \sum_{i=1}^{n_2} (\mathbf{X}_{i.} - \overline{\mathbf{X}})^T \mathbf{S}_Y^{-1}(\mathbf{X}_{i.} - \overline{\mathbf{X}})$$

$$= (126.33 - 130.04, 190.12 - 192.26, 0.20 - 0.26, 6.32 - 10.93,$$
$$98.1 - 98.1, 1.47 - 1.47)$$

$$\times \begin{bmatrix} 72.27 & -44.94 & 0.10 & -5.50 & 0.06 & 0.09 \\ -44.94 & 90.78 & -0.23 & 9.40 & 0.52 & -0.36 \\ 0.10 & -0.23 & 0.005 & -0.047 & -0.008 & 0.008 \\ -5.50 & 9.40 & -0.047 & 3.47 & -0.033 & 0.037 \\ 0.06 & 0.52 & -0.008 & -0.033 & 0.053 & -0.052 \\ 0.09 & -0.36 & 0.008 & 0.037 & -0.052 & 0.055 \end{bmatrix}^{-1} \begin{bmatrix} 126.33 - 130.04 \\ 190.12 - 192.26 \\ 0.20 - 0.26 \\ 6.32 - 10.93 \\ 98.1 - 98.1 \\ 1.47 - 1.47 \end{bmatrix}$$

$+ \cdots$

$$+ (131.72 - 130.04, 191 - 192.26, 0.27 - 0.26, 7.75 - 10.93,$$
$$98.1 - 98.1, 1.47 - 1.47)$$

$$\times \begin{bmatrix} 72.27 & -44.94 & 0.10 & -5.50 & 0.06 & 0.09 \\ -44.94 & 90.78 & -0.23 & 9.40 & 0.52 & -0.36 \\ 0.10 & -0.23 & 0.005 & -0.047 & -0.008 & 0.008 \\ -5.50 & 9.40 & -0.047 & 3.47 & -0.033 & 0.037 \\ 0.06 & 0.52 & -0.008 & -0.033 & 0.053 & -0.052 \\ 0.09 & -0.36 & 0.008 & 0.037 & -0.052 & 0.055 \end{bmatrix}^{-1} \begin{bmatrix} 131.72 - 130.04 \\ 191 - 192.26 \\ 0.27 - 0.26 \\ 7.75 - 10.93 \\ 98.1 - 98.1 \\ 1.47 - 1.47 \end{bmatrix}$$

$$= 44.1788$$

Its upper control limit is

$$\text{UCL} = \chi^2_{a,(n_2-1)p} = \chi^2_{0.01,18} = 34.81$$

Clearly, T_D^2 is also out of control in this example.

10.3.2 Two-phase T^2 multivariate process control for subgroups

In many practical applications of T^2 multivariate process control, the multivariate data are taken in subgroups. In Phase I, we collect a sufficient number of subgroup data during the production run in normal operation condition, and establish control limits based on the data we collect. During Phase I, we examine the data for each subgroup for the Phase I data. If a subgroup is out of control, we remove that subgroup. Eventually, we collect enough in-control subgroups so that corresponding data sets are used to establish the control chart. In Phase II, we use the control limits based on Phase I data to examine the data from current production on a subgroup-by-subgroup basis. Now, we describe this two-phase process for subgroup data in detail.

Phase I: Reference sample preparation. We assume that in Phase I, there are k subgroups of multivariate observations. Each subgroup has n sets of data. All these data are collected in the normal production condition. We use these data to prepare a reference sample.

Example 10.4: Subgroups in Phase I For the reference sample discussed in the Example 10.3, the natural subgroup is a cell. In each cell, we collect four multivariate data sets. So if we subdivide the reference sample by subgroups, as marked by the bold lines in the table in this example, we have five subgroups, that is, $k = 5$, and each subgroup is of size $n = 4$.

NaOH	NaCl	I_1	I_2	Cl_2	O_2
134.89	203	0.05	4	98.37	1.17
129.3	203.1	0.06	1.9	98.37	1.17
145.5	208.6	0.17	6.1	98.23	1.42
143.8	188.1	0.11	0.4	98.44	1.12
146.3	189.1	0.22	0.5	98.44	1.11
141.5	196.19	0.16	3.5	98.26	1.35
157.3	185.3	0.09	2.9	98.23	1.4
141.1	209.1	0.16	0.5	98.69	0.86
131.3	200.8	0.17	3.8	97.95	1.64
156.6	189	0.19	0.5	97.97	1.62
135.6	192.8	0.26	0.5	97.65	1.94
128.39	213.3	0.07	3.6	98.43	1.23
138.1	198.3	0.15	2.7	98.12	1.36
140.4	186.1	0.3	0.3	98.15	1.37
139.3	204	0.25	3.8	98.02	1.54
152.39	176.3	0.19	0.9	98.22	1.3
139.69	186.1	0.15	1.6	98.3	1.25
130.3	190.5	0.23	2.6	98.08	1.37
132.19	198.6	0.09	5.7	98.3	1.16
134.8	196.1	0.17	4.9	97.98	1.5

Phase I consists of the following steps:

Step 1. For each subgroup $i = 1,\ldots,k$, compute
Subgroup averages $\overline{\mathbf{Y}}^{(i)}$ for $i = 1,\ldots,k$
Subgroup covariance matrices $\mathbf{S}^{(i)}$ for $i = 1,\ldots,k$
The grand mean vector $\overline{\overline{\mathbf{Y}}} = \frac{1}{k}\sum_{i=1}^{k}\overline{\mathbf{Y}}^{(i)}$
The pooled covariance matrix $\mathbf{S}_Y = \frac{1}{k}\sum_{i=1}^{k}\mathbf{S}^{(i)}$

Step 2. For each subgroup $i = 1,\ldots,k$, compute

$$T^2_{M^{(i)}} = n(\overline{\mathbf{Y}}^{(i)} - \overline{\overline{\mathbf{Y}}})^T \mathbf{S}_Y^{-1}(\overline{\mathbf{Y}}^{(i)} - \overline{\overline{\mathbf{Y}}}) \qquad (10.19)$$

$$T^2_{D^{(i)}} = \sum_{j=1}^{n}\left(\mathbf{Y}_j^{(i)} - \overline{\mathbf{Y}}^{(i)}\right)^T \mathbf{S}_Y^{-1}\left(\mathbf{Y}_j^{(i)} - \overline{\mathbf{Y}}^{(i)}\right) \qquad (10.20)$$

where $\mathbf{Y}_j^{(i)}$ is the jth multivariate data set in the ith subgroup, $j = 1,\ldots,n$, $i = 1,\ldots,k$.

Step 3. For each subgroup $i = 1,\ldots,k$, compare $T^2_{M^{(i)}}$ with

$$\mathrm{UCL} = \frac{knp - kp - np + p}{kn - k - p + 1} F_{\alpha,p,kn-k-p+1}$$

and $T^2_{D^{(i)}}$ with $\mathrm{UCL} = \chi^2_{\alpha,(n-1)p}$.

If either $T^2_{M^{(i)}} > \mathrm{UCL}$, and/or $T^2_{D^{(i)}} > \mathrm{UCL}$, then the ith subgroup is out of control, and the data in this subgroup should be not be used to calculate the control limits.

Step 4. Delete the data from out of control subgroups, and recalculate $\overline{\overline{\mathbf{Y}}}$ and $\mathbf{S}_Y$ by using the data in in-control subgroups.

After finishing these four steps, we are ready to use $\overline{\overline{\mathbf{Y}}}$ and $\mathbf{S}_Y$ in setting up multivariate process control in Phase II.

Phase II: Process control for new subgroups. In Phase I, we start with k subgroups, and the sample size for each subgroup is n. Because some subgroups might be discarded as out of control, at the beginning of Phase II, $\overline{\overline{\mathbf{Y}}}$ and $\mathbf{S}_Y$ are usually computed by using a smaller number of subgroups. However, k is only a symbol. Without loss of generality, we assume that the initial reference sample after Phase I has k subgroups and each subgroup has n sets of multivariate observations.

The control limits in Phase II are a little bit different from that of Phase I, because future subgroups are assumed to be independent of the current set of subgroups that is used in calculating $\overline{\overline{\mathbf{Y}}}$ and $\mathbf{S}_Y$.

The sample size for the new subgroups is still assumed to be equal to n. In Phase II, for each new subgroup, we calculate the following:

$$T^2_{M(new)} = n(\overline{\mathbf{X}}^{(new)} - \overline{\overline{\mathbf{Y}}})^T \mathbf{S}_Y^{-1}(\overline{\mathbf{X}}^{(new)} - \overline{\overline{\mathbf{Y}}}) \quad (10.21)$$

where $\overline{\mathbf{X}}^{(new)}$ is the vector average of the observations in the new subgroup. The upper control limit for $T^2_{M(new)}$ is

$$\text{UCL} = \frac{p(k-1)(n-1)}{kn-k-p+1} F_{\alpha,p,kn-k-p+1} \quad (10.22)$$

If

$$T^2_{M(new)} > \frac{p(k-1)(n-1)}{kn-k-p+1} F_{\alpha,p,kn-k-p+1}$$

then this new subgroup is out of control for the T^2_M chart, that is, the T^2 chart for mean. This means that this new subgroup's mean is significantly deviated from the mean of the reference sample.

We also compute

$$T^2_{D(new)} = \sum_{j=1}^{n} \left(\mathbf{X}_j^{(new)} - \overline{\mathbf{X}}^{(new)}\right)^T \mathbf{S}_Y^{-1}\left(\mathbf{X}_j^{(new)} - \overline{\mathbf{X}}^{(new)}\right) \quad (10.23)$$

where $\mathbf{X}_j^{(new)}$ is the jth multivariate data set in the new subgroup, $j = 1, \ldots, n$. The upper control limit for $T^2_{D(new)}$ is $\text{UCL} = \chi^2_{\alpha,(n-1)p}$. If $T^2_{D(new)} > \chi^2_{\alpha,(n-1)p}$, then this new subgroup is out of control for T^2_D chart, that is, the T^2 chart for deviation. This means that this new subgroup's internal variation is excessive.

Example 10.5: New Subgroups in Electrolysis Process Assume that for the same electrolysis process, after Phase I, we have $k = 5$, $n = 4$ in the reference sample and we compute that

$$\overline{\overline{\mathbf{Y}}} = (139.94, 195.72, 0.162, 2.535, 98.21, 1.344)^T$$

$$\mathbf{S}_Y = \begin{bmatrix} 72.27 & -44.94 & 0.10 & -5.50 & 0.06 & 0.09 \\ -44.94 & 90.78 & -0.23 & 9.40 & 0.52 & -0.36 \\ 0.10 & -0.23 & 0.005 & -0.047 & -0.008 & 0.008 \\ -5.50 & 9.40 & -0.047 & 3.47 & -0.033 & 0.037 \\ 0.06 & 0.52 & -0.008 & -0.033 & 0.053 & -0.052 \\ 0.09 & -0.36 & 0.008 & 0.037 & -0.052 & 0.055 \end{bmatrix}$$

And we get the following six new subgroups of data from the current production:

NaOH	NaCl	I_1	I_2	Cl_2	O_2
126.33	190.12	0.2	6.32	98.1	1.47
137.83	188.68	0.22	14.28	98.1	1.47
124.3	199.23	0.34	15.56	98.1	1.47
131.72	191	0.27	7.55	98.1	1.47
122.17	191.72	0.14	1.37	98.36	1.21
129.13	193.85	0.13	0.96	98.36	1.21
128.83	188.87	0.15	0.81	98.36	1.21
146.93	184.57	0.23	0.6	98.36	1.21
128.17	192.43	0.12	1.13	98.43	1.12
129.17	191.15	0.06	1.19	98.43	1.12
138.42	180.13	0.18	1.08	98.43	1.12
141.73	176.25	0.2	1.01	98.43	1.12
141.4	197.89	0.11	2.92	97.91	1.64
137.29	201.25	0.18	9.01	97.91	1.64
144	194.03	0.1	0.3	97.91	1.64
139.2	190.21	0.17	5.82	97.91	1.64
118.4	215.24	0.27	21.6	97.82	1.71
144.19	197.75	0.13	0.78	97.82	1.71
128.1	205.33	0.34	10.94	97.82	1.71
128.01	173.79	0.19	0.77	97.82	1.71
136.81	201.74	0.19	3.03	98.18	1.38
133.93	198.98	0.15	2.38	98.18	1.38
140.71	195.19	0.19	0.92	98.18	1.38
140.4	192.03	0.26	0.62	98.18	1.38

Subgroups are separated by the dark lines in the table. If we name the new subgroups as group 1, 2,..., 6, in order from top to bottom, we can compute

$$\overline{X}^{(1)} = (130.04, 192.26, 0.26, 10.93, 98.10, 1.47)$$
$$\overline{X}^{(2)} = (131.77, 189.75, 0.16, 0.94, 98.36, 1.21)$$
$$\overline{X}^{(3)} = (134.37, 184.99, 0.14, 1.10, 98.43, 1.12)$$
$$\overline{X}^{(4)} = (140.47, 195.84, 0.14, 4.51, 97.91, 1.64)$$
$$\overline{X}^{(5)} = (129.68, 198.03, 0.23, 8.52, 97.82, 1.71)$$
$$\overline{X}^{(6)} = (137.96, 196.98, 0.20, 1.74, 98.18, 1.38)$$

By using Eqs. (10.21) and (10.23) for each new subgroup, we have

$$T^2_{M^{(1)}} = 255.9704 \quad T^2_{M^{(2)}} = 32.3648 \quad T^2_{M^{(3)}} = 37.0287$$
$$T^2_{M^{(4)}} = 15.1382 \quad T^2_{M^{(5)}} = 77.6909 \quad T^2_{M^{(6)}} = 2.8721$$

and

$$T^2_{D^{(1)}} = 44.1788 \quad T^2_{D^{(2)}} = 8.9272 \quad T^2_{D^{(3)}} = 8.3150$$
$$T^2_{D^{(4)}} = 32.0262 \quad T^2_{D^{(5)}} = 183.1495 \quad T^2_{D^{(6)}} = 3.3248$$

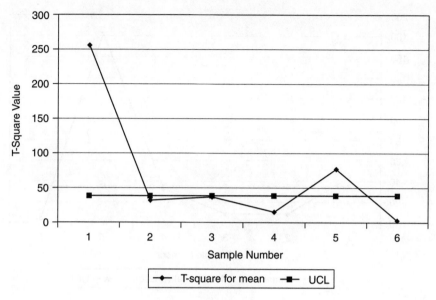

Figure 10.2 T_M^2 control chart plot for Example 10.5.

The upper control limit for the T_M^2 chart is

$$\mathrm{UCL} = \frac{p(k-1)(n-1)}{kn-k-p+1} F_{a,p,kn-k-p+1} = \frac{6 \times 4 \times 3}{5 \times 4 - 5 - 6 + 1} F_{0.01,6,10} = 7.2 \times 5.39 = 38.81$$

The upper control limit for the T_D^2 chart is

$$\mathrm{UCL} = \chi^2_{a,(n-1)p} = \chi^2_{0.01,18} = 34.81$$

Clearly, for subgroups 1 and 5, both T_M^2 and T_D^2 are out of control. We can plot the results in Figs.10.2 and 10.3.

10.4 T^2 Control Chart for Individual Observations

In multivariate process control practice, if the multivariate data are continuously monitored and recorded, that is, one observation at a time, then the sample size of the subgroup is 1. This case is similar to that of an individual control chart, or X chart, in a univariate statistical process control situation. Tracy, Young, and Mason (TYM) (1992) developed a special two-phase multivariate process control procedure based on the T^2 statistic.

In TYM procedure, Phase I is once again to establish a "clean" reference sample. It is very similar to the T^2 control chart procedure for subgroups. In this stage, samples of multivariate data are taken from production runs in normal operation state and are examined by a Phase I T^2 control chart. The out-of-control points will be deleted.

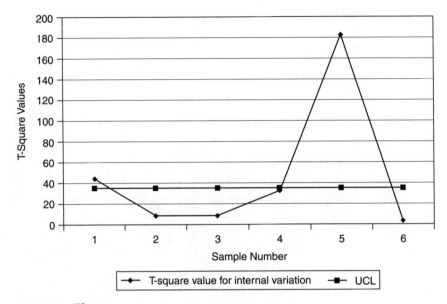

Figure 10.3 T_D^2 control chart plot for Example 10.5.

Then the reduced data set is our clean reference sample. This sample is used to establish the Phase II control chart. Phase II is the process control phase for current production. Now, we describe this two-phase process for individual multivariate data in detail.

10.4.1 Phase I reference sample preparation

In Phase I, assume that there are m pieces of multivariate observations available during a normal state of operation. Specifically, we have m sets of $\mathbf{Y}_{i.} = [y_{i1}, y_{i2}, ..., y_{ip}]^T$ for $i = 1, ..., m$. The mean vector is $\overline{\mathbf{Y}} = [\overline{Y}_{.1}, \overline{Y}_{.2}, ..., \overline{Y}_{.p}]$ where $\overline{Y}_{.j} = (1/m)\sum_{i=1}^{m} y_{ij}$ for $j = 1, ..., p$. The estimated covariance matrix for the reference sample can be computed by

$$\mathbf{S}_Y = \frac{1}{m-1} \sum_{i=1}^{m} (\mathbf{Y}_{i.} - \overline{\mathbf{Y}})(\mathbf{Y}_{i.} - \overline{\mathbf{Y}})^T$$

The T^2 statistic for individual observations is

$$Q_i = (\mathbf{Y}_{i.} - \overline{\mathbf{Y}})^T \mathbf{S}_Y^{-1} (\mathbf{Y}_{i.} - \overline{\mathbf{Y}}) \qquad (10.24)$$

Tracy, Young, and Mason (1992) has shown that Q_i follows a beta distribution, specifically

$$Q_i \sim \frac{(m-1)^2}{m} B(p/2, (m-p-1)/2) \qquad (10.25)$$

where $B(p/2,(m-p-1)/2)$ denotes the beta distribution with parameter $p/2$ and $(m-p-1)/2$. Therefore, by using the beta distribution, we can derive the lower and upper control limits for Q_i as follows:

$$\text{LCL} = \frac{(m-1)^2}{m} B(1-\alpha/2; p/2, (m-p-1)/2) \tag{10.26}$$

$$\text{UCL} = \frac{(m-1)^2}{m} B(\alpha/2; p/2, (m-p-1)/2) \tag{10.27}$$

where $B(\alpha/2; p/2, (m-p-1)/2)$ is the $1-\alpha/2$ percentile of the beta distribution with parameter $p/2$ and $(m-p-1)/2$.

If the table of the beta distribution is not available, we can use the following relationship between the beta distribution and F distribution:

$$B(\alpha; p/2, (m-p-1)/2) = \frac{[p/(m-p-1)]F_{\alpha,p,m-p-1}}{1+[p/(m-p-1)]F_{\alpha,p,m-p-1}} \tag{10.28}$$

Therefore, the LCL and ULC specified by Eqs. (10.26) and (10.27) can be rewritten as

$$\text{LCL} = \frac{(m-1)^2}{m} \frac{[p/(m-p-1)]F_{1-\alpha/2,p,m-p-1}}{1+[p/(m-p-1)]F_{1-\alpha/2,p,m-p-1}} \tag{10.29}$$

$$\text{UCL} = \frac{(m-1)^2}{m} \frac{[p/(m-p-1)]F_{\alpha/2,p,m-p-1}}{1+[p/(m-p-1)]F_{\alpha/2,p,m-p-1}} \tag{10.30}$$

Note that an LCL is stated, unlike the other multivariate control chart procedures given in this section. Although interest will generally be centered at the UCL, a value of Q_i below the LCL should also be investigated, as this could signal problems in data recording. As in the case when subgroups are used, if any of the points plot outside these control limits and special cause(s) that were subsequently removed can be identified, the point(s) would be deleted and the control limits recomputed, making the appropriate adjustments on the degrees of freedom, and retesting the remaining points against the new limits.

Example 10.6: Chemical Process Monitoring (Tracy, Young, and Mason, 1992) There are three variables that are used to monitor a chemical process. They are percentage impurities (X_1), temperature (X_2), and concentration (X_3). The initial sample has 14 observations as listed in Table 10.1

It can be computed that $\overline{Y} = (16.83, 85.19, 43.21)$. The sample covariance is

$$S_Y = \begin{bmatrix} 0.365 & -0.022 & 0.10 \\ -0.022 & 1.036 & -0.245 \\ 0.10 & -0.245 & 0.224 \end{bmatrix}$$

TABLE 10.1   Chemical Process Data

Sample no.	% Impurities X_1	Temperature X_2	Concentration X_3
1	14.92	85.77	42.26
2	16.90	83.77	43.44
3	17.38	84.46	42.74
4	16.90	86.27	43.60
5	16.92	85.23	43.18
6	16.71	83.81	43.72
7	17.07	86.08	43.33
8	16.93	85.85	43.41
9	16.71	85.73	43.28
10	16.88	86.27	42.59
11	16.73	83.46	44.00
12	17.07	85.81	42.78
13	17.60	85.92	43.11
14	16.90	84.23	43.48

Therefore, we can compute

$$Q_1 = (\mathbf{Y}_{1.} - \overline{\mathbf{Y}})^T \mathbf{S}_Y^{-1} (\mathbf{Y}_{1.} - \overline{\mathbf{Y}})$$

$$= (14.92 - 16.83, 85.77 - 85.19, 42.26 - 43.21) \times \begin{bmatrix} 0.365 & -0.022 & 0.10 \\ -0.022 & 1.036 & -0.245 \\ 0.10 & -0.245 & 0.224 \end{bmatrix}^{-1}$$

$$\times \begin{bmatrix} 14.92 - 16.83 \\ 85.77 - 85.19 \\ 42.26 - 43.21 \end{bmatrix} = 10.93$$

Similarly

$Q_2 = 2.04 \quad Q_3 = 5.58 \quad Q_4 = 3.86 \quad Q_5 = 0.04 \quad Q_6 = 2.25 \quad Q_7 = 1.44$

$Q_8 = 1.21 \quad Q_9 = 0.68 \quad Q_{10} = 2.17 \quad Q_{11} = 4.17 \quad Q_{12} = 1.40 \quad Q_{13} = 2.33$

$Q_{14} = 0.90$

The lower and upper control limits for the current data sample are

$$\text{LCL} = \frac{(14-1)^2}{14} \times \frac{[3/(14-3-1)] \times 0.0229}{1 + [3/(14-3-1)] \times 0.0229} = 0.082$$

$$\text{UCL} = \frac{(14-1)^2}{14} \times \frac{[3/(14-3-1)] \times 8.081}{1 + [3/(14-3-1)] \times 8.081} = 8.55$$

Figure 10.4 illustrates the control chart plot based on our calculations. Clearly, the first sample is out of control, and so is sample 5. Later it is found that sample 1 is due to measurement error; and there is no assignable cause for sample 5. So sample 1 is deleted, and sample 5 is retained. Based on the remaining 13 samples, we compute

$$\overline{\mathbf{Y}} = (16.98, 85.14, 43.28)$$

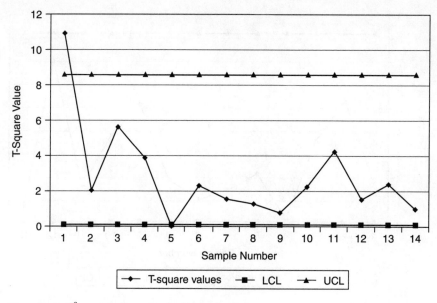

Figure 10.4 T^2 control chart plot for Example 10.6.

The sample covariance is

$$\mathbf{S}_Y = \begin{bmatrix} 0.068 & 0.076 & -0.055 \\ 0.076 & 1.092 & -0.216 \\ -0.055 & -0.216 & 0.163 \end{bmatrix}$$

By using

$$Q_i = (\mathbf{Y}_{i.} - \overline{\mathbf{Y}})^T \mathbf{S}_Y^{-1} (\mathbf{Y}_{i.} - \overline{\mathbf{Y}})$$

We recalculate

$Q_1 = 1.84$ (that corresponds to old Q_2) $Q_2 = 5.33$ $Q_3 = 3.58$ $Q_4 = 0.23$
$Q_5 = 2.17$ $Q_6 = 1.46$ $Q_7 = 1.05$ $Q_8 = 1.91$ $Q_9 = 5.16$ $Q_{10} = 3.84$
$Q_{11} = 1.65$ $Q_{12} = 7.00$ $Q_{13} = 0.77$

We recompute the control limits to be

$$\text{LCL} = \frac{(13-1)^2}{13} \times \frac{[3/(13-3-1)] \times 0.0229}{1 + [3/(13-3-1)] \times 0.0229} = 0.084$$

$$\text{UCL} = \frac{(13-1)^2}{13} \times \frac{[3/(13-3-1)] \times 8.081}{1 + [3/(13-3-1)] \times 8.081} = 8.24$$

Figure 10.5 illustrates the new control chart plot after sample 1 is deleted.

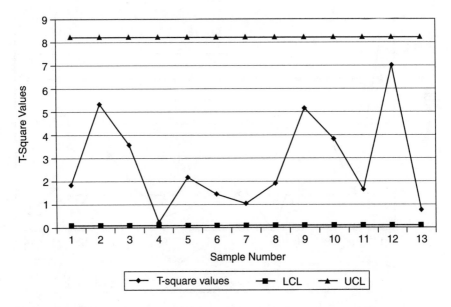

Figure 10.5 T^2 control chart for Example 10.6 after deleting out of control point.

10.4.2 Phase II: Process control for new observations

Phase II of this process control procedure deals with setting up of the control limits for new observations. At this phase, each new observation is denoted by

$$\mathbf{X}^{(new)} = (X_1, X_2, \ldots, X_p)^T$$

The T^2 statistic used in process control for this phase is

$$Q^{(new)} = (\mathbf{X}^{(new)} - \overline{\mathbf{Y}})^T \mathbf{S}_Y^{-1} (\mathbf{X}^{(new)} - \overline{\mathbf{Y}})$$

where $\overline{\mathbf{Y}}$ and $\mathbf{S}_Y$ are based on the reference sample in which out-of-control observations are deleted. It is reasonable to assume that the new observation $\mathbf{X}^{(new)}$ is independent of $\overline{\mathbf{Y}}$ and $\mathbf{S}_Y$, which is different from that of Phase I, where each observation in the reference sample is not independent of $\overline{\mathbf{Y}}$ and $\mathbf{S}_Y$. The control limits for Phase II will be different. Tracy, Young, and Mason (1992) developed the following control limits for Phase II.

$$\text{LCL} = \frac{p(m+1)(m-1)}{m(m-p)} F_{1-\alpha/2; p, m-p} \qquad (10.31)$$

$$\text{UCL} = \frac{p(m+1)(m-1)}{m(m-p)} F_{\alpha/2; p, m-p} \qquad (10.32)$$

Example 10.7: New Observations in Chemical Process Continuing with the situation described in Example 10.6, we assume that we get a new observation:

$$\mathbf{X}^{(new)} = (17.08, 84.08, 43.81)^T$$

$$Q^{(new)} = (\mathbf{X}^{(new)} - \overline{\mathbf{Y}})^T \mathbf{S}_{\overline{\mathbf{Y}}}^{-1} (\mathbf{X}^{(new)} - \overline{\mathbf{Y}})$$

$$= (17.08 - 16.98, 84.08 - 85.14, 43.81 - 43.28) \times \begin{bmatrix} 0.068 & 0.076 & -0.055 \\ 0.076 & 1.092 & -0.216 \\ -0.055 & -0.216 & 0.163 \end{bmatrix}^{-1}$$

$$\times \begin{bmatrix} 17.08 - 16.98 \\ 84.08 - 85.14 \\ 43.81 - 43.28 \end{bmatrix} = 3.52$$

The control limits for the new observations are

$$LCL = \frac{3(13+1)(13-1)}{13(13-3)} 0.0229 = 0.088$$

$$UCL = \frac{3(13+1)(13-1)}{13(13-3)} 8.081 = 31.33$$

Because $Q^{(new)}$ is within control limits, this new observation is in control.

10.5 Principal Component Chart

So far, we have discussed several multivariate process control charts that are based on the T^2 statistic. With the help of computers, these charts are not very difficult to set up and use. They are beginning to be widely accepted by quality engineers and operators. However, there are still difficulties when using T^2-based control charts. First, unlike univariate control charts, the values plotted on the control charts are not original variables themselves, so the meaning of the T^2 statistic is not apparent to online operators. Second, when the T^2 statistic exceeds the control limit, the users will not know which variable(s) caused the out-of-control situation.

We can plot the values of every individual variable on separate charts alongside the T^2 control chart, so that we can monitor the individual variables as well. When T^2 is out of control, we can examine each variable to see if there are excessive variations in some variables. However, some of the out-of-control situations are not caused by the excessive variations of individual variables, but by the change in the covariance/correlation structure. In this case, looking at one variable at a time will not help us detect this situation.

The principal component control chart is an alternative approach or a supplement to the T^2 control chart. In the principal component control chart, usually the principal component on correlation is used, and each principal component is expressed as a linear combination of original standardized variables (the second type of mathematical expression for PCA). Specifically, each principal component Y_i, for $i = 1,\ldots, p$, is represented by a linear combination of standardized variables,

$$Y_i = c_{1i}Z_1 + c_{2i}Z_2 + \cdots + c_{ki}Z_k + \cdots + c_{pi}Z_p$$

where the standardized variable Z_k is computed by subtracting the mean and dividing by its standard deviation.

The principal components have two important advantages:

1. The new variables Y_i's are uncorrelated.
2. Very often, a few (sometimes one or two) principal components may capture most of the variability in the data so that we do not have to use all of the p principal components for control.

The control limits for the principal component control chart are very easy to establish, because the variance of the ith component, that is, Var(Y_i), is equal to the ith eigenvalue λ_i. If we use the standardized principal component score specified in Chap. 5, that is, the regular principal component score divided by $\sqrt{\lambda_i}$, then as we have shown in Chap. 5, the standardized principal component score should follow a standard normal distribution. It is well known that the standard normal distribution has the mean equal to 0 and standard deviation equal to 1. So the 3-Sigma control limit for the principal component control chart is simply ±3 (UCL = 3, LCL = −3, centerline = 0), if the standardized principal component score is used.

Now we describe the step-by-step procedure for setting up and using the principal component control chart.

Step 1. Obtain the initial reference sample from the production, and use the T^2 control chart to eliminate out-of-control observations. So we get a clean reference sample. Then we run a principal component analysis on the correlation matrix on this clean reference sample, to obtain the principal component equations, eigenvalues and percentage variations explained by each principal component.

Step 2. Pick the top k principal components, depending on how much the cumulative percentage variation can be explained by these k components. Usually 80 to 90 percent can be used as a cutoff line. Establish a control chart for each principal component. The center line for the control chart is 0, UCL = 3, LCL = −3.

Step 3. Compute normalized principal component scores for each multivariate observation in the reference sample and for each k component and plot them in the principal component control chart. Study each out-of-control point to find if there is an assignable cause. If there is an assignable cause, this point has to be deleted, and we go to step 1. Otherwise, we go to step 4.

Step 4. Now the control charts established in the first three steps can be used to monitor the current production process. It is very important to figure out what is the plausible interpretation for each principal component. It is desirable to keep both T^2 control charts and principal component charts. Now for each new multivariate observation, plot it in both the T^2 chart and the principal component chart. If this observation is out-of-control in one or several principal component chart(s), then it is caused by excessive variation featured by the corresponding principal component. Because each principal component indicates a special type of correlated variation, an observation with very high principal component score on that component is an extreme case for that type of variation, as discussed in Chap. 5. If an observation is out of control in T^2 but in control in every principal component chart, then the causes of variation for this observation cannot be explained by known principal components and further investigation needs to be made.

Example 10.8: Principal Component Control Chart for the Chemical Process Data in Example 10.6 and 10.7 In Examples 10.6 and 10.7, the initial reference sample has 14 observations. After the Phase I T^2 chart process, observation 1 is out of control and deleted. Now we try to establish principal control charts for the reduced initial sample by running a principal component analysis on correlation matrix in MINITAB. We get

Principal Component Analysis: X_1, X_2, X_3

```
Eigenanalysis of the Correlation Matrix

Eigenvalue    1.8797      0.7184      0.4019
Proportion    0.627       0.239       0.134
Cumulative    0.627       0.866       1.000

Variable         PC1         PC2         PC3
X1            -0.547       0.702      -0.456
X2            -0.544      -0.712      -0.443
X3             0.636      -0.006      -0.772
```

The first two components account for 86.6% of total variation. So it is reasonable to establish the two principal component charts.

From the MINITAB output, we find that the principal component equations for the first two components are

$$Y_1 = -0.547Z_1 - 0.544Z_2 + 0.636Z_3$$

$$Y_2 = 0.702Z_1 - 0.712Z_2 - 0.006Z_3$$

where Z_i's are the normalized X_i's, computed by subtracting the mean and dividing by the standard deviation. The eigenvalues for these two principal components are $\lambda_1 = 1.8797$, $\lambda_2 = 0.7184$.

TABLE 10.2 Normalized Principal Component Scores for the First Two Components

Sample no.	PC1 score	PC2 score
1	0.8229	0.8569
2	−0.9834	1.8495
3	0.0576	−1.1555
4	−0.0611	−0.2484
5	1.4223	0.2123
6	−0.4426	−0.4544
7	−0.0475	−0.7187
8	0.1877	−1.3238
9	−1.0724	−1.2026
10	1.8462	0.5530
11	−0.9722	−0.2282
12	−1.4516	1.3734
13	0.6942	0.4865

The normalized principal component scores for the first two components are computed by using

$$\frac{Y_1}{\sqrt{1.8797}} = \frac{-0.547Z_1 - 0.544Z_2 + 0.636Z_3}{\sqrt{1.8797}}$$

$$\frac{Y_2}{\sqrt{0.7184}} = \frac{0.702Z_1 - 0.712Z_2 - 0.006Z_3}{\sqrt{0.7184}}$$

We list our results in Table 10.2.

These principal component scores can be plotted at the principal component control charts illustrated in Figs. 10.6 and 10.7. Clearly, all points in the control

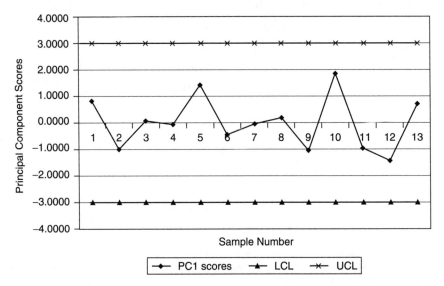

Figure 10.6 Principal component control chart for component 1.

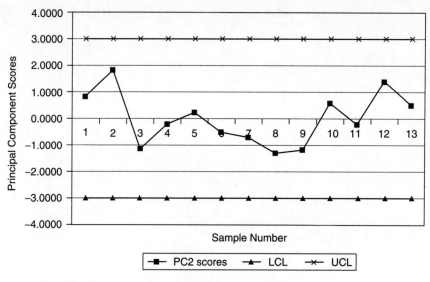

Figure 10.7 Principal component control chart for component 2.

charts are in control. If we have a new observation $\mathbf{X}^{(new)} = (17.08, 84.08, 43.81)^T$, the normalized values for this new observation are

$$Z_1 = \frac{17.08 - 16.98}{\sqrt{0.068}} = 0.385 \quad Z_2 = \frac{84.08 - 85.14}{\sqrt{1.092}} = -1.014 \quad Z_3 = \frac{43.81 - 43.28}{\sqrt{0.163}} = 1.312$$

The normalized principal component score for component 1 is

$$Y_1/\sqrt{1.8797} = (-0.547Z_1 - 0.544Z_2 + 0.636Z_3)/\sqrt{1.8797}$$
$$= [-0.547 \times 0.385 - 0.544 \times (-1.014) + 0.636 \times 1.312]/\sqrt{1.8797}$$
$$= 0.857$$

$$Y_2/\sqrt{0.7184} = (0.702Z_1 - 0.712Z_2 - 0.006Z_3)/\sqrt{0.7184}$$
$$= [0.702 \times 0.385 - 0.712 \times (-1.014) - 0.006 \times 1.312]/\sqrt{0.7184}$$
$$= 1.16$$

Both scores are within $(-3, 3)$, so this new observation is in control for both charts.

Appendix: Probability Distribution Tables

Table A.1 Standard Normal Cumulative Distribution

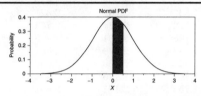

Area under the Normal Curve from 0 to X

X	0.00	0.01	0.02	0.03	0.04	0.05	0.06	0.07	0.08	0.09
0.0	0.00000	0.00399	0.00798	0.01197	0.01595	0.01994	0.02392	0.02790	0.03188	0.03586
0.1	0.03983	0.04380	0.04776	0.05172	0.05567	0.05962	0.06356	0.06749	0.07142	0.07535
0.2	0.07926	0.08317	0.08706	0.09095	0.09483	0.09871	0.10257	0.10642	0.11026	0.11409
0.3	0.11791	0.12172	0.12552	0.12930	0.13307	0.13683	0.14058	0.14431	0.14803	0.15173
0.4	0.15542	0.15910	0.16276	0.16640	0.17003	0.17364	0.17724	0.18082	0.18439	0.18793
0.5	0.19146	0.19497	0.19847	0.20194	0.20540	0.20884	0.21226	0.21566	0.21904	0.22240
0.6	0.22575	0.22907	0.23237	0.23565	0.23891	0.24215	0.24537	0.24857	0.25175	0.25490
0.7	0.25804	0.26115	0.26424	0.26730	0.27035	0.27337	0.27637	0.27935	0.28230	0.28524
0.8	0.28814	0.29103	0.29389	0.29673	0.29955	0.30234	0.30511	0.30785	0.31057	0.31327
0.9	0.31594	0.31859	0.32121	0.32381	0.32639	0.32894	0.33147	0.33398	0.33646	0.33891
1.0	0.34134	0.34375	0.34614	0.34849	0.35083	0.35314	0.35543	0.35769	0.35993	0.36214
1.1	0.36433	0.36650	0.36864	0.37076	0.37286	0.37493	0.37698	0.37900	0.38100	0.38298
1.2	0.38493	0.38686	0.38877	0.39065	0.39251	0.39435	0.39617	0.39796	0.39973	0.40147
1.3	0.40320	0.40490	0.40658	0.40824	0.40988	0.41149	0.41308	0.41466	0.41621	0.41774
1.4	0.41924	0.42073	0.42220	0.42364	0.42507	0.42647	0.42785	0.42922	0.43056	0.43189
1.5	0.43319	0.43448	0.43574	0.43699	0.43822	0.43943	0.44062	0.44179	0.44295	0.44408
1.6	0.44520	0.44630	0.44738	0.44845	0.44950	0.45053	0.45154	0.45254	0.45352	0.45449
1.7	0.45543	0.45637	0.45728	0.45818	0.45907	0.45994	0.46080	0.46164	0.46246	0.46327
1.8	0.46407	0.46485	0.46562	0.46638	0.46712	0.46784	0.46856	0.46926	0.46995	0.47062
1.9	0.47128	0.47193	0.47257	0.47320	0.47381	0.47441	0.47500	0.47558	0.47615	0.47670
2.0	0.47725	0.47778	0.47831	0.47882	0.47932	0.47982	0.48030	0.48077	0.48124	0.48169
2.1	0.48214	0.48257	0.48300	0.48341	0.48382	0.48422	0.48461	0.48500	0.48537	0.48574
2.2	0.48610	0.48645	0.48679	0.48713	0.48745	0.48778	0.48809	0.48840	0.48870	0.48899
2.3	0.48928	0.48956	0.48983	0.49010	0.49036	0.49061	0.49086	0.49111	0.49134	0.49158
2.4	0.49180	0.49202	0.49224	0.49245	0.49266	0.49286	0.49305	0.49324	0.49343	0.49361
2.5	0.49379	0.49396	0.49413	0.49430	0.49446	0.49461	0.49477	0.49492	0.49506	0.49520
2.6	0.49534	0.49547	0.49560	0.49573	0.49585	0.49598	0.49609	0.49621	0.49632	0.49643
2.7	0.49653	0.49664	0.49674	0.49683	0.49693	0.49702	0.49711	0.49720	0.49728	0.49736
2.8	0.49744	0.49752	0.49760	0.49767	0.49774	0.49781	0.49788	0.49795	0.49801	0.49807
2.9	0.49813	0.49819	0.49825	0.49831	0.49836	0.49841	0.49846	0.49851	0.49856	0.49861
3.0	0.49865	0.49869	0.49874	0.49878	0.49882	0.49886	0.49889	0.49893	0.49896	0.49900
3.1	0.49903	0.49906	0.49910	0.49913	0.49916	0.49918	0.49921	0.49924	0.49926	0.49929
3.2	0.49931	0.49934	0.49936	0.49938	0.49940	0.49942	0.49944	0.49946	0.49948	0.49950
3.3	0.49952	0.49953	0.49955	0.49957	0.49958	0.49960	0.49961	0.49962	0.49964	0.49965
3.4	0.49966	0.49968	0.49969	0.49970	0.49971	0.49972	0.49973	0.49974	0.49975	0.49976
3.5	0.49977	0.49978	0.49978	0.49979	0.49980	0.49981	0.49981	0.49982	0.49983	0.49983
3.6	0.49984	0.49985	0.49985	0.49986	0.49986	0.49987	0.49987	0.49988	0.49988	0.49989
3.7	0.49989	0.49990	0.49990	0.49990	0.49991	0.49991	0.49992	0.49992	0.49992	0.49992
3.8	0.49993	0.49993	0.49993	0.49994	0.49994	0.49994	0.49994	0.49995	0.49995	0.49995
3.9	0.49995	0.49995	0.49996	0.49996	0.49996	0.49996	0.49996	0.49996	0.49997	0.49997
4.0	0.49997	0.49997	0.49997	0.49997	0.49997	0.49997	0.49998	0.49998	0.49998	0.49998

Table A.2   Percentage Points of the Student t Distribution

	Probability of exceeding the critical value α					
DF	0.10	0.05	0.025	0.01	0.005	0.001
1.	3.078	6.314	12.706	31.821	63.657	318.313
2.	1.886	2.920	4.303	6.965	9.925	22.327
3.	1.638	2.353	3.182	4.541	5.841	10.215
4.	1.533	2.132	2.776	3.747	4.604	7.173
5.	1.476	2.015	2.571	3.365	4.032	5.893
6.	1.440	1.943	2.447	3.143	3.707	5.208
7.	1.415	1.895	2.365	2.998	3.499	4.782
8.	1.397	1.860	2.306	2.896	3.355	4.499
9.	1.383	1.833	2.262	2.821	3.250	4.296
10.	1.372	1.812	2.228	2.764	3.169	4.143
11.	1.363	1.796	2.201	2.718	3.106	4.024
12.	1.356	1.782	2.179	2.681	3.055	3.929
13.	1.350	1.771	2.160	2.650	3.012	3.852
14.	1.345	1.761	2.145	2.624	2.977	3.787
15.	1.341	1.753	2.131	2.602	2.947	3.733
16.	1.337	1.746	2.120	2.583	2.921	3.686
17.	1.333	1.740	2.110	2.567	2.898	3.646
18.	1.330	1.734	2.101	2.552	2.878	3.610
19.	1.328	1.729	2.093	2.539	2.861	3.579
20.	1.325	1.725	2.086	2.528	2.845	3.552
21.	1.323	1.721	2.080	2.518	2.831	3.527
22.	1.321	1.717	2.074	2.508	2.819	3.505
23.	1.319	1.714	2.069	2.500	2.807	3.485
24.	1.318	1.711	2.064	2.492	2.797	3.467
25.	1.316	1.708	2.060	2.485	2.787	3.450
26.	1.315	1.706	2.056	2.479	2.779	3.435
27.	1.314	1.703	2.052	2.473	2.771	3.421
28.	1.313	1.701	2.048	2.467	2.763	3.408
29.	1.311	1.699	2.045	2.462	2.756	3.396
30.	1.310	1.697	2.042	2.457	2.750	3.385
31.	1.309	1.696	2.040	2.453	2.744	3.375
32.	1.309	1.694	2.037	2.449	2.738	3.365
33.	1.308	1.692	2.035	2.445	2.733	3.356
34.	1.307	1.691	2.032	2.441	2.728	3.348
35.	1.306	1.690	2.030	2.438	2.724	3.340
36.	1.306	1.688	2.028	2.434	2.719	3.333
37.	1.305	1.687	2.026	2.431	2.715	3.326
38.	1.304	1.686	2.024	2.429	2.712	3.319
39.	1.304	1.685	2.023	2.426	2.708	3.313
40.	1.303	1.684	2.021	2.423	2.704	3.307
41.	1.303	1.683	2.020	2.421	2.701	3.301
42.	1.302	1.682	2.018	2.418	2.698	3.296
43.	1.302	1.681	2.017	2.416	2.695	3.291
44.	1.301	1.680	2.015	2.414	2.692	3.286
45.	1.301	1.679	2.014	2.412	2.690	3.281
46.	1.300	1.679	2.013	2.410	2.687	3.277
47.	1.300	1.678	2.012	2.408	2.685	3.273
48.	1.299	1.677	2.011	2.407	2.682	3.269
49.	1.299	1.677	2.010	2.405	2.680	3.265
50.	1.299	1.676	2.009	2.403	2.678	3.261

Table A.2 Percentage Points of the Student *t* Distribution (Continued)

	Probability of exceeding the critical value α					
DF	0.10	0.05	0.025	0.01	0.005	0.001
51.	1.298	1.675	2.008	2.402	2.676	3.258
52.	1.298	1.675	2.007	2.400	2.674	3.255
53.	1.298	1.674	2.006	2.399	2.672	3.251
54.	1.297	1.674	2.005	2.397	2.670	3.248
55.	1.297	1.673	2.004	2.396	2.668	3.245
56.	1.297	1.673	2.003	2.395	2.667	3.242
57.	1.297	1.672	2.002	2.394	2.665	3.239
58.	1.296	1.672	2.002	2.392	2.663	3.237
59.	1.296	1.671	2.001	2.391	2.662	3.234
60.	1.296	1.671	2.000	2.390	2.660	3.232
61.	1.296	1.670	2.000	2.389	2.659	3.229
62.	1.295	1.670	1.999	2.388	2.657	3.227
63.	1.295	1.669	1.998	2.387	2.656	3.225
64.	1.295	1.669	1.998	2.386	2.655	3.223
65.	1.295	1.669	1.997	2.385	2.654	3.220
66.	1.295	1.668	1.997	2.384	2.652	3.218
67.	1.294	1.668	1.996	2.383	2.651	3.216
68.	1.294	1.668	1.995	2.382	2.650	3.214
69.	1.294	1.667	1.995	2.382	2.649	3.213
70.	1.294	1.667	1.994	2.381	2.648	3.211
71.	1.294	1.667	1.994	2.380	2.647	3.209
72.	1.293	1.666	1.993	2.379	2.646	3.207
73.	1.293	1.666	1.993	2.379	2.645	3.206
74.	1.293	1.666	1.993	2.378	2.644	3.204
75.	1.293	1.665	1.992	2.377	2.643	3.202
76.	1.293	1.665	1.992	2.376	2.642	3.201
77.	1.293	1.665	1.991	2.376	2.641	3.199
78.	1.292	1.665	1.991	2.375	2.640	3.198
79.	1.292	1.664	1.990	2.374	2.640	3.197
80.	1.292	1.664	1.990	2.374	2.639	3.195
81.	1.292	1.664	1.990	2.373	2.638	3.194
82.	1.292	1.664	1.989	2.373	2.637	3.193
83.	1.292	1.663	1.989	2.372	2.636	3.191
84.	1.292	1.663	1.989	2.372	2.636	3.190
85.	1.292	1.663	1.988	2.371	2.635	3.189
86.	1.291	1.663	1.988	2.370	2.634	3.188
87.	1.291	1.663	1.988	2.370	2.634	3.187
88.	1.291	1.662	1.987	2.369	2.633	3.185
89.	1.291	1.662	1.987	2.369	2.632	3.184
90.	1.291	1.662	1.987	2.368	2.632	3.183
91.	1.291	1.662	1.986	2.368	2.631	3.182
92.	1.291	1.662	1.986	2.368	2.630	3.181
93.	1.291	1.661	1.986	2.367	2.630	3.180
94.	1.291	1.661	1.986	2.367	2.629	3.179
95.	1.291	1.661	1.985	2.366	2.629	3.178
96.	1.290	1.661	1.985	2.366	2.628	3.177
97.	1.290	1.661	1.985	2.365	2.627	3.176
98.	1.290	1.661	1.984	2.365	2.627	3.175
99.	1.290	1.660	1.984	2.365	2.626	3.175
100.	1.290	1.660	1.984	2.364	2.626	3.174
∞	1.282	1.645	1.960	2.326	2.576	3.090

Table A.3 Percentage Points of the Chi-Square Distribution

DF	\multicolumn{5}{c}{Probability of exceeding the critical value}				
	0.10	0.05	0.025	0.01	0.001
1.	2.706	3.841	5.024	6.635	10.828
2.	4.605	5.991	7.378	9.210	13.816
3.	6.251	7.815	9.348	11.345	16.266
4.	7.779	9.488	11.143	13.277	18.467
5.	9.236	11.070	12.833	15.086	20.515
6.	10.645	12.592	14.449	16.812	22.458
7.	12.017	14.067	16.013	18.475	24.322
8.	13.362	15.507	17.535	20.090	26.125
9.	14.684	16.919	19.023	21.666	27.877
10.	15.987	18.307	20.483	23.209	29.588
11.	17.275	19.675	21.920	24.725	31.264
12.	18.549	21.026	23.337	26.217	32.910
13.	19.812	22.362	24.736	27.688	34.528
14.	21.064	23.685	26.119	29.141	36.123
15.	22.307	24.996	27.488	30.578	37.697
16.	23.542	26.296	28.845	32.000	39.252
17.	24.769	27.587	30.191	33.409	40.790
18.	25.989	28.869	31.526	34.805	42.312
19.	27.204	30.144	32.852	36.191	43.820
20.	28.412	31.410	34.170	37.566	45.315
21.	29.615	32.671	35.479	38.932	46.797
22.	30.813	33.924	36.781	40.289	48.268
23.	32.007	35.172	38.076	41.638	49.728
24.	33.196	36.415	39.364	42.980	51.179
25.	34.382	37.652	40.646	44.314	52.620
26.	35.563	38.885	41.923	45.642	54.052
27.	36.741	40.113	43.195	46.963	55.476
28.	37.916	41.337	44.461	48.278	56.892
29.	39.087	42.557	45.722	49.588	58.301
30.	40.256	43.773	46.979	50.892	59.703
31.	41.422	44.985	48.232	52.191	61.098
32.	42.585	46.194	49.480	53.486	62.487
33.	43.745	47.400	50.725	54.776	63.870
34.	44.903	48.602	51.966	56.061	65.247
35.	46.059	49.802	53.203	57.342	66.619
36.	47.212	50.998	54.437	58.619	67.985
37.	48.363	52.192	55.668	59.893	69.347
38.	49.513	53.384	56.896	61.162	70.703
39.	50.660	54.572	58.120	62.428	72.055
40.	51.805	55.758	59.342	63.691	73.402
41.	52.949	56.942	60.561	64.950	74.745
42.	54.090	58.124	61.777	66.206	76.084
43.	55.230	59.304	62.990	67.459	77.419
44.	56.369	60.481	64.201	68.710	78.750
45.	57.505	61.656	65.410	69.957	80.077
46.	58.641	62.830	66.617	71.201	81.400
47.	59.774	64.001	67.821	72.443	82.720
48.	60.907	65.171	69.023	73.683	84.037
49.	62.038	66.339	70.222	74.919	85.351
50.	63.167	67.505	71.420	76.154	86.661

Table A.3 Percentage Points of the Chi-Square Distribution (Continued)

	Probability of exceeding the critical value				
DF	0.10	0.05	0.025	0.01	0.001
51.	64.295	68.669	72.616	77.386	87.968
52.	65.422	69.832	73.810	78.616	89.272
53.	66.548	70.993	75.002	79.843	90.573
54.	67.673	72.153	76.192	81.069	91.872
55.	68.796	73.311	77.380	82.292	93.168
56.	69.919	74.468	78.567	83.513	94.461
57.	71.040	75.624	79.752	84.733	95.751
58.	72.160	76.778	80.936	85.950	97.039
59.	73.279	77.931	82.117	87.166	98.324
60.	74.397	79.082	83.298	88.379	99.607
61.	75.514	80.232	84.476	89.591	100.888
62.	76.630	81.381	85.654	90.802	102.166
63.	77.745	82.529	86.830	92.010	103.442
64.	78.860	83.675	88.004	93.217	104.716
65.	79.973	84.821	89.177	94.422	105.988
66.	81.085	85.965	90.349	95.626	107.258
67.	82.197	87.108	91.519	96.828	108.526
68.	83.308	88.250	92.689	98.028	109.791
69.	84.418	89.391	93.856	99.228	111.055
70.	85.527	90.531	95.023	100.425	112.317
71.	86.635	91.670	96.189	101.621	113.577
72.	87.743	92.808	97.353	102.816	114.835
73.	88.850	93.945	98.516	104.010	116.092
74.	89.956	95.081	99.678	105.202	117.346
75.	91.061	96.217	100.839	106.393	118.599
76.	92.166	97.351	101.999	107.583	119.850
77.	93.270	98.484	103.158	108.771	121.100
78.	94.374	99.617	104.316	109.958	122.348
79.	95.476	100.749	105.473	111.144	123.594
80.	96.578	101.879	106.629	112.329	124.839
81.	97.680	103.010	107.783	113.512	126.083
82.	98.780	104.139	108.937	114.695	127.324
83.	99.880	105.267	110.090	115.876	128.565
84.	100.980	106.395	111.242	117.057	129.804
85.	102.079	107.522	112.393	118.236	131.041
86.	103.177	108.648	113.544	119.414	132.277
87.	104.275	109.773	114.693	120.591	133.512
88.	105.372	110.898	115.841	121.767	134.746
89.	106.469	112.022	116.989	122.942	135.978
90.	107.565	113.145	118.136	124.116	137.208
91.	108.661	114.268	119.282	125.289	138.438
92.	109.756	115.390	120.427	126.462	139.666
93.	110.850	116.511	121.571	127.633	140.893
94.	111.944	117.632	122.715	128.803	142.119
95.	113.038	118.752	123.858	129.973	143.344
96.	114.131	119.871	125.000	131.141	144.567
97.	115.223	120.990	126.141	132.309	145.789
98.	116.315	122.108	127.282	133.476	147.010
99.	117.407	123.225	128.422	134.642	148.230
100.	118.498	124.342	129.561	135.807	149.449

(*continued*)

Table A.3 Percentage Points of the Chi-Square Distribution (Continued)

	Probability of exceeding the critical value				
DF	0.90	0.95	0.975	0.99	0.999
1.	.016	.004	.001	.000	.000
2.	.211	.103	.051	.020	.002
3.	.584	.352	.216	.115	.024
4.	1.064	.711	.484	.297	.091
5.	1.610	1.145	.831	.554	.210
6.	2.204	1.635	1.237	.872	.381
7.	2.833	2.167	1.690	1.239	.598
8.	3.490	2.733	2.180	1.646	.857
9.	4.168	3.325	2.700	2.088	1.152
10.	4.865	3.940	3.247	2.558	1.479
11.	5.578	4.575	3.816	3.053	1.834
12.	6.304	5.226	4.404	3.571	2.214
13.	7.042	5.892	5.009	4.107	2.617
14.	7.790	6.571	5.629	4.660	3.041
15.	8.547	7.261	6.262	5.229	3.483
16.	9.312	7.962	6.908	5.812	3.942
17.	10.085	8.672	7.564	6.408	4.416
18.	10.865	9.390	8.231	7.015	4.905
19.	11.651	10.117	8.907	7.633	5.407
20.	12.443	10.851	9.591	8.260	5.921
21.	13.240	11.591	10.283	8.897	6.447
22.	14.041	12.338	10.982	9.542	6.983
23.	14.848	13.091	11.689	10.196	7.529
24.	15.659	13.848	12.401	10.856	8.085
25.	16.473	14.611	13.120	11.524	8.649
26.	17.292	15.379	13.844	12.198	9.222
27.	18.114	16.151	14.573	12.879	9.803
28.	18.939	16.928	15.308	13.565	10.391
29.	19.768	17.708	16.047	14.256	10.986
30.	20.599	18.493	16.791	14.953	11.588
31.	21.434	19.281	17.539	15.655	12.196
32.	22.271	20.072	18.291	16.362	12.811
33.	23.110	20.867	19.047	17.074	13.431
34.	23.952	21.664	19.806	17.789	14.057
35.	24.797	22.465	20.569	18.509	14.688
36.	25.643	23.269	21.336	19.233	15.324
37.	26.492	24.075	22.106	19.960	15.965
38.	27.343	24.884	22.878	20.691	16.611
39.	28.196	25.695	23.654	21.426	17.262
40.	29.051	26.509	24.433	22.164	17.916
41.	29.907	27.326	25.215	22.906	18.575
42.	30.765	28.144	25.999	23.650	19.239
43.	31.625	28.965	26.785	24.398	19.906
44.	32.487	29.787	27.575	25.148	20.576
45.	33.350	30.612	28.366	25.901	21.251
46.	34.215	31.439	29.160	26.657	21.929
47.	35.081	32.268	29.956	27.416	22.610
48.	35.949	33.098	30.755	28.177	23.295
49.	36.818	33.930	31.555	28.941	23.983
50.	37.689	34.764	32.357	29.707	24.674

Table A.3 Percentage Points of the Chi-Square Distribution (Continued)

	Probability of exceeding the critical value				
DF	0.90	0.95	0.975	0.99	0.999
51.	38.560	35.600	33.162	30.475	25.368
52.	39.433	36.437	33.968	31.246	26.065
53.	40.308	37.276	34.776	32.018	26.765
54.	41.183	38.116	35.586	32.793	27.468
55.	42.060	38.958	36.398	33.570	28.173
56.	42.937	39.801	37.212	34.350	28.881
57.	43.816	40.646	38.027	35.131	29.592
58.	44.696	41.492	38.844	35.913	30.305
59.	45.577	42.339	39.662	36.698	31.020
60.	46.459	43.188	40.482	37.485	31.738
61.	47.342	44.038	41.303	38.273	32.459
62.	48.226	44.889	42.126	39.063	33.181
63.	49.111	45.741	42.950	39.855	33.906
64.	49.996	46.595	43.776	40.649	34.633
65.	50.883	47.450	44.603	41.444	35.362
66.	51.770	48.305	45.431	42.240	36.093
67.	52.659	49.162	46.261	43.038	36.826
68.	53.548	50.020	47.092	43.838	37.561
69.	54.438	50.879	47.924	44.639	38.298
70.	55.329	51.739	48.758	45.442	39.036
71.	56.221	52.600	49.592	46.246	39.777
72.	57.113	53.462	50.428	47.051	40.519
73.	58.006	54.325	51.265	47.858	41.264
74.	58.900	55.189	52.103	48.666	42.010
75.	59.795	56.054	52.942	49.475	42.757
76.	60.690	56.920	53.782	50.286	43.507
77.	61.586	57.786	54.623	51.097	44.258
78.	62.483	58.654	55.466	51.910	45.010
79.	63.380	59.522	56.309	52.725	45.764
80.	64.278	60.391	57.153	53.540	46.520
81.	65.176	61.261	57.998	54.357	47.277
82.	66.076	62.132	58.845	55.174	48.036
83.	66.976	63.004	59.692	55.993	48.796
84.	67.876	63.876	60.540	56.813	49.557
85.	68.777	64.749	61.389	57.634	50.320
86.	69.679	65.623	62.239	58.456	51.085
87.	70.581	66.498	63.089	59.279	51.850
88.	71.484	67.373	63.941	60.103	52.617
89.	72.387	68.249	64.793	60.928	53.386
90.	73.291	69.126	65.647	61.754	54.155
91.	74.196	70.003	66.501	62.581	54.926
92.	75.100	70.882	67.356	63.409	55.698
93.	76.006	71.760	68.211	64.238	56.472
94.	76.912	72.640	69.068	65.068	57.246
95.	77.818	73.520	69.925	65.898	58.022
96.	78.725	74.401	70.783	66.730	58.799
97.	79.633	75.282	71.642	67.562	59.577
98.	80.541	76.164	72.501	68.396	60.356
99.	81.449	77.046	73.361	69.230	61.137
100.	82.358	77.929	74.222	70.065	61.918

Table A.4 Percentage Points of the F Distribution

Upper Critical Values of the F Distribution for v_1 Numerator Degrees of Freedom and

v_2 \ v_1	1	2	3	4	5	6	7	8	9	10
1	161.448	199.500	215.707	224.583	230.162	233.986	236.768	238.882	240.543	241.882
2	18.513	19.000	19.164	19.247	19.296	19.330	19.353	19.371	19.385	19.396
3	10.128	9.552	9.277	9.117	9.013	8.941	8.887	8.845	8.812	8.786
4	7.709	6.944	6.591	6.388	6.256	6.163	6.094	6.041	5.999	5.964
5	6.608	5.786	5.409	5.192	5.050	4.950	4.876	4.818	4.772	4.735
6	5.987	5.143	4.757	4.534	4.387	4.284	4.207	4.147	4.099	4.060
7	5.591	4.737	4.347	4.120	3.972	3.866	3.787	3.726	3.677	3.637
8	5.318	4.459	4.066	3.838	3.687	3.581	3.500	3.438	3.388	3.347
9	5.117	4.256	3.863	3.633	3.482	3.374	3.293	3.230	3.179	3.137
10	4.965	4.103	3.708	3.478	3.326	3.217	3.135	3.072	3.020	2.978
11	4.844	3.982	3.587	3.357	3.204	3.095	3.012	2.948	2.896	2.854
12	4.747	3.885	3.490	3.259	3.106	2.996	2.913	2.849	2.796	2.753
13	4.667	3.806	3.411	3.179	3.025	2.915	2.832	2.767	2.714	2.671
14	4.600	3.739	3.344	3.112	2.958	2.848	2.764	2.699	2.646	2.602
15	4.543	3.682	3.287	3.056	2.901	2.790	2.707	2.641	2.588	2.544
16	4.494	3.634	3.239	3.007	2.852	2.741	2.657	2.591	2.538	2.494
17	4.451	3.592	3.197	2.965	2.810	2.699	2.614	2.548	2.494	2.450
18	4.414	3.555	3.160	2.928	2.773	2.661	2.577	2.510	2.456	2.412
19	4.381	3.522	3.127	2.895	2.740	2.628	2.544	2.477	2.423	2.378
20	4.351	3.493	3.098	2.866	2.711	2.599	2.514	2.447	2.393	2.348
21	4.325	3.467	3.072	2.840	2.685	2.573	2.488	2.420	2.366	2.321
22	4.301	3.443	3.049	2.817	2.661	2.549	2.464	2.397	2.342	2.297
23	4.279	3.422	3.028	2.796	2.640	2.528	2.442	2.375	2.320	2.275
24	4.260	3.403	3.009	2.776	2.621	2.508	2.423	2.355	2.300	2.255
25	4.242	3.385	2.991	2.759	2.603	2.490	2.405	2.337	2.282	2.236
26	4.225	3.369	2.975	2.743	2.587	2.474	2.388	2.321	2.265	2.220
27	4.210	3.354	2.960	2.728	2.572	2.459	2.373	2.305	2.250	2.204
28	4.196	3.340	2.947	2.714	2.558	2.445	2.359	2.291	2.236	2.190
29	4.183	3.328	2.934	2.701	2.545	2.432	2.346	2.278	2.223	2.177
30	4.171	3.316	2.922	2.690	2.534	2.421	2.334	2.266	2.211	2.165
31	4.160	3.305	2.911	2.679	2.523	2.409	2.323	2.255	2.199	2.153
32	4.149	3.295	2.901	2.668	2.512	2.399	2.313	2.244	2.189	2.142
33	4.139	3.285	2.892	2.659	2.503	2.389	2.303	2.235	2.179	2.133
34	4.130	3.276	2.883	2.650	2.494	2.380	2.294	2.225	2.170	2.123
35	4.121	3.267	2.874	2.641	2.485	2.372	2.285	2.217	2.161	2.114
36	4.113	3.259	2.866	2.634	2.477	2.364	2.277	2.209	2.153	2.106
37	4.105	3.252	2.859	2.626	2.470	2.356	2.270	2.201	2.145	2.098
38	4.098	3.245	2.852	2.619	2.463	2.349	2.262	2.194	2.138	2.091
39	4.091	3.238	2.845	2.612	2.456	2.342	2.255	2.187	2.131	2.084
40	4.085	3.232	2.839	2.606	2.449	2.336	2.249	2.180	2.124	2.077
41	4.079	3.226	2.833	2.600	2.443	2.330	2.243	2.174	2.118	2.071
42	4.073	3.220	2.827	2.594	2.438	2.324	2.237	2.168	2.112	2.065
43	4.067	3.214	2.822	2.589	2.432	2.318	2.232	2.163	2.106	2.059
44	4.062	3.209	2.816	2.584	2.427	2.313	2.226	2.157	2.101	2.054
45	4.057	3.204	2.812	2.579	2.422	2.308	2.221	2.152	2.096	2.049
46	4.052	3.200	2.807	2.574	2.417	2.304	2.216	2.147	2.091	2.044
47	4.047	3.195	2.802	2.570	2.413	2.299	2.212	2.143	2.086	2.039
48	4.043	3.191	2.798	2.565	2.409	2.295	2.207	2.138	2.082	2.035
49	4.038	3.187	2.794	2.561	2.404	2.290	2.203	2.134	2.077	2.030
50	4.034	3.183	2.790	2.557	2.400	2.286	2.199	2.130	2.073	2.026

v_2 Denominator Degrees of Freedom, 5% Significance Level $F_{.05}(v_1, v_2)$

$v_2 \backslash v_1$	11	12	13	14	15	16	17	18	19	20
1	242.983	243.906	244.690	245.364	245.950	246.464	246.918	247.323	247.686	248.013
2	19.405	19.413	19.419	19.424	19.429	19.433	19.437	19.440	19.443	19.446
3	8.763	8.745	8.729	8.715	8.703	8.692	8.683	8.675	8.667	8.660
4	5.936	5.912	5.891	5.873	5.858	5.844	5.832	5.821	5.811	5.803
5	4.704	4.678	4.655	4.636	4.619	4.604	4.590	4.579	4.568	4.558
6	4.027	4.000	3.976	3.956	3.938	3.922	3.908	3.896	3.884	3.874
7	3.603	3.575	3.550	3.529	3.511	3.494	3.480	3.467	3.455	3.445
8	3.313	3.284	3.259	3.237	3.218	3.202	3.187	3.173	3.161	3.150
9	3.102	3.073	3.048	3.025	3.006	2.989	2.974	2.960	2.948	2.936
10	2.943	2.913	2.887	2.865	2.845	2.828	2.812	2.798	2.785	2.774
11	2.818	2.788	2.761	2.739	2.719	2.701	2.685	2.671	2.658	2.646
12	2.717	2.687	2.660	2.637	2.617	2.599	2.583	2.568	2.555	2.544
13	2.635	2.604	2.577	2.554	2.533	2.515	2.499	2.484	2.471	2.459
14	2.565	2.534	2.507	2.484	2.463	2.445	2.428	2.413	2.400	2.388
15	2.507	2.475	2.448	2.424	2.403	2.385	2.368	2.353	2.340	2.328
16	2.456	2.425	2.397	2.373	2.352	2.333	2.317	2.302	2.288	2.276
17	2.413	2.381	2.353	2.329	2.308	2.289	2.272	2.257	2.243	2.230
18	2.374	2.342	2.314	2.290	2.269	2.250	2.233	2.217	2.203	2.191
19	2.340	2.308	2.280	2.256	2.234	2.215	2.198	2.182	2.168	2.155
20	2.310	2.278	2.250	2.225	2.203	2.184	2.167	2.151	2.137	2.124
21	2.283	2.250	2.222	2.197	2.176	2.156	2.139	2.123	2.109	2.096
22	2.259	2.226	2.198	2.173	2.151	2.131	2.114	2.098	2.084	2.071
23	2.236	2.204	2.175	2.150	2.128	2.109	2.091	2.075	2.061	2.048
24	2.216	2.183	2.155	2.130	2.108	2.088	2.070	2.054	2.040	2.027
25	2.198	2.165	2.136	2.111	2.089	2.069	2.051	2.035	2.021	2.007
26	2.181	2.148	2.119	2.094	2.072	2.052	2.034	2.018	2.003	1.990
27	2.166	2.132	2.103	2.078	2.056	2.036	2.018	2.002	1.987	1.974
28	2.151	2.118	2.089	2.064	2.041	2.021	2.003	1.987	1.972	1.959
29	2.138	2.104	2.075	2.050	2.027	2.007	1.989	1.973	1.958	1.945
30	2.126	2.092	2.063	2.037	2.015	1.995	1.976	1.960	1.945	1.932
31	2.114	2.080	2.051	2.026	2.003	1.983	1.965	1.948	1.933	1.920
32	2.103	2.070	2.040	2.015	1.992	1.972	1.953	1.937	1.922	1.908
33	2.093	2.060	2.030	2.004	1.982	1.961	1.943	1.926	1.911	1.898
34	2.084	2.050	2.021	1.995	1.972	1.952	1.933	1.917	1.902	1.888
35	2.075	2.041	2.012	1.986	1.963	1.942	1.924	1.907	1.892	1.878
36	2.067	2.033	2.003	1.977	1.954	1.934	1.915	1.899	1.883	1.870
37	2.059	2.025	1.995	1.969	1.946	1.926	1.907	1.890	1.875	1.861
38	2.051	2.017	1.988	1.962	1.939	1.918	1.899	1.883	1.867	1.853
39	2.044	2.010	1.981	1.954	1.931	1.911	1.892	1.875	1.860	1.846
40	2.038	2.003	1.974	1.948	1.924	1.904	1.885	1.868	1.853	1.839
41	2.031	1.997	1.967	1.941	1.918	1.897	1.879	1.862	1.846	1.832
42	2.025	1.991	1.961	1.935	1.912	1.891	1.872	1.855	1.840	1.826
43	2.020	1.985	1.955	1.929	1.906	1.885	1.866	1.849	1.834	1.820
44	2.014	1.980	1.950	1.924	1.900	1.879	1.861	1.844	1.828	1.814
45	2.009	1.974	1.945	1.918	1.895	1.874	1.855	1.838	1.823	1.808
46	2.004	1.969	1.940	1.913	1.890	1.869	1.850	1.833	1.817	1.803
47	1.999	1.965	1.935	1.908	1.885	1.864	1.845	1.828	1.812	1.798
48	1.995	1.960	1.930	1.904	1.880	1.859	1.840	1.823	1.807	1.793
49	1.990	1.956	1.926	1.899	1.876	1.855	1.836	1.819	1.803	1.789
50	1.986	1.952	1.921	1.895	1.871	1.850	1.831	1.814	1.798	1.784

(continued)

Table A.4 Percentage Points of the F Distribution (Continued)

Upper Critical Values of the F Distribution for v_1 Numerator Degrees of Freedom and

$v_2 \backslash v_1$	1	2	3	4	5	6	7	8	9	10
51	4.030	3.179	2.786	2.553	2.397	2.283	2.195	2.126	2.069	2.022
52	4.027	3.175	2.783	2.550	2.393	2.279	2.192	2.122	2.066	2.018
53	4.023	3.172	2.779	2.546	2.389	2.275	2.188	2.119	2.062	2.015
54	4.020	3.168	2.776	2.543	2.386	2.272	2.185	2.115	2.059	2.011
55	4.016	3.165	2.773	2.540	2.383	2.269	2.181	2.112	2.055	2.008
56	4.013	3.162	2.769	2.537	2.380	2.266	2.178	2.109	2.052	2.005
57	4.010	3.159	2.766	2.534	2.377	2.263	2.175	2.106	2.049	2.001
58	4.007	3.156	2.764	2.531	2.374	2.260	2.172	2.103	2.046	1.998
59	4.004	3.153	2.761	2.528	2.371	2.257	2.169	2.100	2.043	1.995
60	4.001	3.150	2.758	2.525	2.368	2.254	2.167	2.097	2.040	1.993
61	3.998	3.148	2.755	2.523	2.366	2.251	2.164	2.094	2.037	1.990
62	3.996	3.145	2.753	2.520	2.363	2.249	2.161	2.092	2.035	1.987
63	3.993	3.143	2.751	2.518	2.361	2.246	2.159	2.089	2.032	1.985
64	3.991	3.140	2.748	2.515	2.358	2.244	2.156	2.087	2.030	1.982
65	3.989	3.138	2.746	2.513	2.356	2.242	2.154	2.084	2.027	1.980
66	3.986	3.136	2.744	2.511	2.354	2.239	2.152	2.082	2.025	1.977
67	3.984	3.134	2.742	2.509	2.352	2.237	2.150	2.080	2.023	1.975
68	3.982	3.132	2.740	2.507	2.350	2.235	2.148	2.078	2.021	1.973
69	3.980	3.130	2.737	2.505	2.348	2.233	2.145	2.076	2.019	1.971
70	3.978	3.128	2.736	2.503	2.346	2.231	2.143	2.074	2.017	1.969
71	3.976	3.126	2.734	2.501	2.344	2.229	2.142	2.072	2.015	1.967
72	3.974	3.124	2.732	2.499	2.342	2.227	2.140	2.070	2.013	1.965
73	3.972	3.122	2.730	2.497	2.340	2.226	2.138	2.068	2.011	1.963
74	3.970	3.120	2.728	2.495	2.338	2.224	2.136	2.066	2.009	1.961
75	3.968	3.119	2.727	2.494	2.337	2.222	2.134	2.064	2.007	1.959
76	3.967	3.117	2.725	2.492	2.335	2.220	2.133	2.063	2.006	1.958
77	3.965	3.115	2.723	2.490	2.333	2.219	2.131	2.061	2.004	1.956
78	3.963	3.114	2.722	2.489	2.332	2.217	2.129	2.059	2.002	1.954
79	3.962	3.112	2.720	2.487	2.330	2.216	2.128	2.058	2.001	1.953
80	3.960	3.111	2.719	2.486	2.329	2.214	2.126	2.056	1.999	1.951
81	3.959	3.109	2.717	2.484	2.327	2.213	2.125	2.055	1.998	1.950
82	3.957	3.108	2.716	2.483	2.326	2.211	2.123	2.053	1.996	1.948
83	3.956	3.107	2.715	2.482	2.324	2.210	2.122	2.052	1.995	1.947
84	3.955	3.105	2.713	2.480	2.323	2.209	2.121	2.051	1.993	1.945
85	3.953	3.104	2.712	2.479	2.322	2.207	2.119	2.049	1.992	1.944
86	3.952	3.103	2.711	2.478	2.321	2.206	2.118	2.048	1.991	1.943
87	3.951	3.101	2.709	2.476	2.319	2.205	2.117	2.047	1.989	1.941
88	3.949	3.100	2.708	2.475	2.318	2.203	2.115	2.045	1.988	1.940
89	3.948	3.099	2.707	2.474	2.317	2.202	2.114	2.044	1.987	1.939
90	3.947	3.098	2.706	2.473	2.316	2.201	2.113	2.043	1.986	1.938
91	3.946	3.097	2.705	2.472	2.315	2.200	2.112	2.042	1.984	1.936
92	3.945	3.095	2.704	2.471	2.313	2.199	2.111	2.041	1.983	1.935
93	3.943	3.094	2.703	2.470	2.312	2.198	2.110	2.040	1.982	1.934
94	3.942	3.093	2.701	2.469	2.311	2.197	2.109	2.038	1.981	1.933
95	3.941	3.092	2.700	2.467	2.310	2.196	2.108	2.037	1.980	1.932
96	3.940	3.091	2.699	2.466	2.309	2.195	2.106	2.036	1.979	1.931
97	3.939	3.090	2.698	2.465	2.308	2.194	2.105	2.035	1.978	1.930
98	3.938	3.089	2.697	2.465	2.307	2.193	2.104	2.034	1.977	1.929
99	3.937	3.088	2.696	2.464	2.306	2.192	2.103	2.033	1.976	1.928
100	3.936	3.087	2.696	2.463	2.305	2.191	2.103	2.032	1.975	1.927

v_2 Denominator Degrees of Freedom, 5% Significance Level $F_{.05}(v_1, v_2)$

v_2\v_1	11	12	13	14	15	16	17	18	19	20
51	1.982	1.947	1.917	1.891	1.867	1.846	1.827	1.810	1.794	1.780
52	1.978	1.944	1.913	1.887	1.863	1.842	1.823	1.806	1.790	1.776
53	1.975	1.940	1.910	1.883	1.859	1.838	1.819	1.802	1.786	1.772
54	1.971	1.936	1.906	1.879	1.856	1.835	1.816	1.798	1.782	1.768
55	1.968	1.933	1.903	1.876	1.852	1.831	1.812	1.795	1.779	1.764
56	1.964	1.930	1.899	1.873	1.849	1.828	1.809	1.791	1.775	1.761
57	1.961	1.926	1.896	1.869	1.846	1.824	1.805	1.788	1.772	1.757
58	1.958	1.923	1.893	1.866	1.842	1.821	1.802	1.785	1.769	1.754
59	1.955	1.920	1.890	1.863	1.839	1.818	1.799	1.781	1.766	1.751
60	1.952	1.917	1.887	1.860	1.836	1.815	1.796	1.778	1.763	1.748
61	1.949	1.915	1.884	1.857	1.834	1.812	1.793	1.776	1.760	1.745
62	1.947	1.912	1.882	1.855	1.831	1.809	1.790	1.773	1.757	1.742
63	1.944	1.909	1.879	1.852	1.828	1.807	1.787	1.770	1.754	1.739
64	1.942	1.907	1.876	1.849	1.826	1.804	1.785	1.767	1.751	1.737
65	1.939	1.904	1.874	1.847	1.823	1.802	1.782	1.765	1.749	1.734
66	1.937	1.902	1.871	1.845	1.821	1.799	1.780	1.762	1.746	1.732
67	1.935	1.900	1.869	1.842	1.818	1.797	1.777	1.760	1.744	1.729
68	1.932	1.897	1.867	1.840	1.816	1.795	1.775	1.758	1.742	1.727
69	1.930	1.895	1.865	1.838	1.814	1.792	1.773	1.755	1.739	1.725
70	1.928	1.893	1.863	1.836	1.812	1.790	1.771	1.753	1.737	1.722
71	1.926	1.891	1.861	1.834	1.810	1.788	1.769	1.751	1.735	1.720
72	1.924	1.889	1.859	1.832	1.808	1.786	1.767	1.749	1.733	1.718
73	1.922	1.887	1.857	1.830	1.806	1.784	1.765	1.747	1.731	1.716
74	1.921	1.885	1.855	1.828	1.804	1.782	1.763	1.745	1.729	1.714
75	1.919	1.884	1.853	1.826	1.802	1.780	1.761	1.743	1.727	1.712
76	1.917	1.882	1.851	1.824	1.800	1.778	1.759	1.741	1.725	1.710
77	1.915	1.880	1.849	1.822	1.798	1.777	1.757	1.739	1.723	1.708
78	1.914	1.878	1.848	1.821	1.797	1.775	1.755	1.738	1.721	1.707
79	1.912	1.877	1.846	1.819	1.795	1.773	1.754	1.736	1.720	1.705
80	1.910	1.875	1.845	1.817	1.793	1.772	1.752	1.734	1.718	1.703
81	1.909	1.874	1.843	1.816	1.792	1.770	1.750	1.733	1.716	1.702
82	1.907	1.872	1.841	1.814	1.790	1.768	1.749	1.731	1.715	1.700
83	1.906	1.871	1.840	1.813	1.789	1.767	1.747	1.729	1.713	1.698
84	1.905	1.869	1.838	1.811	1.787	1.765	1.746	1.728	1.712	1.697
85	1.903	1.868	1.837	1.810	1.786	1.764	1.744	1.726	1.710	1.695
86	1.902	1.867	1.836	1.808	1.784	1.762	1.743	1.725	1.709	1.694
87	1.900	1.865	1.834	1.807	1.783	1.761	1.741	1.724	1.707	1.692
88	1.899	1.864	1.833	1.806	1.782	1.760	1.740	1.722	1.706	1.691
89	1.898	1.863	1.832	1.804	1.780	1.758	1.739	1.721	1.705	1.690
90	1.897	1.861	1.830	1.803	1.779	1.757	1.737	1.720	1.703	1.688
91	1.895	1.860	1.829	1.802	1.778	1.756	1.736	1.718	1.702	1.687
92	1.894	1.859	1.828	1.801	1.776	1.755	1.735	1.717	1.701	1.686
93	1.893	1.858	1.827	1.800	1.775	1.753	1.734	1.716	1.699	1.684
94	1.892	1.857	1.826	1.798	1.774	1.752	1.733	1.715	1.698	1.683
95	1.891	1.856	1.825	1.797	1.773	1.751	1.731	1.713	1.697	1.682
96	1.890	1.854	1.823	1.796	1.772	1.750	1.730	1.712	1.696	1.681
97	1.889	1.853	1.822	1.795	1.771	1.749	1.729	1.711	1.695	1.680
98	1.888	1.852	1.821	1.794	1.770	1.748	1.728	1.710	1.694	1.679
99	1.887	1.851	1.820	1.793	1.769	1.747	1.727	1.709	1.693	1.678
100	1.886	1.850	1.819	1.792	1.768	1.746	1.726	1.708	1.691	1.676

(continued)

Table A.4 Percentage Points of the F Distribution (Continued)

Upper Critical Values of the F Distribution for v_1 Numerator Degrees of Freedom and

$v_2 \backslash v_1$	1	2	3	4	5	6	7	8	9	10
1	39.863	49.500	53.593	55.833	57.240	58.204	58.906	59.439	59.858	60.195
2	8.526	9.000	9.162	9.243	9.293	9.326	9.349	9.367	9.381	9.392
3	5.538	5.462	5.391	5.343	5.309	5.285	5.266	5.252	5.240	5.230
4	4.545	4.325	4.191	4.107	4.051	4.010	3.979	3.955	3.936	3.920
5	4.060	3.780	3.619	3.520	3.453	3.405	3.368	3.339	3.316	3.297
6	3.776	3.463	3.289	3.181	3.108	3.055	3.014	2.983	2.958	2.937
7	3.589	3.257	3.074	2.961	2.883	2.827	2.785	2.752	2.725	2.703
8	3.458	3.113	2.924	2.806	2.726	2.668	2.624	2.589	2.561	2.538
9	3.360	3.006	2.813	2.693	2.611	2.551	2.505	2.469	2.440	2.416
10	3.285	2.924	2.728	2.605	2.522	2.461	2.414	2.377	2.347	2.323
11	3.225	2.860	2.660	2.536	2.451	2.389	2.342	2.304	2.274	2.248
12	3.177	2.807	2.606	2.480	2.394	2.331	2.283	2.245	2.214	2.188
13	3.136	2.763	2.560	2.434	2.347	2.283	2.234	2.195	2.164	2.138
14	3.102	2.726	2.522	2.395	2.307	2.243	2.193	2.154	2.122	2.095
15	3.073	2.695	2.490	2.361	2.273	2.208	2.158	2.119	2.086	2.059
16	3.048	2.668	2.462	2.333	2.244	2.178	2.128	2.088	2.055	2.028
17	3.026	2.645	2.437	2.308	2.218	2.152	2.102	2.061	2.028	2.001
18	3.007	2.624	2.416	2.286	2.196	2.130	2.079	2.038	2.005	1.977
19	2.990	2.606	2.397	2.266	2.176	2.109	2.058	2.017	1.984	1.956
20	2.975	2.589	2.380	2.249	2.158	2.091	2.040	1.999	1.965	1.937
21	2.961	2.575	2.365	2.233	2.142	2.075	2.023	1.982	1.948	1.920
22	2.949	2.561	2.351	2.219	2.128	2.060	2.008	1.967	1.933	1.904
23	2.937	2.549	2.339	2.207	2.115	2.047	1.995	1.953	1.919	1.890
24	2.927	2.538	2.327	2.195	2.103	2.035	1.983	1.941	1.906	1.877
25	2.918	2.528	2.317	2.184	2.092	2.024	1.971	1.929	1.895	1.866
26	2.909	2.519	2.307	2.174	2.082	2.014	1.961	1.919	1.884	1.855
27	2.901	2.511	2.299	2.165	2.073	2.005	1.952	1.909	1.874	1.845
28	2.894	2.503	2.291	2.157	2.064	1.996	1.943	1.900	1.865	1.836
29	2.887	2.495	2.283	2.149	2.057	1.988	1.935	1.892	1.857	1.827
30	2.881	2.489	2.276	2.142	2.049	1.980	1.927	1.884	1.849	1.819
31	2.875	2.482	2.270	2.136	2.042	1.973	1.920	1.877	1.842	1.812
32	2.869	2.477	2.263	2.129	2.036	1.967	1.913	1.870	1.835	1.805
33	2.864	2.471	2.258	2.123	2.030	1.961	1.907	1.864	1.828	1.799
34	2.859	2.466	2.252	2.118	2.024	1.955	1.901	1.858	1.822	1.793
35	2.855	2.461	2.247	2.113	2.019	1.950	1.896	1.852	1.817	1.787
36	2.850	2.456	2.243	2.108	2.014	1.945	1.891	1.847	1.811	1.781
37	2.846	2.452	2.238	2.103	2.009	1.940	1.886	1.842	1.806	1.776
38	2.842	2.448	2.234	2.099	2.005	1.935	1.881	1.838	1.802	1.772
39	2.839	2.444	2.230	2.095	2.001	1.931	1.877	1.833	1.797	1.767
40	2.835	2.440	2.226	2.091	1.997	1.927	1.873	1.829	1.793	1.763
41	2.832	2.437	2.222	2.087	1.993	1.923	1.869	1.825	1.789	1.759
42	2.829	2.434	2.219	2.084	1.989	1.919	1.865	1.821	1.785	1.755
43	2.826	2.430	2.216	2.080	1.986	1.916	1.861	1.817	1.781	1.751
44	2.823	2.427	2.213	2.077	1.983	1.913	1.858	1.814	1.778	1.747
45	2.820	2.425	2.210	2.074	1.980	1.909	1.855	1.811	1.774	1.744
46	2.818	2.422	2.207	2.071	1.977	1.906	1.852	1.808	1.771	1.741
47	2.815	2.419	2.204	2.068	1.974	1.903	1.849	1.805	1.768	1.738
48	2.813	2.417	2.202	2.066	1.971	1.901	1.846	1.802	1.765	1.735
49	2.811	2.414	2.199	2.063	1.968	1.898	1.843	1.799	1.763	1.732
50	2.809	2.412	2.197	2.061	1.966	1.895	1.840	1.796	1.760	1.729

v_2 Denominator Degrees of Freedom, 10% Significance Level $F_{.10}(v_1, v_2)$

$v_2 \backslash v_1$	11	12	13	14	15	16	17	18	19	20
1	60.473	60.705	60.903	61.073	61.220	61.350	61.464	61.566	61.658	61.740
2	9.401	9.408	9.415	9.420	9.425	9.429	9.433	9.436	9.439	9.441
3	5.222	5.216	5.210	5.205	5.200	5.196	5.193	5.190	5.187	5.184
4	3.907	3.896	3.886	3.878	3.870	3.864	3.858	3.853	3.849	3.844
5	3.282	3.268	3.257	3.247	3.238	3.230	3.223	3.217	3.212	3.207
6	2.920	2.905	2.892	2.881	2.871	2.863	2.855	2.848	2.842	2.836
7	2.684	2.668	2.654	2.643	2.632	2.623	2.615	2.607	2.601	2.595
8	2.519	2.502	2.488	2.475	2.464	2.455	2.446	2.438	2.431	2.425
9	2.396	2.379	2.364	2.351	2.340	2.329	2.320	2.312	2.305	2.298
10	2.302	2.284	2.269	2.255	2.244	2.233	2.224	2.215	2.208	2.201
11	2.227	2.209	2.193	2.179	2.167	2.156	2.147	2.138	2.130	2.123
12	2.166	2.147	2.131	2.117	2.105	2.094	2.084	2.075	2.067	2.060
13	2.116	2.097	2.080	2.066	2.053	2.042	2.032	2.023	2.014	2.007
14	2.073	2.054	2.037	2.022	2.010	1.998	1.988	1.978	1.970	1.962
15	2.037	2.017	2.000	1.985	1.972	1.961	1.950	1.941	1.932	1.924
16	2.005	1.985	1.968	1.953	1.940	1.928	1.917	1.908	1.899	1.891
17	1.978	1.958	1.940	1.925	1.912	1.900	1.889	1.879	1.870	1.862
18	1.954	1.933	1.916	1.900	1.887	1.875	1.864	1.854	1.845	1.837
19	1.932	1.912	1.894	1.878	1.865	1.852	1.841	1.831	1.822	1.814
20	1.913	1.892	1.875	1.859	1.845	1.833	1.821	1.811	1.802	1.794
21	1.896	1.875	1.857	1.841	1.827	1.815	1.803	1.793	1.784	1.776
22	1.880	1.859	1.841	1.825	1.811	1.798	1.787	1.777	1.768	1.759
23	1.866	1.845	1.827	1.811	1.796	1.784	1.772	1.762	1.753	1.744
24	1.853	1.832	1.814	1.797	1.783	1.770	1.759	1.748	1.739	1.730
25	1.841	1.820	1.802	1.785	1.771	1.758	1.746	1.736	1.726	1.718
26	1.830	1.809	1.790	1.774	1.760	1.747	1.735	1.724	1.715	1.706
27	1.820	1.799	1.780	1.764	1.749	1.736	1.724	1.714	1.704	1.695
28	1.811	1.790	1.771	1.754	1.740	1.726	1.715	1.704	1.694	1.685
29	1.802	1.781	1.762	1.745	1.731	1.717	1.705	1.695	1.685	1.676
30	1.794	1.773	1.754	1.737	1.722	1.709	1.697	1.686	1.676	1.667
31	1.787	1.765	1.746	1.729	1.714	1.701	1.689	1.678	1.668	1.659
32	1.780	1.758	1.739	1.722	1.707	1.694	1.682	1.671	1.661	1.652
33	1.773	1.751	1.732	1.715	1.700	1.687	1.675	1.664	1.654	1.645
34	1.767	1.745	1.726	1.709	1.694	1.680	1.668	1.657	1.647	1.638
35	1.761	1.739	1.720	1.703	1.688	1.674	1.662	1.651	1.641	1.632
36	1.756	1.734	1.715	1.697	1.682	1.669	1.656	1.645	1.635	1.626
37	1.751	1.729	1.709	1.692	1.677	1.663	1.651	1.640	1.630	1.620
38	1.746	1.724	1.704	1.687	1.672	1.658	1.646	1.635	1.624	1.615
39	1.741	1.719	1.700	1.682	1.667	1.653	1.641	1.630	1.619	1.610
40	1.737	1.715	1.695	1.678	1.662	1.649	1.636	1.625	1.615	1.605
41	1.733	1.710	1.691	1.673	1.658	1.644	1.632	1.620	1.610	1.601
42	1.729	1.706	1.687	1.669	1.654	1.640	1.628	1.616	1.606	1.596
43	1.725	1.703	1.683	1.665	1.650	1.636	1.624	1.612	1.602	1.592
44	1.721	1.699	1.679	1.662	1.646	1.632	1.620	1.608	1.598	1.588
45	1.718	1.695	1.676	1.658	1.643	1.629	1.616	1.605	1.594	1.585
46	1.715	1.692	1.672	1.655	1.639	1.625	1.613	1.601	1.591	1.581
47	1.712	1.689	1.669	1.652	1.636	1.622	1.609	1.598	1.587	1.578
48	1.709	1.686	1.666	1.648	1.633	1.619	1.606	1.594	1.584	1.574
49	1.706	1.683	1.663	1.645	1.630	1.616	1.603	1.591	1.581	1.571
50	1.703	1.680	1.660	1.643	1.627	1.613	1.600	1.588	1.578	1.568

(continued)

Table A.4 Percentage Points of the F Distribution (Continued)

Upper Critical Values of the F Distribution for v_1 Numerator Degrees of Freedom and

v_2\v_1	1	2	3	4	5	6	7	8	9	10
51	2.807	2.410	2.194	2.058	1.964	1.893	1.838	1.794	1.757	1.727
52	2.805	2.408	2.192	2.056	1.961	1.891	1.836	1.791	1.755	1.724
53	2.803	2.406	2.190	2.054	1.959	1.888	1.833	1.789	1.752	1.722
54	2.801	2.404	2.188	2.052	1.957	1.886	1.831	1.787	1.750	1.719
55	2.799	2.402	2.186	2.050	1.955	1.884	1.829	1.785	1.748	1.717
56	2.797	2.400	2.184	2.048	1.953	1.882	1.827	1.782	1.746	1.715
57	2.796	2.398	2.182	2.046	1.951	1.880	1.825	1.780	1.744	1.713
58	2.794	2.396	2.181	2.044	1.949	1.878	1.823	1.779	1.742	1.711
59	2.793	2.395	2.179	2.043	1.947	1.876	1.821	1.777	1.740	1.709
60	2.791	2.393	2.177	2.041	1.946	1.875	1.819	1.775	1.738	1.707
61	2.790	2.392	2.176	2.039	1.944	1.873	1.818	1.773	1.736	1.705
62	2.788	2.390	2.174	2.038	1.942	1.871	1.816	1.771	1.735	1.703
63	2.787	2.389	2.173	2.036	1.941	1.870	1.814	1.770	1.733	1.702
64	2.786	2.387	2.171	2.035	1.939	1.868	1.813	1.768	1.731	1.700
65	2.784	2.386	2.170	2.033	1.938	1.867	1.811	1.767	1.730	1.699
66	2.783	2.385	2.169	2.032	1.937	1.865	1.810	1.765	1.728	1.697
67	2.782	2.384	2.167	2.031	1.935	1.864	1.808	1.764	1.727	1.696
68	2.781	2.382	2.166	2.029	1.934	1.863	1.807	1.762	1.725	1.694
69	2.780	2.381	2.165	2.028	1.933	1.861	1.806	1.761	1.724	1.693
70	2.779	2.380	2.164	2.027	1.931	1.860	1.804	1.760	1.723	1.691
71	2.778	2.379	2.163	2.026	1.930	1.859	1.803	1.758	1.721	1.690
72	2.777	2.378	2.161	2.025	1.929	1.858	1.802	1.757	1.720	1.689
73	2.776	2.377	2.160	2.024	1.928	1.856	1.801	1.756	1.719	1.687
74	2.775	2.376	2.159	2.022	1.927	1.855	1.800	1.755	1.718	1.686
75	2.774	2.375	2.158	2.021	1.926	1.854	1.798	1.754	1.716	1.685
76	2.773	2.374	2.157	2.020	1.925	1.853	1.797	1.752	1.715	1.684
77	2.772	2.373	2.156	2.019	1.924	1.852	1.796	1.751	1.714	1.683
78	2.771	2.372	2.155	2.018	1.923	1.851	1.795	1.750	1.713	1.682
79	2.770	2.371	2.154	2.017	1.922	1.850	1.794	1.749	1.712	1.681
80	2.769	2.370	2.154	2.016	1.921	1.849	1.793	1.748	1.711	1.680
81	2.769	2.369	2.153	2.016	1.920	1.848	1.792	1.747	1.710	1.679
82	2.768	2.368	2.152	2.015	1.919	1.847	1.791	1.746	1.709	1.678
83	2.767	2.368	2.151	2.014	1.918	1.846	1.790	1.745	1.708	1.677
84	2.766	2.367	2.150	2.013	1.917	1.845	1.790	1.744	1.707	1.676
85	2.765	2.366	2.149	2.012	1.916	1.845	1.789	1.744	1.706	1.675
86	2.765	2.365	2.149	2.011	1.915	1.844	1.788	1.743	1.705	1.674
87	2.764	2.365	2.148	2.011	1.915	1.843	1.787	1.742	1.705	1.673
88	2.763	2.364	2.147	2.010	1.914	1.842	1.786	1.741	1.704	1.672
89	2.763	2.363	2.146	2.009	1.913	1.841	1.785	1.740	1.703	1.671
90	2.762	2.363	2.146	2.008	1.912	1.841	1.785	1.739	1.702	1.670
91	2.761	2.362	2.145	2.008	1.912	1.840	1.784	1.739	1.701	1.670
92	2.761	2.361	2.144	2.007	1.911	1.839	1.783	1.738	1.701	1.669
93	2.760	2.361	2.144	2.006	1.910	1.838	1.782	1.737	1.700	1.668
94	2.760	2.360	2.143	2.006	1.910	1.838	1.782	1.736	1.699	1.667
95	2.759	2.359	2.142	2.005	1.909	1.837	1.781	1.736	1.698	1.667
96	2.759	2.359	2.142	2.004	1.908	1.836	1.780	1.735	1.698	1.666
97	2.758	2.358	2.141	2.004	1.908	1.836	1.780	1.734	1.697	1.665
98	2.757	2.358	2.141	2.003	1.907	1.835	1.779	1.734	1.696	1.665
99	2.757	2.357	2.140	2.003	1.906	1.835	1.778	1.733	1.696	1.664
100	2.756	2.356	2.139	2.002	1.906	1.834	1.778	1.732	1.695	1.663

v_2 Denominator Degrees of Freedom, 5% Significance Level $F_{.10}(v_1, v_2)$

$v_2 \backslash v_1$	11	12	13	14	15	16	17	18	19	20
51	1.700	1.677	1.658	1.640	1.624	1.610	1.597	1.586	1.575	1.565
52	1.698	1.675	1.655	1.637	1.621	1.607	1.594	1.583	1.572	1.562
53	1.695	1.672	1.652	1.635	1.619	1.605	1.592	1.580	1.570	1.560
54	1.693	1.670	1.650	1.632	1.616	1.602	1.589	1.578	1.567	1.557
55	1.691	1.668	1.648	1.630	1.614	1.600	1.587	1.575	1.564	1.555
56	1.688	1.666	1.645	1.628	1.612	1.597	1.585	1.573	1.562	1.552
57	1.686	1.663	1.643	1.625	1.610	1.595	1.582	1.571	1.560	1.550
58	1.684	1.661	1.641	1.623	1.607	1.593	1.580	1.568	1.558	1.548
59	1.682	1.659	1.639	1.621	1.605	1.591	1.578	1.566	1.555	1.546
60	1.680	1.657	1.637	1.619	1.603	1.589	1.576	1.564	1.553	1.543
61	1.679	1.656	1.635	1.617	1.601	1.587	1.574	1.562	1.551	1.541
62	1.677	1.654	1.634	1.616	1.600	1.585	1.572	1.560	1.549	1.540
63	1.675	1.652	1.632	1.614	1.598	1.583	1.570	1.558	1.548	1.538
64	1.673	1.650	1.630	1.612	1.596	1.582	1.569	1.557	1.546	1.536
65	1.672	1.649	1.628	1.610	1.594	1.580	1.567	1.555	1.544	1.534
66	1.670	1.647	1.627	1.609	1.593	1.578	1.565	1.553	1.542	1.532
67	1.669	1.646	1.625	1.607	1.591	1.577	1.564	1.552	1.541	1.531
68	1.667	1.644	1.624	1.606	1.590	1.575	1.562	1.550	1.539	1.529
69	1.666	1.643	1.622	1.604	1.588	1.574	1.560	1.548	1.538	1.527
70	1.665	1.641	1.621	1.603	1.587	1.572	1.559	1.547	1.536	1.526
71	1.663	1.640	1.619	1.601	1.585	1.571	1.557	1.545	1.535	1.524
72	1.662	1.639	1.618	1.600	1.584	1.569	1.556	1.544	1.533	1.523
73	1.661	1.637	1.617	1.599	1.583	1.568	1.555	1.543	1.532	1.522
74	1.659	1.636	1.616	1.597	1.581	1.567	1.553	1.541	1.530	1.520
75	1.658	1.635	1.614	1.596	1.580	1.565	1.552	1.540	1.529	1.519
76	1.657	1.634	1.613	1.595	1.579	1.564	1.551	1.539	1.528	1.518
77	1.656	1.632	1.612	1.594	1.578	1.563	1.550	1.538	1.527	1.516
78	1.655	1.631	1.611	1.593	1.576	1.562	1.548	1.536	1.525	1.515
79	1.654	1.630	1.610	1.592	1.575	1.561	1.547	1.535	1.524	1.514
80	1.653	1.629	1.609	1.590	1.574	1.559	1.546	1.534	1.523	1.513
81	1.652	1.628	1.608	1.589	1.573	1.558	1.545	1.533	1.522	1.512
82	1.651	1.627	1.607	1.588	1.572	1.557	1.544	1.532	1.521	1.511
83	1.650	1.626	1.606	1.587	1.571	1.556	1.543	1.531	1.520	1.509
84	1.649	1.625	1.605	1.586	1.570	1.555	1.542	1.530	1.519	1.508
85	1.648	1.624	1.604	1.585	1.569	1.554	1.541	1.529	1.518	1.507
86	1.647	1.623	1.603	1.584	1.568	1.553	1.540	1.528	1.517	1.506
87	1.646	1.622	1.602	1.583	1.567	1.552	1.539	1.527	1.516	1.505
88	1.645	1.622	1.601	1.583	1.566	1.551	1.538	1.526	1.515	1.504
89	1.644	1.621	1.600	1.582	1.565	1.550	1.537	1.525	1.514	1.503
90	1.643	1.620	1.599	1.581	1.564	1.550	1.536	1.524	1.513	1.503
91	1.643	1.619	1.598	1.580	1.564	1.549	1.535	1.523	1.512	1.502
92	1.642	1.618	1.598	1.579	1.563	1.548	1.534	1.522	1.511	1.501
93	1.641	1.617	1.597	1.578	1.562	1.547	1.534	1.521	1.510	1.500
94	1.640	1.617	1.596	1.578	1.561	1.546	1.533	1.521	1.509	1.499
95	1.640	1.616	1.595	1.577	1.560	1.545	1.532	1.520	1.509	1.498
96	1.639	1.615	1.594	1.576	1.560	1.545	1.531	1.519	1.508	1.497
97	1.638	1.614	1.594	1.575	1.559	1.544	1.530	1.518	1.507	1.497
98	1.637	1.614	1.593	1.575	1.558	1.543	1.530	1.517	1.506	1.496
99	1.637	1.613	1.592	1.574	1.557	1.542	1.529	1.517	1.505	1.495
100	1.636	1.612	1.592	1.573	1.557	1.542	1.528	1.516	1.505	1.494

(continued)

Table A.4 Percentage Points of the F Distribution (Continued)

Upper Critical Values of the F Distribution for v_1 Numerator Degrees of Freedom and v_2

$v_2 \backslash v_1$	1	2	3	4	5	6	7	8	9	10
1	4052.19	4999.52	5403.34	5624.62	5763.65	5858.97	5928.33	5981.10	6022.50	6055.85
2	98.502	99.000	99.166	99.249	99.300	99.333	99.356	99.374	99.388	99.399
3	34.116	30.816	29.457	28.710	28.237	27.911	27.672	27.489	27.345	27.229
4	21.198	18.000	16.694	15.977	15.522	15.207	14.976	14.799	14.659	14.546
5	16.258	13.274	12.060	11.392	10.967	10.672	10.456	10.289	10.158	10.051
6	13.745	10.925	9.780	9.148	8.746	8.466	8.260	8.102	7.976	7.874
7	12.246	9.547	8.451	7.847	7.460	7.191	6.993	6.840	6.719	6.620
8	11.259	8.649	7.591	7.006	6.632	6.371	6.178	6.029	5.911	5.814
9	10.561	8.022	6.992	6.422	6.057	5.802	5.613	5.467	5.351	5.257
10	10.044	7.559	6.552	5.994	5.636	5.386	5.200	5.057	4.942	4.849
11	9.646	7.206	6.217	5.668	5.316	5.069	4.886	4.744	4.632	4.539
12	9.330	6.927	5.953	5.412	5.064	4.821	4.640	4.499	4.388	4.296
13	9.074	6.701	5.739	5.205	4.862	4.620	4.441	4.302	4.191	4.100
14	8.862	6.515	5.564	5.035	4.695	4.456	4.278	4.140	4.030	3.939
15	8.683	6.359	5.417	4.893	4.556	4.318	4.142	4.004	3.895	3.805
16	8.531	6.226	5.292	4.773	4.437	4.202	4.026	3.890	3.780	3.691
17	8.400	6.112	5.185	4.669	4.336	4.102	3.927	3.791	3.682	3.593
18	8.285	6.013	5.092	4.579	4.248	4.015	3.841	3.705	3.597	3.508
19	8.185	5.926	5.010	4.500	4.171	3.939	3.765	3.631	3.523	3.434
20	8.096	5.849	4.938	4.431	4.103	3.871	3.699	3.564	3.457	3.368
21	8.017	5.780	4.874	4.369	4.042	3.812	3.640	3.506	3.398	3.310
22	7.945	5.719	4.817	4.313	3.988	3.758	3.587	3.453	3.346	3.258
23	7.881	5.664	4.765	4.264	3.939	3.710	3.539	3.406	3.299	3.211
24	7.823	5.614	4.718	4.218	3.895	3.667	3.496	3.363	3.256	3.168
25	7.770	5.568	4.675	4.177	3.855	3.627	3.457	3.324	3.217	3.129
26	7.721	5.526	4.637	4.140	3.818	3.591	3.421	3.288	3.182	3.094
27	7.677	5.488	4.601	4.106	3.785	3.558	3.388	3.256	3.149	3.062
28	7.636	5.453	4.568	4.074	3.754	3.528	3.358	3.226	3.120	3.032
29	7.598	5.420	4.538	4.045	3.725	3.499	3.330	3.198	3.092	3.005
30	7.562	5.390	4.510	4.018	3.699	3.473	3.305	3.173	3.067	2.979
31	7.530	5.362	4.484	3.993	3.675	3.449	3.281	3.149	3.043	2.955
32	7.499	5.336	4.459	3.969	3.652	3.427	3.258	3.127	3.021	2.934
33	7.471	5.312	4.437	3.948	3.630	3.406	3.238	3.106	3.000	2.913
34	7.444	5.289	4.416	3.927	3.611	3.386	3.218	3.087	2.981	2.894
35	7.419	5.268	4.396	3.908	3.592	3.368	3.200	3.069	2.963	2.876
36	7.396	5.248	4.377	3.890	3.574	3.351	3.183	3.052	2.946	2.859
37	7.373	5.229	4.360	3.873	3.558	3.334	3.167	3.036	2.930	2.843
38	7.353	5.211	4.343	3.858	3.542	3.319	3.152	3.021	2.915	2.828
39	7.333	5.194	4.327	3.843	3.528	3.305	3.137	3.006	2.901	2.814
40	7.314	5.179	4.313	3.828	3.514	3.291	3.124	2.993	2.888	2.801
41	7.296	5.163	4.299	3.815	3.501	3.278	3.111	2.980	2.875	2.788
42	7.280	5.149	4.285	3.802	3.488	3.266	3.099	2.968	2.863	2.776
43	7.264	5.136	4.273	3.790	3.476	3.254	3.087	2.957	2.851	2.764
44	7.248	5.123	4.261	3.778	3.465	3.243	3.076	2.946	2.840	2.754
45	7.234	5.110	4.249	3.767	3.454	3.232	3.066	2.935	2.830	2.743
46	7.220	5.099	4.238	3.757	3.444	3.222	3.056	2.925	2.820	2.733
47	7.207	5.087	4.228	3.747	3.434	3.213	3.046	2.916	2.811	2.724
48	7.194	5.077	4.218	3.737	3.425	3.204	3.037	2.907	2.802	2.715
49	7.182	5.066	4.208	3.728	3.416	3.195	3.028	2.898	2.793	2.706
50	7.171	5.057	4.199	3.720	3.408	3.186	3.020	2.890	2.785	2.698

Denominator Degrees of Freedom, 1% Significance Level $F_{.10}$ (v_1, v_2)

v_2\\v_1	11	12	13	14	15	16	17	18	19	20
1.	6083.35	6106.35	6125.86	6142.70	6157.28	6170.12	6181.42	6191.52	6200.58	6208.74
2.	99.408	99.416	99.422	99.428	99.432	99.437	99.440	99.444	99.447	99.449
3.	27.133	27.052	26.983	26.924	26.872	26.827	26.787	26.751	26.719	26.690
4.	14.452	14.374	14.307	14.249	14.198	14.154	14.115	14.080	14.048	14.020
5.	9.963	9.888	9.825	9.770	9.722	9.680	9.643	9.610	9.580	9.553
6.	7.790	7.718	7.657	7.605	7.559	7.519	7.483	7.451	7.422	7.396
7.	6.538	6.469	6.410	6.359	6.314	6.275	6.240	6.209	6.181	6.155
8.	5.734	5.667	5.609	5.559	5.515	5.477	5.442	5.412	5.384	5.359
9.	5.178	5.111	5.055	5.005	4.962	4.924	4.890	4.860	4.833	4.808
10.	4.772	4.706	4.650	4.601	4.558	4.520	4.487	4.457	4.430	4.405
11.	4.462	4.397	4.342	4.293	4.251	4.213	4.180	4.150	4.123	4.099
12.	4.220	4.155	4.100	4.052	4.010	3.972	3.939	3.909	3.883	3.858
13.	4.025	3.960	3.905	3.857	3.815	3.778	3.745	3.716	3.689	3.665
14.	3.864	3.800	3.745	3.698	3.656	3.619	3.586	3.556	3.529	3.505
15.	3.730	3.666	3.612	3.564	3.522	3.485	3.452	3.423	3.396	3.372
16.	3.616	3.553	3.498	3.451	3.409	3.372	3.339	3.310	3.283	3.259
17.	3.519	3.455	3.401	3.353	3.312	3.275	3.242	3.212	3.186	3.162
18.	3.434	3.371	3.316	3.269	3.227	3.190	3.158	3.128	3.101	3.077
19.	3.360	3.297	3.242	3.195	3.153	3.116	3.084	3.054	3.027	3.003
20.	3.294	3.231	3.177	3.130	3.088	3.051	3.018	2.989	2.962	2.938
21.	3.236	3.173	3.119	3.072	3.030	2.993	2.960	2.931	2.904	2.880
22.	3.184	3.121	3.067	3.019	2.978	2.941	2.908	2.879	2.852	2.827
23.	3.137	3.074	3.020	2.973	2.931	2.894	2.861	2.832	2.805	2.781
24.	3.094	3.032	2.977	2.930	2.889	2.852	2.819	2.789	2.762	2.738
25.	3.056	2.993	2.939	2.892	2.850	2.813	2.780	2.751	2.724	2.699
26.	3.021	2.958	2.904	2.857	2.815	2.778	2.745	2.715	2.688	2.664
27.	2.988	2.926	2.871	2.824	2.783	2.746	2.713	2.683	2.656	2.632
28.	2.959	2.896	2.842	2.795	2.753	2.716	2.683	2.653	2.626	2.602
29.	2.931	2.868	2.814	2.767	2.726	2.689	2.656	2.626	2.599	2.574
30.	2.906	2.843	2.789	2.742	2.700	2.663	2.630	2.600	2.573	2.549
31.	2.882	2.820	2.765	2.718	2.677	2.640	2.606	2.577	2.550	2.525
32.	2.860	2.798	2.744	2.696	2.655	2.618	2.584	2.555	2.527	2.503
33.	2.840	2.777	2.723	2.676	2.634	2.597	2.564	2.534	2.507	2.482
34.	2.821	2.758	2.704	2.657	2.615	2.578	2.545	2.515	2.488	2.463
35.	2.803	2.740	2.686	2.639	2.597	2.560	2.527	2.497	2.470	2.445
36.	2.786	2.723	2.669	2.622	2.580	2.543	2.510	2.480	2.453	2.428
37.	2.770	2.707	2.653	2.606	2.564	2.527	2.494	2.464	2.437	2.412
38.	2.755	2.692	2.638	2.591	2.549	2.512	2.479	2.449	2.421	2.397
39.	2.741	2.678	2.624	2.577	2.535	2.498	2.465	2.434	2.407	2.382
40.	2.727	2.665	2.611	2.563	2.522	2.484	2.451	2.421	2.394	2.369
41.	2.715	2.652	2.598	2.551	2.509	2.472	2.438	2.408	2.381	2.356
42.	2.703	2.640	2.586	2.539	2.497	2.460	2.426	2.396	2.369	2.344
43.	2.691	2.629	2.575	2.527	2.485	2.448	2.415	2.385	2.357	2.332
44.	2.680	2.618	2.564	2.516	2.475	2.437	2.404	2.374	2.346	2.321
45.	2.670	2.608	2.553	2.506	2.464	2.427	2.393	2.363	2.336	2.311
46.	2.660	2.598	2.544	2.496	2.454	2.417	2.384	2.353	2.326	2.301
47.	2.651	2.588	2.534	2.487	2.445	2.408	2.374	2.344	2.316	2.291
48.	2.642	2.579	2.525	2.478	2.436	2.399	2.365	2.335	2.307	2.282
49.	2.633	2.571	2.517	2.469	2.427	2.390	2.356	2.326	2.299	2.274
50.	2.625	2.562	2.508	2.461	2.419	2.382	2.348	2.318	2.290	2.265

(continued)

Table A.4 Percentage Points of the F Distribution (Continued)

Upper Critical Values of the F Distribution for v_1 Numerator Degrees of Freedom and

$v_2\backslash v_1$	1	2	3	4	5	6	7	8	9	10
51	7.159	5.047	4.191	3.711	3.400	3.178	3.012	2.882	2.777	2.690
52	7.149	5.038	4.182	3.703	3.392	3.171	3.005	2.874	2.769	2.683
53	7.139	5.030	4.174	3.695	3.384	3.163	2.997	2.867	2.762	2.675
54	7.129	5.021	4.167	3.688	3.377	3.156	2.990	2.860	2.755	2.668
55	7.119	5.013	4.159	3.681	3.370	3.149	2.983	2.853	2.748	2.662
56	7.110	5.006	4.152	3.674	3.363	3.143	2.977	2.847	2.742	2.655
57	7.102	4.998	4.145	3.667	3.357	3.136	2.971	2.841	2.736	2.649
58	7.093	4.991	4.138	3.661	3.351	3.130	2.965	2.835	2.730	2.643
59	7.085	4.984	4.132	3.655	3.345	3.124	2.959	2.829	2.724	2.637
60	7.077	4.977	4.126	3.649	3.339	3.119	2.953	2.823	2.718	2.632
61	7.070	4.971	4.120	3.643	3.333	3.113	2.948	2.818	2.713	2.626
62	7.062	4.965	4.114	3.638	3.328	3.108	2.942	2.813	2.708	2.621
63	7.055	4.959	4.109	3.632	3.323	3.103	2.937	2.808	2.703	2.616
64	7.048	4.953	4.103	3.627	3.318	3.098	2.932	2.803	2.698	2.611
65	7.042	4.947	4.098	3.622	3.313	3.093	2.928	2.798	2.693	2.607
66	7.035	4.942	4.093	3.618	3.308	3.088	2.923	2.793	2.689	2.602
67	7.029	4.937	4.088	3.613	3.304	3.084	2.919	2.789	2.684	2.598
68	7.023	4.932	4.083	3.608	3.299	3.080	2.914	2.785	2.680	2.593
69	7.017	4.927	4.079	3.604	3.295	3.075	2.910	2.781	2.676	2.589
70	7.011	4.922	4.074	3.600	3.291	3.071	2.906	2.777	2.672	2.585
71	7.006	4.917	4.070	3.596	3.287	3.067	2.902	2.773	2.668	2.581
72	7.001	4.913	4.066	3.591	3.283	3.063	2.898	2.769	2.664	2.578
73	6.995	4.908	4.062	3.588	3.279	3.060	2.895	2.765	2.660	2.574
74	6.990	4.904	4.058	3.584	3.275	3.056	2.891	2.762	2.657	2.570
75	6.985	4.900	4.054	3.580	3.272	3.052	2.887	2.758	2.653	2.567
76	6.981	4.896	4.050	3.577	3.268	3.049	2.884	2.755	2.650	2.563
77	6.976	4.892	4.047	3.573	3.265	3.046	2.881	2.751	2.647	2.560
78	6.971	4.888	4.043	3.570	3.261	3.042	2.877	2.748	2.644	2.557
79	6.967	4.884	4.040	3.566	3.258	3.039	2.874	2.745	2.640	2.554
80	6.963	4.881	4.036	3.563	3.255	3.036	2.871	2.742	2.637	2.551
81	6.958	4.877	4.033	3.560	3.252	3.033	2.868	2.739	2.634	2.548
82	6.954	4.874	4.030	3.557	3.249	3.030	2.865	2.736	2.632	2.545
83	6.950	4.870	4.027	3.554	3.246	3.027	2.863	2.733	2.629	2.542
84	6.947	4.867	4.024	3.551	3.243	3.025	2.860	2.731	2.626	2.539
85	6.943	4.864	4.021	3.548	3.240	3.022	2.857	2.728	2.623	2.537
86	6.939	4.861	4.018	3.545	3.238	3.019	2.854	2.725	2.621	2.534
87	6.935	4.858	4.015	3.543	3.235	3.017	2.852	2.723	2.618	2.532
88	6.932	4.855	4.012	3.540	3.233	3.014	2.849	2.720	2.616	2.529
89	6.928	4.852	4.010	3.538	3.230	3.012	2.847	2.718	2.613	2.527
90	6.925	4.849	4.007	3.535	3.228	3.009	2.845	2.715	2.611	2.524
91	6.922	4.846	4.004	3.533	3.225	3.007	2.842	2.713	2.609	2.522
92	6.919	4.844	4.002	3.530	3.223	3.004	2.840	2.711	2.606	2.520
93	6.915	4.841	3.999	3.528	3.221	3.002	2.838	2.709	2.604	2.518
94	6.912	4.838	3.997	3.525	3.218	3.000	2.835	2.706	2.602	2.515
95	6.909	4.836	3.995	3.523	3.216	2.998	2.833	2.704	2.600	2.513
96	6.906	4.833	3.992	3.521	3.214	2.996	2.831	2.702	2.598	2.511
97	6.904	4.831	3.990	3.519	3.212	2.994	2.829	2.700	2.596	2.509
98	6.901	4.829	3.988	3.517	3.210	2.992	2.827	2.698	2.594	2.507
99	6.898	4.826	3.986	3.515	3.208	2.990	2.825	2.696	2.592	2.505
100	6.895	4.824	3.984	3.513	3.206	2.988	2.823	2.694	2.590	2.503

v_2 Denominator Degrees of Freedom, 1% Significance Level $F_{.01}(v_1, v_2)$

$v_2\backslash v_1$	11	12	13	14	15	16	17	18	19	20
51.	2.617	2.555	2.500	2.453	2.411	2.374	2.340	2.310	2.282	2.257
52.	2.610	2.547	2.493	2.445	2.403	2.366	2.333	2.302	2.275	2.250
53.	2.602	2.540	2.486	2.438	2.396	2.359	2.325	2.295	2.267	2.242
54.	2.595	2.533	2.479	2.431	2.389	2.352	2.318	2.288	2.260	2.235
55.	2.589	2.526	2.472	2.424	2.382	2.345	2.311	2.281	2.253	2.228
56.	2.582	2.520	2.465	2.418	2.376	2.339	2.305	2.275	2.247	2.222
57.	2.576	2.513	2.459	2.412	2.370	2.332	2.299	2.268	2.241	2.215
58.	2.570	2.507	2.453	2.406	2.364	2.326	2.293	2.262	2.235	2.209
59.	2.564	2.502	2.447	2.400	2.358	2.320	2.287	2.256	2.229	2.203
60.	2.559	2.496	2.442	2.394	2.352	2.315	2.281	2.251	2.223	2.198
61.	2.553	2.491	2.436	2.389	2.347	2.309	2.276	2.245	2.218	2.192
62.	2.548	2.486	2.431	2.384	2.342	2.304	2.270	2.240	2.212	2.187
63.	2.543	2.481	2.426	2.379	2.337	2.299	2.265	2.235	2.207	2.182
64.	2.538	2.476	2.421	2.374	2.332	2.294	2.260	2.230	2.202	2.177
65.	2.534	2.471	2.417	2.369	2.327	2.289	2.256	2.225	2.198	2.172
66.	2.529	2.466	2.412	2.365	2.322	2.285	2.251	2.221	2.193	2.168
67.	2.525	2.462	2.408	2.360	2.318	2.280	2.247	2.216	2.188	2.163
68.	2.520	2.458	2.403	2.356	2.314	2.276	2.242	2.212	2.184	2.159
69.	2.516	2.454	2.399	2.352	2.310	2.272	2.238	2.208	2.180	2.155
70.	2.512	2.450	2.395	2.348	2.306	2.268	2.234	2.204	2.176	2.150
71.	2.508	2.446	2.391	2.344	2.302	2.264	2.230	2.200	2.172	2.146
72.	2.504	2.442	2.388	2.340	2.298	2.260	2.226	2.196	2.168	2.143
73.	2.501	2.438	2.384	2.336	2.294	2.256	2.223	2.192	2.164	2.139
74.	2.497	2.435	2.380	2.333	2.290	2.253	2.219	2.188	2.161	2.135
75.	2.494	2.431	2.377	2.329	2.287	2.249	2.215	2.185	2.157	2.132
76.	2.490	2.428	2.373	2.326	2.284	2.246	2.212	2.181	2.154	2.128
77.	2.487	2.424	2.370	2.322	2.280	2.243	2.209	2.178	2.150	2.125
78.	2.484	2.421	2.367	2.319	2.277	2.239	2.206	2.175	2.147	2.122
79.	2.481	2.418	2.364	2.316	2.274	2.236	2.202	2.172	2.144	2.118
80.	2.478	2.415	2.361	2.313	2.271	2.233	2.199	2.169	2.141	2.115
81.	2.475	2.412	2.358	2.310	2.268	2.230	2.196	2.166	2.138	2.112
82.	2.472	2.409	2.355	2.307	2.265	2.227	2.193	2.163	2.135	2.109
83.	2.469	2.406	2.352	2.304	2.262	2.224	2.191	2.160	2.132	2.106
84.	2.466	2.404	2.349	2.302	2.259	2.222	2.188	2.157	2.129	2.104
85.	2.464	2.401	2.347	2.299	2.257	2.219	2.185	2.154	2.126	2.101
86.	2.461	2.398	2.344	2.296	2.254	2.216	2.182	2.152	2.124	2.098
87.	2.459	2.396	2.342	2.294	2.252	2.214	2.180	2.149	2.121	2.096
88.	2.456	2.393	2.339	2.291	2.249	2.211	2.177	2.147	2.119	2.093
89.	2.454	2.391	2.337	2.289	2.247	2.209	2.175	2.144	2.116	2.091
90.	2.451	2.389	2.334	2.286	2.244	2.206	2.172	2.142	2.114	2.088
91.	2.449	2.386	2.332	2.284	2.242	2.204	2.170	2.139	2.111	2.086
92.	2.447	2.384	2.330	2.282	2.240	2.202	2.168	2.137	2.109	2.083
93.	2.444	2.382	2.327	2.280	2.237	2.200	2.166	2.135	2.107	2.081
94.	2.442	2.380	2.325	2.277	2.235	2.197	2.163	2.133	2.105	2.079
95.	2.440	2.378	2.323	2.275	2.233	2.195	2.161	2.130	2.102	2.077
96.	2.438	2.375	2.321	2.273	2.231	2.193	2.159	2.128	2.100	2.075
97.	2.436	2.373	2.319	2.271	2.229	2.191	2.157	2.126	2.098	2.073
98.	2.434	2.371	2.317	2.269	2.227	2.189	2.155	2.124	2.096	2.071
99.	2.432	2.369	2.315	2.267	2.225	2.187	2.153	2.122	2.094	2.069
100.	2.430	2.368	2.313	2.265	2.223	2.185	2.151	2.120	2.092	2.067

References

Aldenderfer, M.S., and Blashfield R.K., *Cluster Analysis*, Sage Publications, 1985.
Alt, F.B., "Multivariate Quality Control: State of Art," *ASQC Tech. Conf. Transactions*, 1982, pp. 886–893.
Anderson, T.W., *An Introduction to Multivariate Statistical Analysis*, 2nd ed., Wiley, New York, 1984.
Arbuckle, J.L., *AMOS 3.5 for Windows*, SmallWaters Corporation for Statistical Modeling, Chicago, 1995.
Bartlett, M.S., "Multivariate Analysis," *J. Royal Stat. Soc. Suppl. (B)* **9**:176–197 (1974).
Bartlett, M.S., "A Note on the Multiplying Factor from Various X^2 Approximations," *J. Royal Stat. Soc. Suppl. (B)* **16**:296–298 (1954).
Bartlett, M.S., "Properties of Sufficiency and Statistical Tests," *Proc. Royal Soc. London (A)* **160**:268–282 (1937).
Bentler, P.M., and Wu, E.J.C. *EQS/Windows: User's Guide*, BMDP Statistical Software, Los Angeles, 1937.
Berry, M., and Linoff, G.S., *Master Data Mining*, Willey, New York, 2000.
Bewig, K.M., Clarke, A.D., Roberts, C., and Unklesbay, N., "Discriminant Analysis of Vegetable Oils by Near-Infrared Reflectance Spectroscopy," *J. Am. Oil Chem. Soc.* **71**(2):195–200 (1994).
Bickel, P.J., and Doksum, K.A., *Mathematical Statistics: Basic Ideas and Selected Topics*, Holden Day, San Francisco, 1977.
Breyfogle, III, F.W., *Implementing Six Sigma: Smart Solutions using Statistical Methods*, Wiley, Hoboken, New Jersey 2003.
Bruckner, L.A., "On Chernoff Faces," *Graphical Representation of Multivariate Data, Proc. Symp. Graphical Representation of Multivariate Data*, P.C. Wang (ed.), Academic Press, 1978, pp. 93–121.
Ceglarek, D.J., "Knowledge-based Diagnosis for Automotive Body Assembly: Methodology and Implementation," Ph.D. dissertation, University of Michigan, Ann Arbor, 1994.
Ceglarek, D.J., "Knowledge-Based Diagnosis for Automotive Body Assembly: Methodology and Implementation," Ph.D. dissertation, University of Michigan, Ann Arbor, Michigan, 1994.
Chernoff, H., "Graphical Representation as a Discipline," *Graphical Representation of Multivariate Data, Proc. Symp. Graphical Representation of Multivariate Data*, P.C. Wang (ed.), Academic Press, 1978, pp. 1–11.
Chernoff, H., "The Use of Faces to Represent Points in n-Dimensional Space Graphically," Tech. Rep. 71, Department of Statistics, Stanford University, Stanford, CA, 1971.
Dillion, W.R., and Goldstein, M., *Multivariate Analysis*, 2nd ed., Wiley, New York, 1984.
Doganaksoy, N., Fulton, J., and Tucker, W.T., "Identification of Out of Control Quality Characteristics in a Multivariate Manufacturing Environment," *Commun. Stat. A* **20**:2775–2790 (1991).
Duncan, O.D., "Path Analysis: Sociology Examples," *Am. J. Sociol.* **72**:1–16 (1966).
Edelstein, H.A., *Introduction to Data Mining and Knowledge Discovery*, 3rd ed., Potomac, Maryland, Two Crows Corporation, 1999.
Everitt, B.S., Landau, S., and Leese, M., *Cluster Analysis*, 4th ed., Edward Arnold, 2001.
Fisher, R.A., "The Use of Multiple Measurements in Taxonomic Problems," *Ann. Eugenics* **7**:179–188 (1936).
Fortner, B., *The Data Handbook: A guide to Understanding the Organization and Visualization of Technical Data*, Springer-Verlag, New York, 1995.

Gerth, R.J., and Hancock, W.M. "Output Variation Reduction of Production Systems Involving a Large Number of Subprocesses," *Quality Engineering* **8**(1):145–163 (1995).

Girshick, M.A., "On the Sampling Theory of the Roots of Determinental Equations," *Ann. Math. Stat.* **10**:203–224 (1939).

Girshick, M.A., "Principal Components," *J. Am. Stat. Assoc.* **31**:519–528 (1936).

Grant, E.L., and Leavenworth, R.S., *Statistical Quality Control*, 5th ed., McGraw-Hill, New York, 1980.

Hamade, H.M., "A Systematic Approach to Engineering Design Evaluation Using Multivariate Statistics," Ph.D. dissertation, Industrial and Manufacturing Engineering Wayne State University, Detroit, Michigan, 1996.

Hand, D.J., *Discriminant Analysis*, Willey, New York, 1981.

Harris, R.J., *A Primer of Multivariate Statistics*, Academic Press, New York, 1975.

Hawkins, D.M., "A Review and Analysis of the Mahalanobis-Taguchi System," *Technometrics* **45**(1):25–29 (2003).

Hines, W., and Montgomery, D., *Probability and Statistics in Engineering and Management Science*, 3rd ed., Wiley, New York, 1990.

Hotelling, H., "Multivariate Quality Control," in *Techniques of Statistical Analysis*, C. Eisenhart, M.W. Hastay, and W.A. Wallis (eds.), McGraw-Hill, New York, 1947, pp. 111–184.

Hotelling, H., "Analysis of a Complex of Statistical Variables into Principal Components," *J. Education Psychol.*, **24**:417–441, 498–520 (1933).

Hu, S. Jack, and Wu, S.M., "Impact of 100% In-Process Measurement on Statistical Process (SPC) in Automobile Body Assembly," *ASME WAM* **44**:433–448 (1990).

Hu, S. Jack, and Wu, S.M., "Identifying Sources of Variation in Automobile Body Assembly Using Principal Component Analysis," *Transactions of NAMRI/SME* **20**:311–316 (1992).

Huberty, C.J., *Applied Discriminant Analysis*, Willey, New York, 1994.

Jackson, J.E., "Quality Control Methods for Several Related Variables," *Technometrics* **1**:359–377 (1959).

Jackson, J.E., "Multivariate Quality Control," *Commun. Stat. A* **14**:2657–2688 (1985).

Jackson, J.E., *A User's Guide to Principal Components*, Wiley, New York 1991.

Johnson, R.A., and Wichern, D.W., *Applied Multivariate Statistical Analysis*, 5th ed., Prentice Hall, New Jersey, 2002.

Johnson, R.A., and Wichern, D.W., *Applied Multivariate Statistical Analysis*, Prentice Hall, New Jersey, 1982.

Jolicoeur, P., and Mosimann, J.E., "Size and Shape Variation in the Painted Turtle: A Principal Component Analysis," *Growth* **24**:339–354 (1960).

Jolliffe, I.T., *Principal Component Analysis*, Spriner-Verlag, New York, 1986.

Joreskog, K.G., *Factor Analysis by Least-Squares and Maximum-Likelihood Methods*, Statistical Methods for Digital Computers, vol III, K. Enslein et al. (eds.), Wiley, New York, 1977a, pp. 125–135.

Joreskog, K.G., "Structural Equation Models in Social Sciences: Specification, Estimation, and Testing," *Application of Statistics*, North-Holland, Amsterdam, 1977b, pp. 265–287.

Joreskog, K.G., and Sorbom, D., *LISREL & Aguide to the Programs and Applications*, 2nd ed., SPSS Publication, Chicago, 1989.

Kaiser, H.F., "The Varimax Criterion for Analytic Rotation in Factor Analysis," *Psychometrika* **23**:187–200, 1958.

Kang, I.J., Matsumura, Y., and Mori, T., "Characterization of Texture and Mechanical Properties of Heat-Induced Soy Protein Gels," *J. Am. Oil Chem. Soc.* **68**(5):339–345 (1991).

Kenett, R.S., "The Mahalanobis-Taguchi Strategy," *Quality Progress* **36**(8):98 (2003).

Lachenbruch, P.A., *Discriminant Analysis*, Hafner Press, New York, 1975.

Lowry, C.A., and Montgomery, D.C., "A Review of Multivariate Control Charts," *IIE Transactions* **27**:800–810 (1995).

Magnus, J.R., and Neudecker, H., *Matrix Differential Calculus with Applications in Statistics and Econometrics*, Willey, New York, 1999.

Mason, R.L., and Young, J.C., *Multivariate Statistical Process Control with Industrial Applications*, ASA-SIAM, 2002.
Montgomery, D.C., *Introduction to Statistical Quality Control*, 4th ed., Wiley, New York, 2000.
Morrison, D.F., *Multivariate Statistical Methods*, 2nd ed., McGraw-Hill, 1976.
Pande, P.S., Neuman, R.P., and Cavanagh, R.R., *The Six Sigma Way: How GE, Motorola and Other Top Companies are Honing Their Performance*, McGraw-Hill, New York, 2000.
Park, S., Hartley, J., and Wilson, D., "Quality Management Practices and Their Relationship to Buyer's Supplier Ratings: A Study in the Korean Automotive Industry," *J. Operation Management* **19**:695–712 (2001).
Pearson, K., "On Lines and Planes of Closest Fit to Systems of Points in Space," *Phil. Mag. Ser. B* **2**:559–572 (1901).
Post, F.H., Nielson, G.M., and Bonneau, G., *Data Visualization, The State of Art*, Kluwer Academic Publishers, Norwell, MA, 2003.
Press S.J., and Wilson S., "Choosing Between Logistic Regression and Discriminant Analysis," *J. Am. Stat. Assoc.* **73**:699–705 (1978).
Price, B., and Barth, B., "A Structural Model Relating Process Inputs and Final Product Characteristics," *Quality Engng.* **7**(4):693–704 (1995).
Rao, C.R., "The Use and Interpretation of Principal Component Analysis in Applied Research," *Sankhya A* **26**:329–358 (1964).
Roan, C., Hu, S.J., and Wu, S.M., "Computer Aided Identification of Root Causes of Variation in Automotive Body Assembly," *J. Manufacturing Sci. Engng* **64**:391–400 (1993).
Rosenblum, L., et al., *Scientific Visualization, Advances and Challenges*, Academic Press, London, UK, 1994.
Ryan, T.P., *Statistical Methods for Quality Improvement*, Wiley, New York, 1989.
Schmid, Calvin F., *Handbook of Graphical Presentation*, Ronal Press, New York, 1954.
Seder, L.A., "Diagnosis with Diagrams—Part I," *Industrial Quality Control* **4**:11–19 (1950).
Srivastava, M.S., *Methods of Multivariate Statistics*, Wiley, New York, 2002.
Srivastava, M.S., and Carter, E.M., *An Introduction to Applied Multivariate Statistics*, North-Holland, Elsiver, 1983.
Taguchi, G., *Taguchi on Robust Technology Development*, ASME, New York, 1993.
Taguchi, G., *Introduction to Quality Engineering*, UNIPUB/Kraus International Publications, White Plains, NY, 1986.
Taguchi, G., Chowdhury, S., and Wu, Y., *The Mahalanobis Taguchi System*, McGraw-Hill, New York, 2001.
Taguchi, G., Elsayed, E., and Hsiang, T., *Quality Engineering in Production Systems*, McGraw-Hill, New York, 1989.
Taguchi G., and Jugulum, R., *The Mahalanobis Taguchi Strategy: A Patten Technology System*, Wiley, New York, 2002.
Taguchi, G., and Wu, Y., *Introduction to Off-line Quality Control*, Magaya, Japan, Central Japan Quality Control Association, 1979.
Thurstone, L.L., *Multiple Factor Analysis*, University of Chicago Press, Chicago, 1947.
Tiku, M.L., and Balakrishnan, N., "Testing the Equality of Variance-Covariance Matrices the Robust Way," *Commun. Stat. – Theory and Method* **14**(9):985–1001 (1985).
Tracy, N.D., Young, J.C., and Mason, R.L., "Multivariate Quality Control Charts for Individual Observations," *J. Quality Technol.* **24**:88–95 (1992).
Wilks, S.S., "Certain Generalizations in the Analysis of Variance," *Biometrica* **24**:471–494 (1932).
Woodall, W.H., et al., "A Review and Analysis of the Mahalanobis-Taguchi System," *Technometrics* **45**(1):1–15 (2003).
Workman, J.J., Mobley, P.R., Kowalski, B.R., and Bro, R., "Review of Chemometrics Applied to Spectroscopy: 1985–95," *Appl. Spectrosc. Rev.* **31**:73–124 (1996).
Yang, K., "Improving Automotive Dimensional Quality by Using Principle Component Analysis," *Quality Reliability Engng. Int.* **12**:401–409 (1996).

Yang, K., and El-Haik, B., *Design for Six Sigma: A Roadmap for Product Development*, McGraw-Hill, New York, 2003.

Yang, K., and Yang, G.B., "Performance Degradation Analysis for IRLEDs," *Quality Engng.* **13**(1):27–34 (2000).

Yates, A., *Multivariate Exploratory Data Analysis: A Perspective on Exploratory Factor Analysis*, State University of New York Press, 1988.

Young, F.W., and Sarle, W.S., *SAS Views: Exploratory Data Analysis*, 1982 ed., SAS Institute, Cary, NC, 1983.

Index

AMOS, 225, 235–239
animation of principal component analysis, 121–122
average link method, 187–188

between cluster distance, 185–187
Bonferroni simultaneous confidence intervals, 69–70

casual correlation relationship, 225
Cauchy-Schwarz inequality, 164
causation in path analysis, 225
centroid, 188–189, 194
characteristic root, 77
characteristic vector, 77
Chernoff faces, 31–33
circular profile, 27
classification criteria, 163
classification rule, 163–170
cluster algorithm, 183
cluster analysis, 181–200
cluster centroids, 195
CMM, 126
common factor analysis, 142–148
communality, 143–144
complete linkage method, 187
component score, 111–113
confidence interval, 67–73
confidence region, 67–73
control charts, 243–269
control limits, 243–269
correlation coefficient, 48–50
correlation matrix, 50, 54
covariance matrix, 48–53
covariance, 48–53
covary relationship, 225
cutoff score, 164–165, 168
cutoff similarity score, 198

data animation, 38
data mining, 2, 11
data reduction, 2, 98
data stratification, 2, 25, 38–46
data visualization and animation, 33–38
data visualization scheme, 42
dendogram, 190–195
determinant, 55, 77
diagonal matrix, 74
dimensional quality, 124–127
dimensional variability, 125–127
discriminant analysis, 6, 161–180
discriminant classification rule, 163–172
distance and similarity measure, 183–190
distance between clusters, 185–189
distance matrix, 184, 186, 190–191, 195
DMAIC, 20

eigenvalues, 77, 99–105, 117–119, 266
eigenvectors, 77, 99–105, 117–119, 266
EQS, 225, 235
Euclidean distance, 183, 189

factor analysis, 5, 141–160
farthest neighbor method, 187

geometrical dimensional variations, 116–121
geometrical variation modes, 117–121
graphical analysis of multivariate variation pattern, 41–46
graphical analysis of variance, 3, 39–45
graphical multivariate data display, 3, 25–39
graphical templates for multivariate data, 26–33

hierarchical clustering method, 190–195
Hotelling's T^2, 59–60, 244–265
hypothesis testing, 61–73

interpiece variability, 41
intrapiece variability, 41
isosurfaces, 37

K-mean method, 195–197

large sample multivariate hypothesis testing, 63, 66
Lawley-Hotelling statistic, 92
line charts, 27
linear discriminant analysis, 163–170
linear discriminant classification, 163–170
linkage method, 185–189
LISREL, 225, 235
lower triangular matrix, 74

Mahalanobis distance, 163, 166, 171, 178, 201–222
Mahalanobis space creation, 202
Mahalanobis squared distance, 4, 166, 170
Mahalanobis Taguchi system, 7, 201–222
Manhattan distance, 184
MANOVA model, 81–96
MANOVA table, 92
MANOVA, 81–96
matrix rank, 77
misclassification error, 166
misclassification probability, 166, 168
MTS, 201
multicolinearity, 224
multidimensional data, 3
multi-vari chart, 31, 39, 40–46
multivariate analysis of variance, 5, 81–96
multivariate calibration, 12
multivariate control charts, 10, 243–270
multivariate data-based diagnosis system, 201
multivariate data-based pattern recognition, 201, 203
multivariate data display, 25–46
multivariate data scatter plots, 55
multivariate data sets, 25, 51–55
multivariate data stratification, 38–46
multivariate data visualization, 33–38
multivariate descriptive statistics, 51–55
multivariate monitoring system, 214

multivariate normal distribution, 3, 48–56
multivariate normal, 48–56
multivariate population covariance matrix, 56
multivariate population mean vector, 56
multivariate process control, 243–270
multivariate random variables, 47–50
multivariate random vector, 47–50
multivariate sample data sets, 50
multivariate sampling distribution, 3, 57–60
multivariate statistical inferences, 60–73
multivariate statistical process monitoring and control, 12, 243–270
multivariate variation patterns, 42–45
nonhierarchical clustering method, 195–197

nonsingular matrix, 77
normalized eigenvector, 79
null matrix, 74

optical coordinate measurement machines, 5
orthogonal matrix, 76
orthomax rotation, 150
overall T^2, 249

path analysis, 223–242
path diagram, 9, 227–235
pattern recognition, 12
PCA on correlation matrix, 98, 109–114
PCA on covariance matrix, 98, 99–108
PCA, 97–140
P-diagram, 223
performance degradation analysis, 131
piece-to-piece variation, 127
Pillai's statistic, 92
population covariance matrix, 54, 99–103
principal component analysis, 5, 97–140
principal component chart, 265–270
principal component control chart, 245, 265–270
principal component equations, 266
procrustes factor analysis, 154–159
procrustes rotation, 152
quadratic discriminant function, 170
quality loss function, 203, 214
quantitative structure-activity modeling, 12
quartimax rotation, 150

regular Mahalanobis distance, 205
robust engineering, 201
rotated factor matrix, 142
rotation matrix, 149

sample correlation matrix, 54, 108–111
sample covariance matrix, 52, 54, 103–105
sample mean vector, 52
scaled Mahalanobis distance, 204
SEM, 236
Shanin method, 22
simultaneous confidence intervals, 68–72
single linkage method, 186
Six Sigma, 18, 223
small sample multivariate hypothesis testing, 62, 64
spectroscopy, 12
square matrix, 74
standardized Euclidean distance, 183, 191
standardized Manhattan distance, 184
standardized principal component score, 266
standardized structural mode, 228
statistical process control, 243–270
structure-activity, 12
subgroup covariance matrices, 256
symmetric matrix, 53, 75

templates for displaying multivariate data, 29–33
three-cluster partition, 194
three-dimensional animation, 131
three-dimensional matrix data, 33–35
trace, 76
two-dimensional matrix data, 33–35
two-phase T^2 control chart, 245
two-phase T^2 multivariate control chart, 251–259
two-group discriminant analysis, 163–170

unrotated factor matrix, 142
upper control limit, 247
upper triangular matrix, 73

variation extent limit, 119–120
variation mode chart, 118–120
variation mode decomposition, 117
varimax rotation, 150
VEL, 119
visual display of principal component analysis, 121–122
volumetric visualization, 37

Wilks' Lambda, 90
Wishart distribution, 58
within cluster distance, 185

zero matrix, 74

ABOUT THE AUTHORS

KAI YANG, PH.D., has consulted extensively in many areas of quality and reliability engineering. He is Associate Professor of Industrial and Manufacturing Engineering at Wayne State University, Detroit, Michigan. He lives in West Bloomfield, Michigan.

JAYANT TREWN, PH.D, is a research faculty member at Beaumont Hospital in Royal Oak Michigan. He is responsible for implementing cutting edge industrial engineering tools in hospital and health care management. Dr. Trewn was a Director of Quality and Productivity Improvement at Vetri Systems, a Lason Company. He was responsible for business process design and improvement in the global business environment.